Revised and Updated 2nd Edition

From Photon to Pixel

The Digital Camera Handbook

Henri Maître

iSTE

WILEY

First edition published 2015 in Great Britain and the United States by ISTE Ltd and John Wiley & Sons, Inc. © ISTE Ltd 2015.

First published 2017 in Great Britain and the United States by ISTE Ltd and John Wiley & Sons, Inc.

ISTE Ltd
27-37 St George's Road
London SW19 4EU
UK

www.iste.co.uk

John Wiley & Sons, Inc.
111 River Street
Hoboken, NJ 07030
USA

www.wiley.com

Library of Congress Control Number: 2016961650

British Library Cataloguing-in-Publication Data
A CIP record for this book is available from the British Library
ISBN 978-1-78630-137-6

Contents

Acknowledgements

I would like to thank here the large number of people without whom this book could not have been written. First, my colleagues from Telecom ParisTech who often mastered some of the areas described here much better than myself: Andrès Almansa, Isabelle Bloch, Hans Brettel, Yann Gousseau, Saïd Ladjal and Yohann Tendero. I thank those who were at Telecom ParisTech for a short time before going elsewhere: Cécilia Aguerrebere, Julie Delon and Edoardo Provenzi. With all of them, I have had the opportunity to exchange ideas about many chapters and I have largely been inspired by their work, their results and that of their students.

I also generously thank the reviewers who have kindly toiled over some chapters and improved them with their judicious observations: Pierre, Chavel, Cyril Concolato, Jacques Debize, Frédéric Dufaux, Claude Maître, Alain Maruani, Yves Matthieu, Pierre Chavel, Michel Rousseau, and Alain Maruani.

I would finally like to thank the students of Telecom ParisTech, zealous of signal processing, image processing, multimedia, data research, social networks or online services, to whom I dedicate these lines. I hope that they will find the answers to the questions that are not often addressed in classes but that close off many of the problems from the moment the image generated by a modern photographic camera must be automatically processed.

1

First Contact

It could be said of photography what Hegel said of philosophy: "No other art, no other science is exposed to this ultimate degree of contempt based on the belief that one can take possession of them all at once[a]" [BOU 65]

a. Preface to the Principle of the Philosophy of Right, G.W.F. Hegel, 1820.

1.1. Toward a society of the image

To say that, over the last 30 years, a real revolution has taken the world of photography by storm and deeply modified the multiple technical, economic, industrial and societal aspects in which it develops would be an understatement.

From a technical perspective, the replacement of analog silver film by solid digital sensors, tentatively began 40 years ago, emulating a transition from analog to digital that is found in many other fields (the telephone, television, printing, etc.), could have certainly been no more than a significant progress, but in the end a natural one and of little impact for the user (is the user really conscious of the transition to digital terrestrial television or of the phototypesetting of newspapers?). However, it has profoundly modified the concept of photography itself, bringing forward several types of original devices; first, the point-and-shoot that fits in a pocket and that can be forgotten, then the mobile phone or the tablet, which the photographic industry would gladly repudiate as illegitimate and somewhat degenerate children if they did not hold the promise of an inexhaustible market.

The consequences of this technical mutation have proved to be devastating for the established economy of photography. Major players (Kodak, Agfa, Fuji, Ilford, Minolta, etc.) which seemed to be ruling empires were forced to fall back on uncertain niches, to regroup and sometimes disappear. Newcomers have settled down, whose photographic culture was often thin (Sony and Samsung), sometimes non-existent (Nokia, Apple, etc.). Other players are building kingdoms from scratch around online image services or their social networks, but these are people with Internet, computing and telecoms backgrounds and not photographic ones. Whereas the chemical industries that produced films and derived products have naturally strongly suffered, processing laboratories, if they had not disappeared entirely, had to be totally reconverted and distribution has begun a profound transformation whether it relates to materials or to services.

The reconfigurations of the industrial fabric, far from being completed, continue today, involving more and more closely players who ignored each other, some coming from imaging, but many who prospered without being related to photography professions: on-board electronics experts, creators of integrated circuits, software developers, network and mobile phone key players have been transformed into designers of camera bodies and objectives.

Society's activities are themselves deeply affected by these reconfigurations. Have they accompanied the industrial mutation or are they one of the causes? It is not up to us to say. But the societal mutations of photography are equally significant and irreversible. They reveal themselves in several ways, but first of all it is the generalization of the use of photography that is striking. Generalization within the family: photography is no longer the prerogative of the *pater familias* as at the beginning of the previous century, of adults and of old adolescents as at the end of the century. It is now a personal attribute that every one uses, children as well as grandparents. Society as a whole is also exposed to it and there is no niche regardless of the level of wealth or education, that escapes from it, no population, regardless of its location, city or rural, that does not participate in it. And its distribution is remarkably diffuse over the globe, no country is excluded, at least in the component of its population exposed to modern life.

At this point, we need to recall, by rereading the famous analysis that Bourdieu and his collaborators made fifty years ago, the achievements made for half a century [BOU 65]. While there is no evidence that photography is no longer an "average art" as they deplored at the time, it is obvious that the family, the cultural and the socio-professional categories that then allowed the identification of typical behaviors with regard to photography are now completely blurred. Attitudes are surprisingly similar between a class of Parisian students on a field trip, a bus of older Japanese tourists in Capri, or the crowd gathered for the return of a *rock-star* at Wembley Stadium. Photography is ubiquitous and permanent, it is spontaneous, individual and collective, frantic and viral, intimate and shared and very few can escape from it.

This universal infatuation with photography profoundly affects its usages. Certainly whole swathes of the photographic "culture" remain almost unchanged (art photography, event photography (family or public), professional newspapers, catalogues and advertising photography, scientific photography, etc.), and the reader would find word for word, for these areas, the relevance of Bourdieu's analysis. But old facets tend to rise or become distorted, new facets emerge: leisure or travel photography now free from "old fashion" conformity stereotypes (at the expense of a "young" conformity?), everyday-life photography, detailed, anecdotal, sensational, observational, unusual photography, self portraiture alone or with others, microcosm, microcommunity photography and its culmination into the narcissic *selfie*. These new forms combine an often simplified manner of photography and modern means of instantaneous, remote and mass communication; it relies on a formidable technique, both hidden and exhibited, following the rules of the best marketing. MMS[1] and the Internet are the natural extensions of the photographic image; social networks work as their magnifier unless they are the objective. It remains undecided nowadays whether YouTube, Facebook, Twitter, Instagram, Tumblr, Picasa and Flickr must be included or not in "photography products".

The figures say everything about this unprecedented evolution, but they do not explain anything[2]. It would be naive to imagine that it owes everything to the advent of digital image and of its natural creation tool, the still camera.

1 MMS = *multimedia messaging services* = the extension to documents (including images) of SMS telephonic messages (*short message services*).

2 In bulk, and on the basis of the only statements of the owners, 150 million Instagram users in 2013, 400 million in 2015, 6 billion pictures registered on Flickr and a 100 billion on Facebook in 2011, 32 million photo galleries on jAlbum and more than 200 million blogs on Tumblr, 70 million landscapes and travel images on Panoramio, a 100 million images visited every day on Facebook and in the order of a billion on all the exchange and network sites, Instagram bought for one billion dollars in 2012, with a growth of 23% in 2013, a digital photographic market (products and services) estimated at 70 billion dollars in 2014, a growth of 5 to 6% per year probably until at least 2020.

It is however doubtful that such a development never would have occurred to such an extent without a simple and universal image acquisition system having been made available, fully compatible with modern communication and information processing means. Contrary to film cameras, digital cameras play this key role[3].

The objective of this book is not to extend this sociological study, which is worthy nevertheless, but to explain how the digital camera works by examining in detail each of the components that constitute it.

1.1.1. *A bit of vocabulary in the form of zoology*

The generic term we will subsequently use the most is the term "camera", standing in short for "photographic camera". This term covers the various forms of the devices, with film or digital and in its various current versions. What are these forms today? We have illustrated them in Figures 1.1 and 1.2. The proposed naming ensues from their functionalities. However, denominations are not fixed. Because of many hybrid configurations, these denominations may be attributed to items rather different from those presented here as prototypes.

The SLR: SLR stands for single lens reflex expressing that the same lens is used for viewing and for the image as opposed to other cameras that use two different optical paths. SLR is the long-standing reference of the film photography camera, especially in the historical 24 × 36 mm format. The optical path is made conventionally, either toward the sensor, or toward the viewfinder, by using a moving mirror and a prism with a pentagonal cross-section. It has interchangeable lenses. Its digital version appeared very early on the market (from the 1990s), but with small-sized sensors, generally less than 24 × 36 mm. Since around 2010, it has been available with a sensor of size 24 × 36 mm.

The compact camera: this is a complete photography system of small size, which may be slipped into a pocket or a small bag so that it can be carried everywhere without any discomfort. Its lens is fixed, usually with a variable

---

3 The numbers above will be compared to those that Bourdieu estimated to be considerable at the time: "*8,135,000 cameras in working condition, (...), and 845,000 cameras sold each year,*" ([BOU, p. 23]) which are several orders of magnitude below the current figures (for example, the annual market forecasts of last generation photoscopes are largely above a hundred million units).

and retractable focal length. The viewing is done with a screen and not through an eyepiece. The smaller compacts are the size of a credit card. Many compacts are very simple and very intuitive to operate, but the range also offers almost professional compacts which present all the functionalities of an SLR with a very reduced size and allow, for example, work to be prepared that will eventually be continued later with a more sophisticated camera. The compact is the first digital camera to have conquered a place on the market, in the early 1980s.

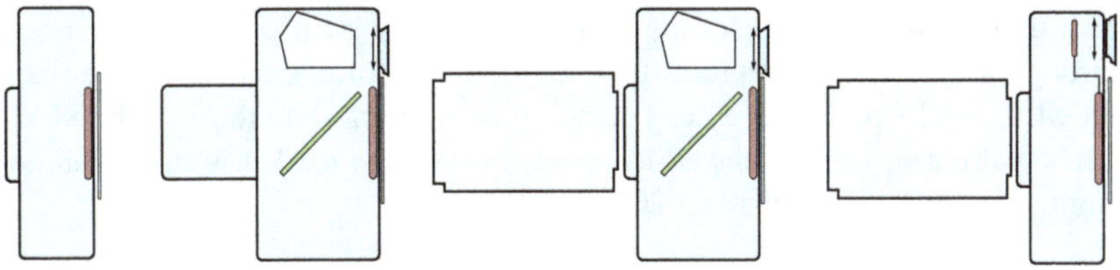

Figure 1.1. *The four main architectures of general public digital cameras. From left to right:*
- the compact: fixed objective, no eyepiece, viewing through the back screen;
- the bridge: fixed objective, viewing through an eyepiece in an optical path returned by a tilting mirror, display on the back screen after shooting;
- the single lens reflex (SLR): interchangeable objective, viewing through an eyepiece in an optical path returned by a tilting mirror, display on the back screen after shooting;
- the hybrid: interchangeable objective, viewing through an eyepiece in an electronic path or on the back screen, no tilting mirror.
Intermediate solutions exist (for example using the live view *function of SLRs that allows them to display on the back screen the image during focusing adjustments)*

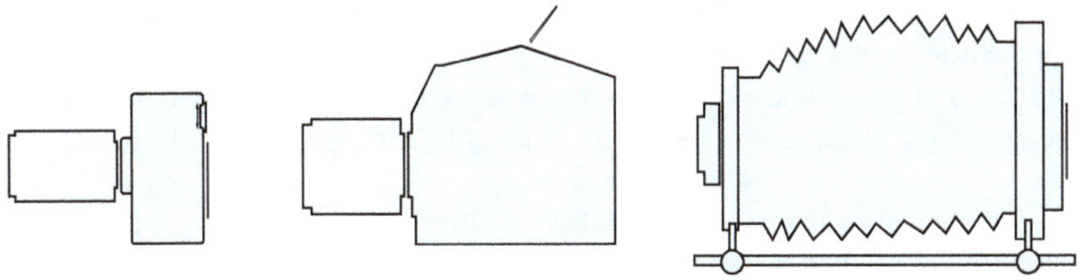

Figure 1.2. *Professional cameras: on the left, 24 ×36 mm format SLR. In the center: medium-format camera. On the right: view camera. The diagrams are not to scale, the sizes are indicative*

The bridge: this also has a fixed lens, usually a zoom, but a body, a build and an optical path similar to those of an SLR. It typically uses a prism and a moving mirror in the optical circuit allowing a reflex viewing. Its name comes from its intermediary positioning between compact and reflex. It appeared on the market in 1995 and has suffered a strong decline in its distribution in the 2010s.

The hybrid camera: this looks like an SLR because of its interchangeable objectives and its often advanced functionalities, but it does not use a prism or a moving reflex mirror in the optical path, the viewing being carried out through an electronic eyepiece. Its body therefore offers a smaller size than the SLR, but its performance and its usage are very close to those of the SLR. Technical reasons have delayed its appearance on the market where it had no significant presence until about 2010.

The medium format: this is a camera whose sensor (traditionally film) is larger than that of the 24 × 36 mm. In its film version, it uses sensitive surfaces in spools and takes pictures of 4 × 4 cm, 6 × 6 cm, or 6 × 7 cm in size. A number of digital medium formats have been available on the market for a few years, but with generally high costs which designate them for professional or semi-professional purposes.

View cameras: for formats beyond those of the medium format, cameras are referred to as view cameras (formats from 9 × 12 cm to 20 × 25 cm), which make use of plates or individually packaged film sheets. View cameras are reserved for professional applications: architecture, fashion, works of art, etc. As of 2015, there have actually been no digital sensors available on the market adapted to view cameras. The very large-dimension sensors which can be adapted to view cameras are used especially in the scientific field, in microelectronics, astronomy or particle physics, and remote sensing. They are still often prototypes made of mosaics of juxtaposed sensors. Moreover, for applications that allow it, very large images (typically 50,000 × 50,000 pixels and beyond) are obtained by the movements of a sensor (linear or matrix) using robotic mechanisms such as in biology or for capturing works of art.

Photoscopes: we will also mention in this book sensors that perform the photographic function of computers, tablets as well as that of mobile phones. These devices are very similar in their architecture and in their design to smaller compacts. They differ from them, on the one hand by automating most of the functions, on the other hand by the intensive use of communication and computing functions. They thus appear ahead in numerous technical aspects compared to their cousins solely dedicated to photography. Although the limited quality of the images they provide and the small freedom they afford to the photographer exposes them to the condescension of part of the community of photographers, they have gradually become the most important source of pictures of the huge market that we have described above. As such, they receive the utmost attention of all the manufacturers, the components and software developers, and this attention is bearing fruit. They now achieve amazing performances. We will be especially looking at all the innovations they propose; these are good markers of trends in photography. We will refer to them consequently either as photoscopes, or as mobile phones.

Among the new terms and along with photoscope, the acronym "DC" is often found, which is generically used to refer to digital cameras in all its forms.

Finally, it should be noted that none of the above terms, either in French or in English, are included in the recent *Vocabulaire Technique de la Photographie*, [CAR 08], which reflects rather well the gap which remains within the world of photography between those who design cameras and those who use them.

In the English vocabulary the term camera is universally recognized. It covers any device that allows to capture a picture (either still or moving). In the case of the new electronic cameras, many more concise forms are proposed, making use of the letters D (*digital*) or E (*electronic*) associated with acronyms not always very explicit, such as digital still camera (DSC), electronic still picture camera (*ESPC*), Electronic still picture imaging (*ESPI*), digital single lens reflex (*DSLR*).

This standardized vocabulary for photography is the subject of a recently completed ISO norm [ISO 12], but still seldom followed.

1.1.2. *A brief history of photography*

> *As we move forward, we have entered the videosphere, a technical and moral revolution which does not mark the peak of the "society of the spectacle " but its end.* [DEB 92]

The technical components allowing the capture of an image were all available at the beginning of the 19th Century, some for a long time previously: the *camera obscura* which constitutes the body of the view camera had been known since antiquity, in Europe and in Asia as well, and was particularly familiar to the artists of the Renaissance. The lens which canalises channels the captured luminous flux can be traced back several millennia before our era but has only been really useful in the formation of images since the 12th century, the photosensitive components, either in negative (such as silver chloride), or in positive (as the bitumen of Judea, mixture of natural hydrocarbons) were familiar to chemists at the end of 17th century. In addition, the laws of propagation and the mysteries of light and of color have been correctly mastered for two hundred years for the former and fifty for the latter.

The first photography tests, which can be dated to 1812, were by Nicéphore Niépce. However, while an image could then correctly be captured, at the expense of very long exposure times, it was not stable and disappeared too quickly. Efforts were therefore being made in these two directions: on the one hand, by improving the sensitivity of receptors, on the other hand, above all, by maintaining the image after its formation.

The first photograph from life was made in 1826 by Nicéphore Niépce of his suroundings: "View from the Window at Le Gras". It was achieved on a pewter plate coated with bitumen and has required an exposure time of 8 h.

Seeking to improve his process, Nicéphore Niépce tested a very large number of media and developers, the best being silver plates and iodine vapors[4]. He entered into a partnership with Louis Daguerre, in 1829 to develop what would become an industrial process: the daguerreotype. He used silver salts on a copper plate and develop them with iodine vapors. Niépce died in 1833. In 1839, the daguerreotype was presented to the public and Arago officially presents photography at the Academy of Sciences. This was the beginning of the immediate commercial success of the daguerreotype, and of photography among the general public.

In Italy and in England, William Fox Talbot was working in parallel on photographic recording processes focusing in particular on the reproduction on paper (paper coated with chloride sodium and with silver nitrate fixed with potassium salts). He achieved his first photos in 1835. In 1840 he developed the paper negative, which allows the reproduction of several photographs from a single original. He generalized the use of hyposulphite of soda as a fixer and patented an original process in 1841: the calotype.

4 Iodine was discovered in 1811 by Bernard Courtois.

It was nevertheless the daguerreotype, rights-free, which was widespread, more than the calotype, penalized by its patent. The calotype will take its revenge later, since the principle of the photographic negative has been, for 100 years at least, at the heart of the photographic industry.

Another major breakthrough was made significantly later by George Eastman, in the United States. In 1884, he proposed photographic media not on glass but on flexible celluloid cut into strips: film is born. In the wake of this discovery (1888), he proposed a very compact camera body using this film that could take 100 pictures in a row. Individual photography was ready for the general public who no longer wanted the inconvience of bulky camera bodies, tripods and boxes of heavy and fragile photographic plates

This evolution was confirmed by the release of the Kodak Brownie in 1900, a camera priced at $1.00 that made it possible to take 20 photos, each with a cost of 25 cents. The market shifted from the camera to film: the consumable is the engine of the market.

However, for most cameras, especially those of quality, sensitive surfaces were still quite large, with sides in the order of 10 cm. It was not until the Leica in 1925 that the small format became widespread, popularizing for the occasion the 24×36 mm.

However, color photography remained to be improved. The Auguste and Louis Lumière brothers' "autochrome" process, patented in 1903 and commercialized in 1907, was the starting point. The autochrome used potato starch whose colorful grains mixed within lampblack acted as a sensitive support. Nonetheless, the sensitivity was very low (equivalent today to a few ISO) and the development process was complex. The quality of the image was nevertheless exceptional and would have nothing to envy from our best emulsions such as can be seen in the pictures still available a hundred years later.

The color emulsions Kodacolor (1935) then Agfacolor (1936) would appear much later on the market and would stand out in turn by the simplicity of their use and the quality of the images they provide.

In 1948, Edwin Land conceived a process that enabled the instantaneous development of a positive, the Polaroid, whose color version appeared in 1963. Despite its success, it never really competed with the emulsions developed in laboratories which had an extremely dense network of distributors.

With regard to cinema, it was Thomas Edison who filmed the first movies from 1891 to 1894 with the kinetograph, which used the flexible film recently invented by Eastman to which he added side perforations for transporting (he thus imposed the 35 mm-sided format which would provide the basis for the success of the 24 × 36 mm). As early as 1895, the Lumière brothers had improved the kinetograph with a very large number of patents and gave it the well known popular momentum.

In the field of image transmission (the forerunner of video), the German engineer Paul Nipkow was the first to study the analysis process of a picture with a perforated disk. His work which began in 1884 would, however, only be presented to the public in 1928. Edouard Belin introduced a process of transmitting photographs, first by cable in 1908, and then by telephone in 1920: the belinograph. Meanwhile, the Russian Vladimir Zworykin filed a patent for the iconoscope in 1923, based on the principle of the Nipkow disk. In 1925, John Logie Baird performed the first experimental transmission of animated pictures in London.

The first digital cameras were created at the beginning of 1975 within the Kodak laboratories. Broadcast first in a very confidential manner and rather as laboratory instruments, they were first professionally distributed in areas very different from professional photography: insurance companies, real estate agencies, etc. They were not really competing with cameras which gave much better quality images. The consumer market was however very quickly conquered, and followed gradually by the professional market. From the 1990s, the market of digital cameras was more significant than the analog market and companies relying on film disappeared one after another (AgfaPhoto went bankrupt in 2005, Eastman Kodak had to file for bankruptcy in 2012) or convert to digital (Fujifilm).

1.2. The reason for this book

In recent years, the evolution of digital photographic equipment has been considerable. It has affected the most important components of cameras (such as sensors, for example) and the specialized press has widely reported the progresses obtained. However, the evolution has also affected aspects much less accessible to the public, either because they are too technical or because they appear as accessories in a first analysis, or often because they are hidden within products and reveal manufacturers' secrets. These advances relate to very varied scientific fields: optics, electronics, mechanics, science materials, computer science and as a result do not fully find their place in specialized technical journals. In order to be be used by photographers, they often require long recontextualizations that explain their role and the principles of their functioning. It is in this spirit that this text has been written which reviews all the functions essential to the proper functioning of the camera and explains the proposed solutions to solve them.

But before addressing these major features of the camera, we will in global terms situate photography in the field of science, present a few key features, particularly important for the formation of the image and introduce a bit of vocabulary and of formalism that will accompany us throughout the book. We will use this quick description to give an overview of the various chapters that follow and to indicate the manner in which we have chosen to address each problem.

1.3. Physical principle of image formation

Photography is a matter of light and as such we will have to speak of optics, the science of light. There are numerous books covering this area, and often excellent despite being a bit old [BOR 70, PER 94, FOW 90]. A few important elements should be remembered concerning light, its nature and its propagation that will allow us to place photography within the major chapters of physics.

1.3.1. *Light*

Light is an electromagnetic radiation in a narrow window of frequencies. Today, it is likely to be addressed by formalisms of classical physics or by quantum or semi-quantum approaches. Photography is overwhelmingly explained using traditional approaches: image formation is very well described in terms of geometrical optics, fine phenomena concerning the resolution are explained by the theory of diffraction, most often in a scalar, and eventually in a vector representation. Diffraction theory takes its rightful place when addressing the ultimate limits of resolution, one of the key problems of photography. The polarization of light occurs in a few specific points that we will take care to point out. The concept of coherence is one of the finest subtleties of photography, especially in the field of microphotography.

Only the basic phenomenon of the transformation of the photon into an electron within the photodetector relies on more advanced theories since it is based on the photoelectric effect[5] which can only be explained with the help of quantum theory[6]. However, we will not need quantum theory in this book, once the fundamental result of the photoelectric effect is admitted: the exchange of energy between a photon particle and an electron, if the photon has at least the energy necessary to change the state of the electron.

1.3.2. *Electromagnetic radiation: wave and particle*

Light is electromagnetic radiation perceived by the human visual system. It covers a range of wavelengths from 400 to around 800 nm[7] and therefore a frequency range of 7.5×10^{14} to 3.75×10^{14} Hz, which corresponds to a transmission window of the atmosphere, on the one hand, and to a maximum of the solar emission, on the other hand.

In its wave representation, light is a superposition A of monochromatic, ideal, plane-waves characterized by their frequency ν, their direction of propagation $\mathbf{k}$ and their phase ϕ [PER 94]. We will represent it thereafter, at point $\mathbf{x}$ and time t, by a formula such as:

$$A(\mathbf{x}, t) = \sum_i a_i \cos(2\pi\nu_i t - \mathbf{k}_i.\mathbf{x} + \phi_i) \qquad [1.1]$$

In its particle representation, a photon of frequency ν (and therefore of wavelength $\lambda = c/\nu$, with c = 299,800,000 $m.s^{-1}$, the speed of light) carries an energy $h\nu$, where h is Planck's constant ($h = 6.626 \times 10^{-34}$ J.s), either between 2.5×10^{-19} and 5×10^{-19} J, or, in a more convenient unit, between about 1.55 eV and 3.1 eV [BOR 70, NEA 11].

5 The photoelectric effect, explained by A. Einstein in 1905, earned him the Nobel Prize in 1921.

6 As a matter of fact, it is not the only case, because as we will quickly mention it in section 5.1.1, many aspects of the appearance of the objects, and in particular their color, could not be justified without a quantum interpretation of the light–matter interaction.

7 There is no consensus about the extent of the visible spectrum. We have chosen here a range (from 400 to 800 nm) to simplify the figures. But different, more narrow limits can often be found in the literature: 420–750 or 450–720. If these exact values are not very important for the human observer since the response of the sensor is almost non-existent at the terminals, this is not the case for cameras whose sensor is potentially sensitive beyond.

A photon therefore carries a very low energy and a normally lit scene involves a very large number of photons[8]. It is important for image formation to consider what relationship these photons have relative to each other: these are aspects of coherence which will be discussed in section 2.6. It should be considered now that photography is first and foremost concerned with incoherent optics. We will also examine in section 9.7 the polarization properties of waves, but these properties are only marginally involved in photography.

Color is one of the most striking manifestations of the diversity of the frequency content of electromagnetic waves. It is also one of the most complex challenges of photography. Chapter 5 will be dedicated to it, where we will have to introduce numerous notions of physiological optics to account for the richness of human perception that *in fine* governs the choices concerning image quality.

The corpuscular aspects of the photon will be at the heart of the chapter dedicated to the photodetector (Chapter 3) as well as at that concerning noise affecting the image signal (Chapter 7). The wave aspects will be used to address the properties of propagation through the lens (Chapter 2), of image quality (Chapter 6) and of very prospective aspects related to the improvement of the images, which we will see in Chapter 10.

The wave is either emitted, or reflected by the object of interest to the photographer. It then travels freely in space, and then in a guided fashion by the optical system which turns it into an image on the sensor. The simplest instrument to create an image is the pinhole camera (or dark room), known since antiquity as a (discrete) observation system, which does not use lenses, but a simple hole by way of an image formation system. We will examine the image constructed by the pinhole camera because it is a very convenient model which allows the simple processing of numerous computer vision problems and remains widely used today.

1.3.3. *The pinhole*

The dark room (or *camera obscura*, or pinhole) is a box perforated with a small hole (of diameter d) placed at the center of one of the sides and with a focusing screen[9] on the opposite side (Figure 1.3). The principle of the pinhole

8 However, we will come to process imaging at a very low luminous level, single-photon photography being today the issue of numerous research projects.

9 Focusing screen that can be replaced by film or a digital photographic sensor.

is well explained in geometric optics approximation. Detailed studies of the pinhole are available in [LAV 03, CAR 05]. In this approximation, light travels in a straight line without diffraction. An inverted image of an object placed in front of the camera is formed on the screen, at the scale of the ratio between the depth p' of the box and the distance p of the object to the hole. If the object is at a great distance p with regard to the depth p' of the box, each of its points generates a small homothetic spot of the hole (of diameter $\varepsilon = (p + p')d/p \approx d$). It is thus advantageous to keep this hole as small as possible to ensure a sharp image. But the energy that hits the screen is directly proportional to the area of the hole and in order to have a bright pinhole camera it would therefore be helpful that this hole be wide. It should be noted that any object placed in the "object space" will be imaged in an identical manner regardless of its distance to the pierced side as soon as $p \gg p'$. As a consequence, there is no notion of focusing with a pinhole camera except for the very close points ($p' \simeq p$) which will be proportionally affected with a greater blur.

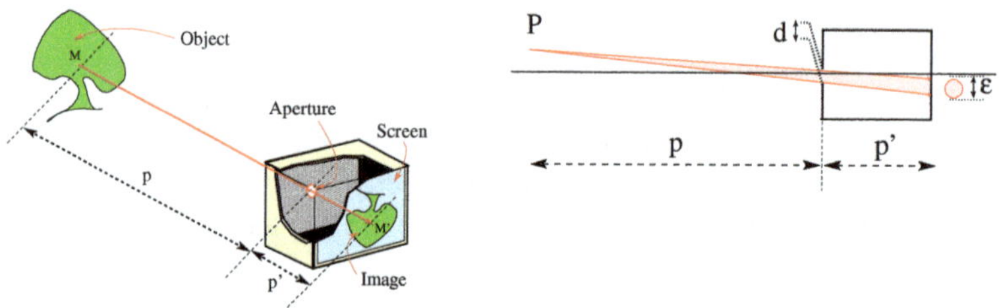

Figure 1.3. *Pinhole – On the left: diagram of the formation of the image in a pinhole camera. On the right: if the object point M is far away from the camera ($p' \ll p$), its image M' is a circle of diameter approximately equal to that of the opening: $\varepsilon = d$*

Under what conditions can the rectilinear propagation of light from geometrical optics be applied to the pinhole camera? It is important that diffraction phenomena, which are observable as soon as the dimensions of the openings are small compared to the wavelength, be negligible. Let us calculate, for a wavelength λ, the diameter of the diffraction figure of an circular opening d observed on the screen (therefore at distance p'). This diameter ε', limited to the first ring of the Airy disk (diffraction figure of a circular aperture, this point will be examined in detail in section 2.6) is equal to $\varepsilon' = 1.22\,\lambda d/p'$ and the diffraction figure will have the same extension as the geometric spot previously calculated if $d = \sqrt{1.22\,\lambda p'}$. For a green wavelength ($\lambda = 0.5 \times 10^{-6}$ m) and a camera with a depth of $p = 10$ cm, the diffraction gives a spot equal to the geometric optics for a hole in the order of

0.2 mm, well below what is used in practice. We are thus generally entitled to ignore the diffraction in the analysis of a pinhole image[10].

1.3.4. *From pinholes to photo cameras*

1.3.4.1. *The role of the lens*

In order to increase the amount of light received on the screen, we replace the opening in the pinhole camera by a lens. We build a camera. Between conjugate planes of the lens, all the rays issued from an object point M converge at an image point M' (approximated stigmatism of the lenses in the paraxial approximation). This solves the problem of the energy since the whole aperture of the lens is now used to capture the light issued from M. This gain comes at a price with the selectivity of the object planes that are clearly seen: only those that are close to the plane of the screen are sharp, the others will be more or less blurred. Consequently, it is important to provide a mechanism to ensure this focusing.

1.3.4.2. *Image formation*

The focal plane in a camera is determined by translating the lens on its optical axis such that to vary its distance to the image plane. This movement can be automated if the camera has a telemetry function that measures the distance to the object (section 9.5). Points M and M' are conjugate if they verify the conjugation relations. Let us recall them. These are Snell–Descartes relations (with origin at the optical centre) or Newton's (with origin in focal points) [FOW 90][11] . If $s = p - f$ and $s' = p' - f$ (Figure 1.5):

$$1/p' + 1/p = 1/f \quad \text{(Snell – Descartes)} \qquad ss' = f^2 \quad \text{(Newton)} \qquad [1.2]$$

---

10 The reader will find in [CAR 05] more accurate derivations to determine the pinhole resolution. In this reference, instead of using infinite distance diffraction as we did, finite distance diffraction is used.

11 In photography, object and image are by nature always real and never virtual, see [FOW 90], the object is thus "on the left" of the object focus and the image "on the right" of the image focus. In addition, the object medium and the image medium are (usually) air, of index $n = 1$. To simplify the equations, we will give p, p', σ and σ' positive values and adapt accordingly the conjugation equations: $p + 1/p' = 1/f$.

Also we should recall the magnification relations. The transversal magnification G and the longitudinal magnification Γ of the image, which will be useful later, are equal to:

$$G = -f/s = -s'/f, \quad \Gamma = p'/p = G^2 \qquad [1.3]$$

The sign "-" reminds us that the image is reversed. The other planes become increasingly more blurry when gradually moving away from P, upstream as well as downstream. It is a defect well known by any photographer (see Figure 1.4).

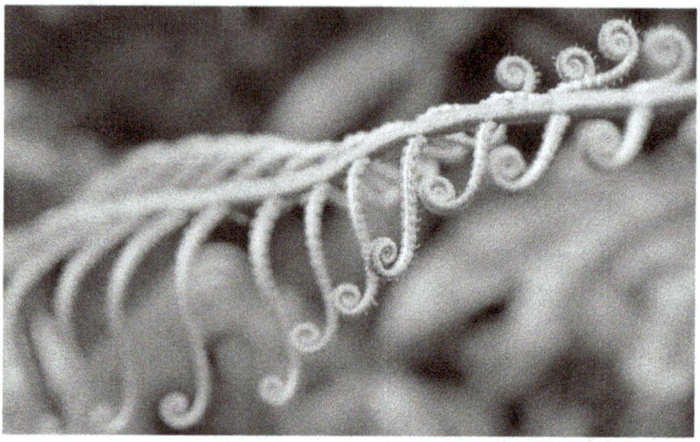

Figure 1.4. *The depth of field is limited: the sharpness of details decreases regularly upstream and downstream of the plane chosen for focusing*

We can see in Figure 1.5, on the left, what happens for a point Q that is located in a plane at distance $q \neq p$ from the lens. Its image converges in Q', in a plane Θ different from Π. The image spot of Q on Π is much greater if the aperture is significant and if the point Q is far from the focus plane. The opening of the lens is in practice controlled by a diaphragm (Figure 1.5 on the right).

The calculation of the depth of field of a photographic lens can be found in section 2.2[12]. If with a lens of focal length f, fitted with a diaphragm of diameter D, observing an object at distance p, (see Figure 1.5), an image spot

12 The depth of field is the distance along the optical axis that separates the closest object considered as focused from the most distant object also considered focused.

of diameter ε can be tolerated, then noting Δ the depth of field, distance between the two extreme focusing points, upstream and downstream of P, Q_1 and Q_2, and in the event of a distant object ($p \gg f$) and a low blur (ε small), it yields:

$$\Delta \approx \frac{2\varepsilon p^2}{fD} \qquad [1.4]$$

expressing that the depth of field varies inversely to the aperture of the diaphragm D and to the focal length f, but grows very fast when the object moves backwards to infinity (p large).

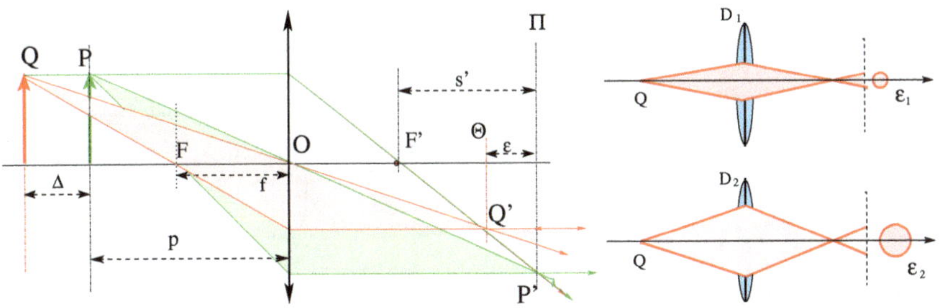

Figure 1.5. *Camera – On the left: The focus is done on the image of P, in the plane Π by varying the "shift" of the lens (s'). A point Q, gives a sharp image Q' in a plane Θ but in the plane Π it gives a blurred image. On the right: the amount of blur depends directly on the size of the diaphragm which limits the inclination of the rays on the axis. A small aperture can decrease the blurring*

1.3.4.3. *Object and image position*

In practice, the purpose is to obtain the image of an object at the distance L of the observer with a camera of focal length f. Two equations are thus available: $s + s' + 2f = L$ and $ss' = f^2$ which lead to a second degree equation, and under the constraint (image and real objects) that $L > 4f$, with two solutions:

$$\begin{aligned} s_1 &= 1/2.(L - 2f + \sqrt{L^2 - 4fL}) \\ s_2 &= 1/2.(L - 2f - \sqrt{L^2 - 4fL}) \end{aligned} \qquad [1.5]$$

These two solutions correspond to symmetric dispositions where the object of one of them becomes the image of the other and vice versa (Figure 1.6). Most often in photography, a focal length f (of a few centimeters) is chosen with a small value compared to L (of a few meters) and a distant object is observed. Then the solution s_1 is adopted which results in a magnification significantly smaller than 1: $G = f/s_1$. The object is then at the position $s \approx L$, the image plane being roughly merged with the image focal plane: $p' \approx f$.

In the case of macro-photography (photography which consists of dramatically enlarging a small object), and such that the magnification $G = s'/f$ be maximum, it requires approximating the object to the object focal point in such manner that s' be the largest possible. The configuration on the right of Figure 1.6 is then adopted.

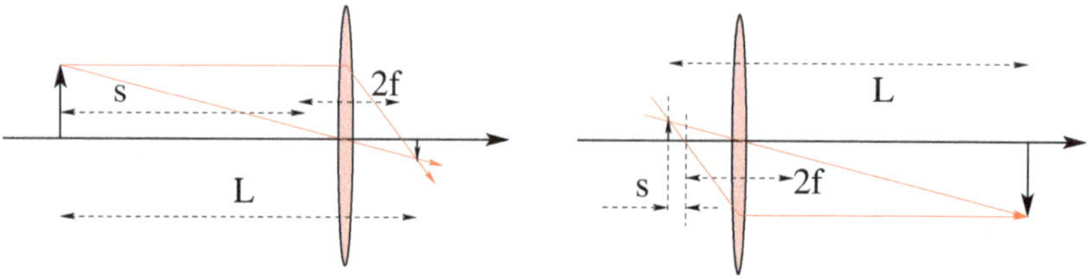

Figure 1.6. *The two solutions to combine two planes at a distance L with a lens of focal length f are symmetrical. When $L \gg f$, we are faced with an ordinary photography situation, the image plane is practically in the focal plane (situation on the left). The other solution (situation on the right) is the case of macrophotography where a large magnification for small objects is achieved*

1.3.4.4. *Lens aperture*

The role of the diaphragm D in the quality of the image is clearly evident in relation [1.4], despite being introduced to control the energy flow. It is therefore a key element in the adjustment of a photographic system, generally controlled by means of a ring on the objective (see Figure 1.7).

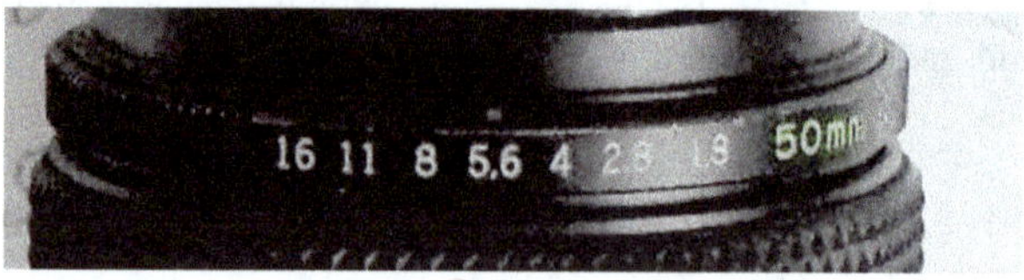

Figure 1.7. *The diaphragm ring on a camera lens*

It is usual to apply the terms lens aperture to the diameter D of the diaphragm and f-number N to the ratio between the focal distance f and the physical diameter D: $N = f/D$. Thus a $50\,mm$ lens, with an f-number $N = 4$, has an open diaphragm with a diameter of $D = 12.5\,mm$. Therefore an f-number N or an aperture f/N refer to the same concept.

Moreover, the diaphragm controls the energy received by the sensor. It is proportional to the free area of the diaphragm, thus to the square of D. It is therefore convenient to propose a range of apertures following a geometric progression with a common ratio $\sqrt{2}$, in order to divide by 2 the energy received at each step.

The most commonly proposed apertures are thus:

f/1 *f*/1.4 *f*/2 *f*/2.8 *f*/4 *f*/5.6 *f*/8 *f*/11 *f*/16 *f*/22 *f*/32 *f*/45 *f*/64

The largest apertures (1/1, 1/1.4) are reserved for high-quality objectives because of they operate very far from the optical axis, and they ought to compensate for various aberrations (see section 2.8). The smallest apertures (1/32 and 1/64) can only be used in very strong brightness conditions or with very long exposure times.

We will have the opportunity to revisit these definitions of the aperture in section 2.4.4.2, where we will bring some clarification to these simple forms.

1.3.4.5. *Photography and energy*

The aperture of the diaphragm D controls the luminous flux arriving at the photodetector. The relationship between the received energy and the aperture is quadratic. The shutter also intervenes to determine the amount of energy that builds the image by controlling the duration of the exposure. In order to make an energy appraisal, the quality of the optical components that constitute the lenses should also be taken into account, but their contribution to the flow of energy is often negligible because objectives are nowadays carefully coated in such a manner that only very few stray reflections can occur.

Finally, the last item that should be taken into account to transcribe the relation between the incident energy and the value of the resulting image is the process of the conversion of photons into signal. This process is very complex and will be discussed in detail in Chapter 4. We should here very quickly recall this problem.

In film photography, the process of the conversion of photons into image on film, undergoes the various stages of the photochemistry of the exposure first, followed by those of developing and of fixing. In order to reduce the numerous parameters of these stages, standard processing conditions were made available which allowed for a common usage, the association of a fixed single number to a commercial product, (a specific film), and the definition of what fully described the result of the conversion: the film sensitivity.

Although the conversion process is now very different in digital photography, since it involves stages of electronic amplification that we will examine in detail in Chapter 3, the notion of sensitivity, very similar to that defined for photographic film, has been recovered for solid sensors. Here again, it expresses the capability of the sensor to generate a signal as a function of the flow of incident light, and as for film photography, a greater sensitivity will *a priori* be possible at the expense of lower image quality.

This sensitivity, such as defined by the standardization body, the ISO[13], varies generally between the values of 25 and 3,000 ISO (i.e. 0.4 to 0.003 lux.second) for film and between values from 100 to 100,000 ISO in the case of solid sensors.

We will denote N the aperture f-number, τ the exposure time and $\mathcal{S}$ the sensitivity of the receiver.

Relying on the three variables above, the formula which ensures a correct energy evaluation is of the form:

$$\frac{\mathcal{S}\tau}{N^2} = \frac{\mathcal{S}\tau f^2}{D^2} = cste \qquad [1.6]$$

Whereas the three terms $\mathcal{S}$,τ and N seem to compensate for each other, they however control in a particular manner the image quality:

– the sensitivity $\mathcal{S}$, as we have said, is notably responsible for the noise that affects the image and it will be advantageous to choose it as low as possible;

– the exposure time τ controls motion blurring and will have to be adapted to both the movement of the objects in the scene and to the movements of the camera;

– the f-number N controls the depth of field: if N is low (a wide open aperture), the depth of field will be small (equation [1.4]).

13 ISO = International Standard Organization, the standardization of sensitivity is the subject of the document [ISO 06a].

It is the art of the photographer to balance these terms in order to translate a particular effect that he wishes to achieve. Rules of use and of good sense, associated with a rough classification of the scene expressed by a choice in terms of priority ("aperture" or "speed" priority) or in terms of thematics ("Sport", or "Landscape" or "Portrait", for example), allow settings to be offered from the measured conditions (average brightness, focus distance) in more or less automatic modes.

1.3.4.6. *Further, color*

It would not be possible to mention photography without addressing the sensitive issue of color, halfway between physics on the one hand and human perception on the other hand. We will do so in Chapter 5. We will show the complexity of any representation of a chromatic signal and the difficulty of defining color spaces that are simultaneously accurate with regard to the observed stimuli and sufficiently broad in order for any image to be exchanged and retouched without betraying the perception that the photographer had of it. We will see that technology offers several solutions which are not at all equivalent and that lead each to difficult compromises. We will also see why the photographer must imperatively worry about white balance.

1.4. Camera block diagram

Nonetheless, the photographic camera cannot be reduced to these major areas of physics, important as they are. The camera is a very complex technological tool which assembles multiple elementary components responsible for specific functions: to measure the distance to the target, to measure the distribution of energy, to ensure the conversion of the optical signal, to select the chromatic components, to archive the signal in memory, to stabilize the sensor during shooting, etc. It is through these features that one should also study the camera and it is according to this schema that part of this book is organized.

The elementary functions are grouped in the diagram of Figure 1.8. It can be recognized that:

– the optical block, a lens or most often an objective consisting of several lenses. It is responsible for achieving the reduced image of the scene. Its role is essential for the quality of the image and all the functions of the camera aim to make the best use of its capabilities. We will examine its function in Chapter 2;

– a sensor, the heart of the acquisition. In analog photography technology, it was film, but for this book it will be of course a CCD or CMOS semiconductor matrix. Chapter 3 will be dedicated to it;

– the range finder measures the distance of the objects in the scene. From its measurements the focus distance is derived. It has significantly evolved since the era of photography film. It will be examined in section 2.2, and we will come back to this subject in section 9.5;

– the photometer: it measures the received brightness of the scene (section 4.4.1). It is sometimes confused with the rangefinder because in some systems these two sensors work in a coupled manner (the contrast is measured there where the focus is achieved). It has similarly evolved in recent years;

– the shutter adjusts the duration of the exposure, which will be mechanical, electromechanical or electronic. It is a functionally essential accessory, but not widely publicized. We will discuss this point in section 9.4;

– the diaphragm associated with the shutter controls the amount of light received by the sensor during exposure. Its role is fundamental, but it has hardly evolved in recent years;

– finally, a moving device ensures the optical conjugation between the plane of the target object and that of the image on the sensor in accordance with the instructions provided by the range finder (section 2.2).

Other accessories are often associated with the camera, accessories that are often useful. We will also engage in their study. We will consider:

– optical filters (infrared and ultraviolet) and anti-aliasing, placed over the sensor (section 3.3.2);

– additional objectives or accessories that modify the properties of the lens to suit a particular purpose: converters/extenders, macro photography, etc. (section 9.7.1).

This functional description, viewed through the eyes of the photographer, must be supplemented with new components introduced by digital sensors. First of all, we will concentrate on the processor which controls all the features of the camera. On the one hand, it manages the information coming from the various sensors (range finders, photometers), on the other hand, it controls the settings: focus, aperture, exposure time. Finally, it ensures the proper functioning of the sensor and the dialog with the user. Its major function is to recover the signal originating from the measurement, to give it a shape as an image and in particular ensuring its compatibility with storage, transmission and archiving systems. The digital processor will be the subject of section 9.1.

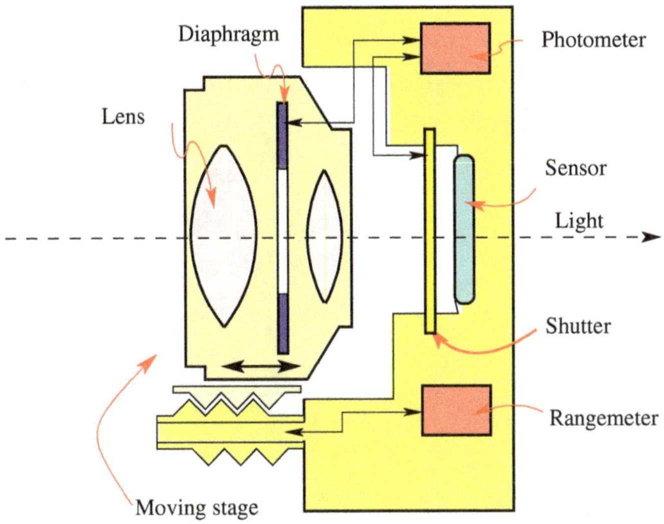

Figure 1.8. *Schematic layout of a camera, whether it be analog (captured by film) or digital (solid sensor)*

Along with the processor, numerous auxiliary features deserve our attention: the energy source, the display screen, and the memory. They will also find their place in Chapter 9.

Finally, we could not conclude this book without mentioning the algorithms and the software that can implement images. We will do so in Chapter 10. We will examine those that create images within the body of the camera, but also those who are transposed to the host computer and that allow the quality of images to be improved or the functionalities of the camera to be increased. This chapter gives an important place to very prospective developments that explore new domains of application for photography, domains of application made possible by the intensive usage of calculations to obtain images that sensors only are not capable of delivering: images with extended resolution or dynamics, images with perfect focus at any distance, images correcting the effects of camera movement. This chapter is an open door to the new field of computational imaging which prefigures the camera of the future.

2

The Photographic Objective Lens

The objective lens has always been considered to be the most important camera component. Certainly, the advent of digital cameras and the emergence of photographic features within systems dedicated to other roles (telephones and tablet computers) have drawn attention to other components, in particular to the sensor (that we will encounter again in Chapter 3) as well as to the numerous accessories of its automatism (that we will detail in Chapter 9). However, we will see in this chapter, that the key role of the lens cannot be denied. It is a complex component whose technical description requires complementary angles of analysis: we will first discuss image formation, in the geometric optics approach, the most relevant in photography, and according to the thin lens model which allows a large number of simple developments (sections 2.1–2.3), then briefly in the case of heterogeneous objectives (*fisheyes* that do not verify the Gauss conditions (paraxial approximation)) and in the case of thick centered systems, closer to real objective lenses (section 2.4). In section 2.6, we will then examine the role of diffraction to establish the limitations of the geometric approach. We will present the relations that can be established between the position of a point in the image and the position of the object which has given rise to this image in the observed world, and for this purpose we will discuss the basic calibration techniques (section 2.7). Finally, we will examine the aberrations that affect photographic lenses (section 2.8).

This chapter has a natural extension in Chapter 6 which covers image quality, since of course, the objective lens (and its settings) is largely responsible for it. It also extends to Chapter 9, dedicated to the various components that complement the camera, in which accessories that can be added to the objective will be presented (lenses and filters), or those that control its functioning (range finder, photometer, diaphragm, etc.).

2.1. Focusing

In this section we will examine some of the properties associated with the construction of images. First, we will examine the geometric elements that constitute the quality of the focusing, then those related to the depth of field. Then, we will examine the devices that ensure this focusing and certain consequences of their usage.

2.1.1. *From focusing to blurring*

We are faced with the approximation of a simple lens under Gauss optics (thus, in the case where the rays make small angles with the optical axis) with a focal length f and a diameter D, therefore, with an f-number $N = f/D$. P is the object upon which the focusing is exactly made. It is located at distance $p - f$ of the focal point F or at the distance p of the optical center O (Figure 2.1).

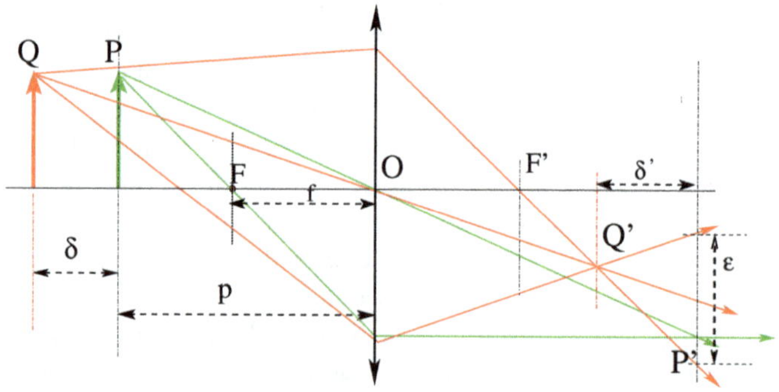

Figure 2.1. *Construction of an image of a sharp object, P, and a blurred object, Q. The point P is focused in P'. If it moves to Q, by δ its image will be formed in Q', at the distance δ', with $\delta' = \Gamma\delta$ (Γ being the longitudinal magnification, see equation [1.3]). The blur in the image plane of Q' will have a diameter ε*

Let us consider a second point Q at distance δ of P. Its image is formed by Q' in a plane at distance δ' of the plane containing P'. Its image is, therefore, blurred. In the geometric optics approximation (thus, in the absence of diffraction), the image in Q' is a disk of diameter ε (in the small angle approximation, the light cone intersection being based on the diaphragm).

Using the conjugation equations (equations [1.2]), yields:

$$\varepsilon = \frac{\delta f D}{(p+\delta)(p-f)} = \frac{\delta f^2}{N(p+\delta)(p-f)} \qquad [2.1]$$

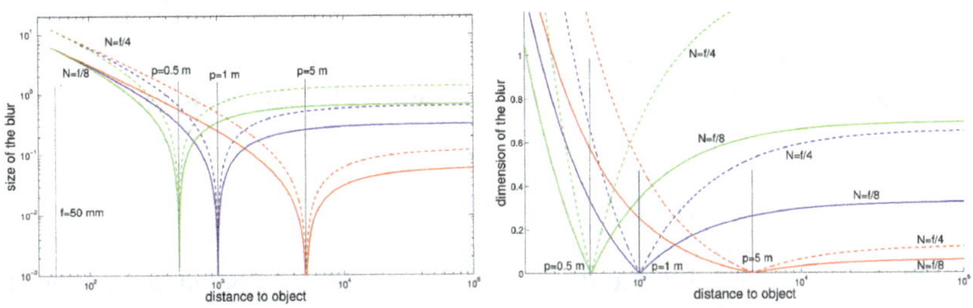

Figure 2.2. *Change in the size of the blur spot relative to the displacement along the optical axis. The scales on the two axes are in millimeters. The focal length is fixed at 50 mm. Three focusing distances are studied: 0.5, 1 and 5 m, as well as two apertures of the diaphragm: $N = f/4$ and $N = f/8$. On the left, in a log-log plot of the focal point F at a distance of 100 m, On the right, in a semi-log plot. For a color version of this figure, see www.iste.co.uk/maitre/pixel.zip*

Behaviors that are significant for any photographer can be observed (which are illustrated in Figure 2.2):

– starting at a point Q at infinity, the size of its blurred image is fixed by the focal length f, focus p and aperture N: $\varepsilon_\infty = fD/(p-f)$; this is the maximum extent of a blur of a point beyond the focal point;

– when it approximates the focal point, δ tends toward 0, the blur decreases (to 0 in the geometric approximation, but in practice toward the diffraction figure of the diaphragm, see section 2.6);

– when it comes closer still, it passes beyond the focal point position. The focusing decreases and the image of Q grows until it occupies the object focal plane. Its image in the focusing plane then has the dimension of the diaphragm $\varepsilon_f = D$ (the ray originated from Q is parallel in the image half-space). This is of course a case with no interest for the photographer, but this case defines the left limit ε_f of the curves 2.2, generally much higher than ε_∞, since $p \gg f$.

The main points of these findings relate to the central role of the focal length as well as of the numerical aperture. Thus, it can be seen that in ordinary photography (that is to say outside micrography), the important blurs are those located in front of the focal point, because those that are behind have a more limited extension. Examining Figure 2.2 on the right, which more accurately represents what is happening in the vicinity of the focal points, it can be seen that the behavior is almost linear on both sides of the location of the exact focal point, but that the variation is of greater magnitude when the object is closer.

We are going to examine in the next section, in more detail, the appearance of an image consisting of multiple objects, of which some are not focused.

2.1.2. *Focusing complex scenes*

Phenomena related to the depth of field are the daily bread and butter of any photographer who can take advantage of them to best suit his intentions. It is a phenomenon whose whys and wherefores are well understood and which, as we have seen, lends itself pretty well to a mathematical expression in simple situations.

The same cannot be said of complex situations, and in the following, we will examine how some situations, where certain objects are simultaneously sharp and others blurry (as that shown in Figure 2.3, on the left), can be interpreted.

We will choose a spatial coordinate system $\{x, y, z\}$ whose origin is at the optical center and where z is along the optical axis. In the small angle hypothesis, the planes at $z = cste$ are fronto-parallel planes (Figure 2.4). The variable z plays, therefore, the role of variables p or q of previous sections. The space object is characterized by its brightness in the direction of the camera (we will refer to it as $i(x, y, z)$) (see Chapter 4). After the formation of the image, the sensor receives a variable illumination which will form the image: $i'(x', y')$. The lens will be considered as perfect, that is without aberrations or losses, and we will ignore diffraction when we will not need it. The transversal magnification of the lens is G ($G = -f/(f + p)$) (see equation [1.3]), p being the object focus distance; G is a number usually much smaller than 1 in photography, generally negative, expressing that the image is inverted relative to the object.

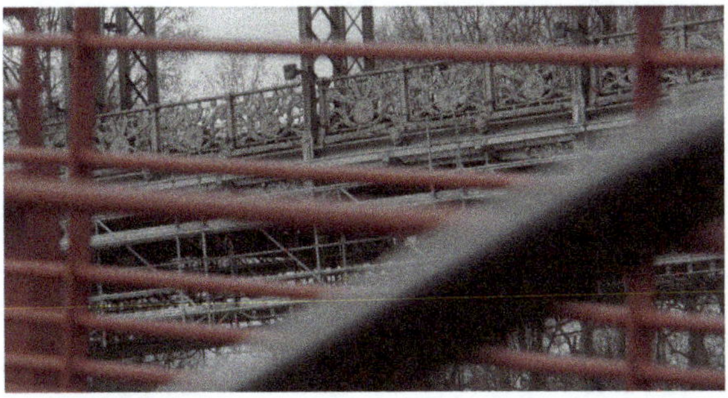

Figure 2.3. *Image formation is complex in the presence of blurring and partial masking. The foreground, if it is blurry, and only partially hides the background. Thus, in this image taken from a scaffolding, the tubes at middle distance are partially visible inside the blur of the black tube in the foreground and themselves do not completely hide the structures of the background by their blurring*

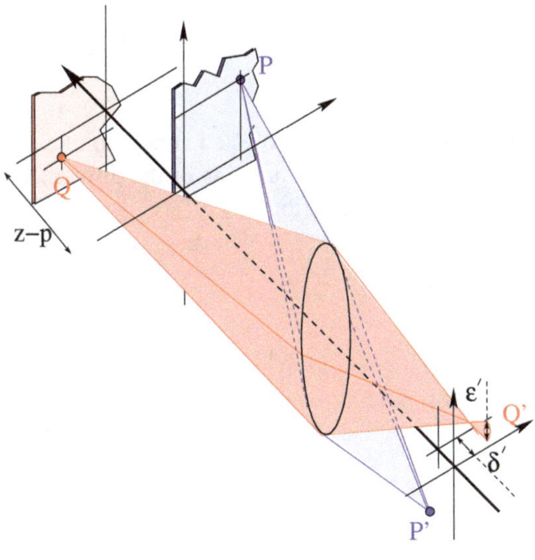

Figure 2.4. *Geometry of rays originating from 2 points, one P in the focusing plane, the other Q in a plane beyond. P' is the object in the image plane. In this plane, Q' is a homothetic spot of the entrance pupil, whose size is a function of the value $z - p$ between the two fronto-parallel planes containing P and Q*

First, we should recall the laws that govern the formation of an image $i'(x,y)$ in the case where all the points of the object are correctly focused (thus, such as $\forall(x,y), i(x,y,z) = i(x,y,p))$:

$$\begin{aligned} i'(x',y',p') &= i(Gx,Gy,p) * h(x',y') \\ &= \iint i(x'',y'',p)h(\frac{x'}{G} - x'', \frac{y'}{G} - y'')dx''dy'' \end{aligned} \qquad [2.2]$$

where h is the impulse response of the lens.

If we ignore the diffraction and lenses defects, and in the case of a perfect focusing, we have: $h(x,y) = \delta(x,y)$ and : $i'(x',y',p') = i(\frac{x'}{G}, \frac{y'}{G}, p)$, the image is just a version reduced by a factor G of what is happening in the object plane where the focusing is carried out $z = p$. Furthermore, we will observe the image only in the conjugate plane of p and therefore omit p' when expressing i'.

2.1.2.1. *Case of an object in a fronto-parallel plane*

We have extensively explained the case of the unfocused source point ($z \neq p$) in the previous sections.

A single source point $\delta(x,y,z)$ projected at position (x',y') gives for the image a spot whose diameter ε' is a function of the difference between z and the exact focusing distance p (equation [2.1]):

$$i'(x',y') = \frac{4}{\pi\varepsilon'^2}\text{circle}(\rho'/\varepsilon') \qquad [2.3]$$

where $\rho'^2 = x'^2 + y'^2$ and the multiplying coefficient expresses that the energy contained in point i is spread all over the spot. The diameter ε' is a function of the aperture N, focal length f, focus distance p and distance of the plane z, (equation [2.1]):

$$\varepsilon' = \frac{|z-p|f^2}{Nz(p-f)} \qquad [2.4]$$

The case where the entire object i is in this plane z, derived immediately from the previous case by convolution of all the points of the object that share the same impulse response. If the field $\mathcal{D}$ of the sensor has a size $[-X', X'; -Y', Y']$:

$$i'(x', y') = \frac{4}{\pi\varepsilon'^2}\, i(Gx, Gy, z) * \text{circle}(\rho'/\varepsilon')$$
$$= \frac{4}{\pi\varepsilon'^2} \int_{-X'}^{X'} \int_{-Y'}^{Y'} i(x'', y'', z)\, \text{circle}\left[(\frac{\rho'}{G} - \rho'')/\varepsilon'\right] dx''dy''$$

which is also explicitly written as:

$$i'(x', y') = \frac{4}{\pi}\left[\frac{Nz(p-f)}{(z-p)f^2}\right]^2$$
$$\times \iint_{\mathcal{D}} \text{circle}\left[\frac{Nz(p-f)\sqrt{(\frac{x'}{G} - x'')^2 + (\frac{y'}{G} - y'')^2}}{|z-p|f^2}\right]$$
$$\times i(x'', y'', z)dx''dy'' \qquad [2.5]$$

2.1.2.2. *Case of a scene without masking*

A scene composed of opaque objects is without masking, if any ray originated from a point of the scene in the direction of the field diaphragm of the sensor does not meet any other object of the scene.

In a scene without masking, all the points are thus seen at the widest aperture. Each contributes, therefore, to the construction of the image under the form of a spot whose dimension is dependent of its position (see Figure 2.5). Equation [2.5] is thus still valid, but the distance z is a function of x and y (that we clarify by denoting as $z_{x,y}$):

$$i'(x', y') = \frac{4}{\pi} \iint_{\mathcal{D}} \left[\frac{Nz_{x,y}(p-f)}{(z_{x,y}-p)f^2}\right]^2 i(x'', y'', z_{x,y})$$
$$\text{circle}\left[\frac{Nz_{x,y}(p-f)\sqrt{(\frac{x'}{G} - x'')^2 + (\frac{y'}{G} - y'')^2}}{|z_{x,y}-p|f^2}\right] dx''dy'' \qquad [2.6]$$

This new equation is thus no longer a convolution, since the impulse response is no longer constant for the whole image. It can no longer be calculated by fast Fourier transform algorithms. It cannot be inversed by linear filtering algorithms anymore. But, the exact calculation of the image can still be achieved explicitly if, for any point of the scene, its amplitude and distance from the optical center are known.

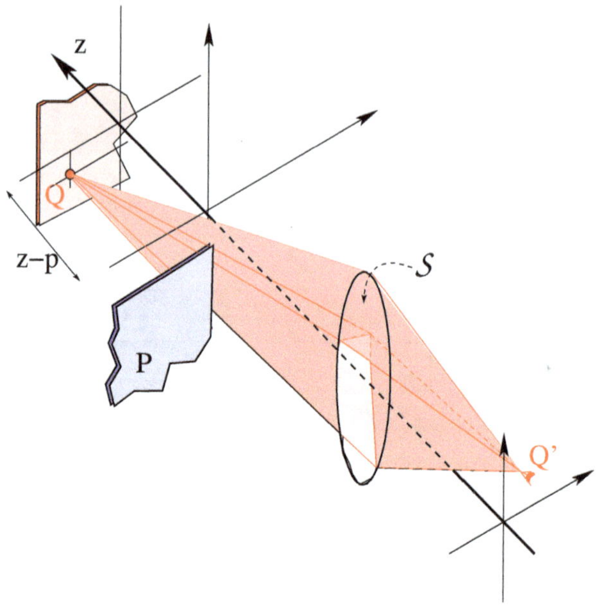

Figure 2.5. *Case of masking by a foreground object. The source point Q only sees part of the entrance pupil. It is this part that will determine the shape of the blurred image of Q*

2.1.2.3. *Case of a scene with masking*

This case is that of ordinary photography. It led to a much more complex formulation than the previous one.

As a result, it is important to consider for each point $P(x, y, z)$ of the scene, all the rays that originate therefrom, and to examine whether or not they meet an obstacle before traversing the field diaphragm:

– if no ray enters the objective, the point P is not seen and does not contribute to the image;

– if no ray meets any obstacle, the contribution of the point to the image is that which we have just considered in the previous section: it is centered around the image point which is convolved by a spot of diameter ε' and intensity proportional to $4/\pi\varepsilon'^2$;

– if certain rays are stopped by objects placed between the point and objective, the contribution of this point should consider it:

- rays that traverse the lens cutoff a subpupil $\mathcal{S}_{x,y}$ on the entrance pupil,

- the image of P, always in the geometric optics approximation, is then no longer a circle of diameter $\varepsilon'(z)$ as in equation [2.4] (image of the diaphragm D), but the homothetic form $\mathcal{S}'_{x',y'}$ of $\mathcal{S}_{x,y}$ but inverted, in the ratio $\varepsilon'(z)/D$,

- the amplitude of this image fraction will always be affected by the attenuation $4/\pi\varepsilon'^2$, although only a portion of this energy arrives at the sensor.

All the points P of the field add their contributions at a point (x', y') of the sensor if these contributions are not zero, that is if their image extends to (x', y'). Thus, if focusing is achieved at a distance p, the objects such as $z < p$ are affected by blurring. Their borders are particularly unclear and the most distant objects, very close to this edge, will be visible through the blur of the close object, be they clear or blurred. This explains the transparency areas of Figure 2.3. Note, however, that if a foreground object is sharp, then any object located behind this object will not be able to contribute to its image. In effect, by the principle of the inverse return of light, all the rays issued from the image point will converge to the object point without meeting an obstacle on the way.

We can see that it is now necessary to know for each point in the field, not only its distance to the camera lens, but also the subpupil with which it sees the objective in order to carry out an exact calculation. In image processing, where these elements are not known, these calculations could not be performed to explain the image. In image synthesis, it is on quite similar schemes that rendering engines operate, aiming for high-image quality. Instead of analytically determining the subpupil $\mathcal{S}'_{x',y'}$, they instead carry out either a fine meshing of the entrance pupil or send rays that statistically make it possible to determine a reasonable area of this subpupil [SIL 94, COH 93, PHA 10].

Among the consequences that should be expected from these partial maskings, one of the most surprising is the possible presence in the image of colors that neither exists in the background object, or in that of the mask, but resulting from the combination of their luminous flux at the location where they are superimposed.

2.2. Depth of field

These lines about the focusing of complex scenes raise as a matter of course a number of questions about the depth of field of the optical system, that is the distance between the first sharp points and the last sharp points along the optical axis, from the camera to infinity. To this end, a focusing criterion is chosen. This is, for example, a threshold value of the dimension ε of the figure of a point that we have used in previous sections.

We will encounter again here some rules used by photographers. We will have to distinguish between cases where the object is distant and the case of macrophotography. Finally, we will introduce widely used hyperfocal expressions.

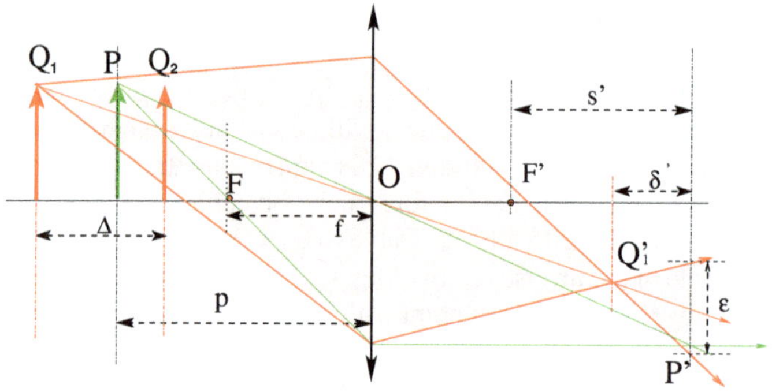

Figure 2.6. *Depth of field: the focus is made on the image of P. A point Q_1 gives an image Q_1' in a plane at distance δ' of the blurred image plane, and in this image plane a spot of diameter ε, similarly gives a point Q_2 on the other side of P*

By referring to Figure 2.6, the aim is, therefore, to determine the distance denoted by Δ between the points Q_1 and Q_2, one upstream of the focal point P, the other downstream, such that the blurs on Q_1 and Q_2 are equal to the threshold value ε.

Using the formulas of Newton, it yields:

$$PQ_1 = f^2(1/s' - 1/s_1') = \frac{f^2\delta'}{s'(s' - \delta')}$$

$$PQ_2 = f^2(1/s_2' - 1/s') = \frac{f^2\delta'}{s'(s' + \delta')}$$

Therefore, it can be derived that:

$$\Delta = \frac{2f^2\delta'}{s'^2 - \delta'^2}$$

where δ' is related to the aperture D of the diaphragm and to the blur ε by: $\delta' = \frac{\varepsilon q'}{D}$. Moreover:

$$p' = \frac{pf}{p-f} \qquad \text{and} \qquad s' = \frac{f^2}{s} = \frac{f^2}{p-f}$$

By expressing Δ from the f-number of the lens $N = f/D$ and the only variable p, it gives:

$$\Delta = \frac{2\varepsilon Dpf(p-f)}{D^2f^2 - \varepsilon^2(p-f)^2} = \frac{2\varepsilon Npf^2(p-f)}{f^4 - \varepsilon^2(p-f)^2N^2} \qquad [2.7]$$

The positions of Q_1 and Q_2 are also given by:

$$q_1 = p + \frac{\varepsilon N(p-f)p}{f^2 - \varepsilon(p-f)N} \qquad \text{and} \qquad q_2 = p - \frac{\varepsilon N(p-f)p}{f^2 + \varepsilon(p-f)N} \qquad [2.8]$$

These equations are subject to the only hypothesis that Gauss conditions be verified.

If faced with the conditions of a very small blur ($\delta' \ll s'$), then equation [2.7] becomes:

$$\Delta = \frac{2\varepsilon p(p-f)}{Df} \qquad \text{or} \qquad \Delta = \frac{2\varepsilon Np(p-f)}{f^2} \qquad [2.9]$$

2.2.1. *Long-distance photography*

In conventional photography, it is usual to work with large values of p ($p \gg f$). Consequently, the conventional formula emerges again (reported in section 1.3.4):

$$\Delta = \frac{2\varepsilon Np^2}{f^2} \qquad [2.10]$$

that certainly recalls that we must work at small numerical apertures for the depth of field to increase. Photographic lenses with fixed focal length generally offer a scale that makes it possible to determine the focal range corresponding to an aperture and a fixed viewing distance (see Figure 2.7). This depth of field is calculated from an arbitrary value of ε, often determined from the maximal angular resolution of the eye (1/1,000 approximately), applied to a 24×36 mm^2 negative.

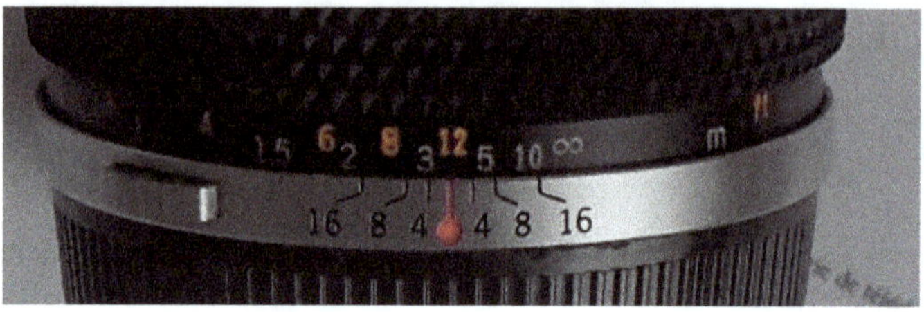

Figure 2.7. *On fixed focal length lenses, a scale allows the depth of field to be determined for a given focus distance and a given aperture*

It can be noted that P is not the middle of Q_1Q_2 (what we had already set out in Chapter 1), but it verifies: $p^2 = q_1q_2$.

2.2.2. *Macrophotography*

The field of photography concerning small objects is very cryptic. Experts distinguish:

– microscopic photography or photomicrography which provides magnifications greater than 10 using microscope frames coupled to photo cameras;

– macrophotography, for magnifications between 1 and 10 that make use of cameras bodies, fitted with bellows or extension tubes (see section 9.7.1);

– proxiphotography, for magnifications comprised between 1/10 and 1.

We will use the same term: macrophotography for these last two areas that share the same schemes of image formation.

In macrophotography, we work with the object very close to the focal plane ($p \approx f$) and the position of the object is adjusted in order to define the magnification. Knowing that the radial magnification is equal to: $G = s/f$, equation [2.9] becomes:

$$\Delta = \frac{2\varepsilon N}{G} \qquad [2.11]$$

Here again, an experimentally well-known fact can be observed: high magnification is accompanied by a shallow depth of field. It can also be recalled that formula [2.10], which is often given without conditions, is no longer applicable.

2.2.3. *Hyperfocal*

The focusing distance (denoted by p_H) is called hyperfocal for which the depth of field extends to infinity. It actually depends on the aperture of the lens N and on the tolerance ε that is imposed on the blur. From equations [2.8], by letting q_1 tend to infinity, it yields:

$$p_H = f + \frac{f^2}{\varepsilon N} \qquad [2.12]$$

When the focus distance p is fixed at p_H, the focus expands, within a tolerance ε, from Q_2 to infinity. It yields Q_2 with (equation [2.8]):

$$q_2 = \frac{1}{2}\left(f + \frac{f^2}{\varepsilon N}\right) = \frac{p_H}{2}$$

This relationship expresses that when focusing is set to hyperfocal, the depth of field extends from infinity to half of the hyperfocal.

This relation is also used the other way around: knowing that the aim is to ensure focusing from a distance x to infinity, the focus will be at $2x$, and from the product εN we can derive how to choose the f-number N to ensure a given blur ε.

Lenses with short focal length have a very small hyperfocal. Some cheap cameras take advantage of this to propose lenses adjusted in the factory to the hyperfocal, but without focusing. This was the case of very popular cameras such as the Kodak Brownie; it has become once again fairly common, with compacts, mobile phones or tablets.

2.3. Angle of view

The angle of view is the extension, in degrees, of the scene recorded by the sensor. It is usually characterized by a single number, the largest of the two horizontal or vertical angles, except for very wide angle objectives (*fisheye*) which are sometimes described by the angle associated with the diagonal. To this end, it is necessary to know the format of the sensor since sensors with very different formats can be found on the market (we can refer, for example, to Figure 3.1 to illustrate this variety).

We have seen that in practice measures were taken for the field aperture to completely cover the sensor in order to avoid vignetting. Under these conditions, it is, therefore, the sensor which determines the angle of view. If the large dimension of the sensor is l and the focal length f, the half angle of view is equal to θ such that:

$$\operatorname{tg}\theta = l\,/\,2f \qquad [2.13]$$

This equation is only valid for objectives following the small-angle hypothesis of Gauss optics. It is still verifiable wide field objectives of ordinary usage which are very correctly corrected in order to remain in this projection model known as equidistant, but it is at fault for very wide angle objectives (*fisheyes*) as we will see later.

2.3.1. *Angle of view and human visual system*

It is sometimes mentioned that a film camera using a 135 film (therefore, such that the image field measures $24\,mm \times 36\,mm$), equipped with a 50 mm lens, is by means of its angle of view equivalent to human vision and this assertion is sometimes brought forward to justify the standard role of this format. This equivalence is actually not simple and deserves a discussion.

By means of equation [2.13], a 24×36 camera with a 50 mm lens offers a total angle of view of $27° \times 40°$, either in "landscape" format (the large horizontal axis), or in "portrait" format.

What is the human angle of view? This is a delicate issue discussed in detail in physiological optics works (for example, [LEG 67]), which has no simple answer since visual functions are complex. The total angle of view is in the order of 150° horizontally and 120° vertically, but it is strongly asymmetrical (stronger on the temporal side than on the nasal side and stronger downward than upward), it is very heterogeneous (its performance in resolution as in color discrimination decreases when moving away from the fovea) and the right-left superposition area is smaller than the full angle. The high-resolution area of the eye, that on which we form the image of the objects that we examine (the fovea) is on the contrary much smaller, its angle not exceeding 3–5°. Then how do we qualify an equivalence? In practice, it refers to a visual comfort zone during the recognition of shapes that, for example, allows symbols to be read. This angle of view covers approximately 40° as it can be seen in Figure 2.8. These are the angles of view that appear as very "natural" to a human observer, but it is clear that nothing is really rigorous.

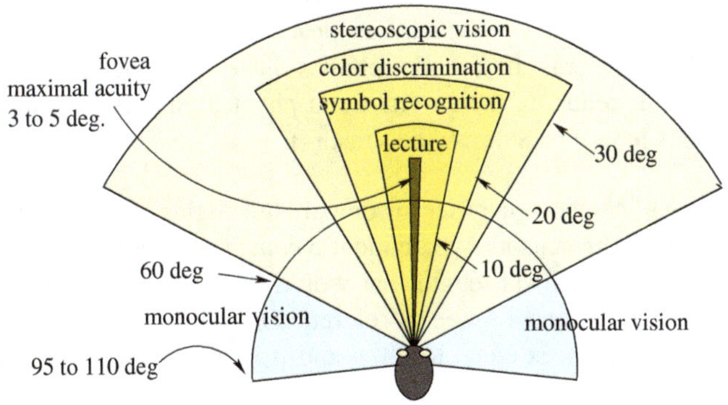

Figure 2.8. *The various angles of view of human vision (©Wikipedia.fr, article "champ de vision")*

2.3.2. *Angle of view and focal length*

If an objective with a focal length of 18 mm is mounted on a camera ordinarily fitted with a 50 mm objective, a much wider angle will be obtained in the image given by the sensor. It can be said that the objective is "wide-angle". Conversely, the use of a long focal length: 150 or 300 mm (telephoto lens) will give a narrow field of vision. The widest angles, with the objectives known as *fisheyes*, exceed 150°, whereas telephoto lenses with very long focal lengths have angles reduced to a few degrees.

The size of the sensor is a very important parameter of any camera. It directly affects the choice of the lenses that are used. The question of the resolution of the sensor which, combined with its size, allows the number of pixels to be defined in the image will be addressed further. However, an important difference between film-based and electronic sensors should be noted here. It is not very difficult to spread a photographic emulsion on large surfaces, for example, on 120-format film (6×6 cm-sized images), and even on "4×5 inch" film ($10\times$ 12 cm images); the gain in image quality, when the quality of the lenses allows it, is then directly related to the size of the film and professionals have, therefore, often used large formats when there are no cluttering problems. On the contrary, the manufacturing of large-sized solid sensors provides great technological difficulties because these monoblock components require extremely accurate microelectronics design methods (photocomposition, masking, implantation, etc.). It is not surprising that the industry of electronic imaging has very largely favored small formats (smaller than the standard 24×36) which has promoted the emergence of shorter focal lengths. Cameras in mobile phones have focal lengths of several millimeters, thus reducing their weight and physical size (for example, most mobile phones have 4 or 8 mm-wide sensors).

In the case of cameras, in order to compare the angles of view of the images obtained with certain sensors, we define the *conversion factor* which makes it possible to match the focal length that would give an equivalent angle with a $24\ mm \times 36\ mm$ sensor taken as reference to a lens f used with a sensor of width d. This factor is equal to $36/d$ and it is often comprised between 1 and 2.

Thus, for a camera A whose sensor measures 16 mm $\times$ 24 mm, the conversion factor is 1.5 and a 28–300 mm zoom mounted on this camera is the equivalent of a 42–450 mm mounted on a film lens reflex (SLR) B. Two images, one taken with A and a focal length of 33.3 mm, the other with B and a focal length of 50 mm will present two images exactly superposable: same framing, same magnification and same geometry within the image. If the numerical apertures are the same (for example, f/4), exposure times (e.g. $1/100th$) and identical sensitivities (for example, ISO 200), then the images will only differ by:

– the fine structure of the image (photographic grain in one case and pixels in the other);

– defects due to lenses (chromatic and geometric aberrations, and internal reflections);

– the reproduction of contrasts (linear for the digital sensor and nonlinear for film);

– noise affecting the gray levels.

2.4. Centered systems

The optical system of a photographic camera is not limited to a single lens as we have assumed so far (except for very cheap cameras), but consists of a *"centered system"*, an assembly of lenses sharing the same axis. This arrangement provides an improved image quality by reducing a good number of defects of thin lenses, in particular geometric aberrations and achromaticity (see section 2.8). In addition, these assemblies sometimes make it possible to vary the focal length f and thus to achieve zoom functions. Centered systems are said to be "dioptric" if they only make use of lenses. If they also make use of mirrors, they are called "catadioptric".

In photography, the vast majority of objective lenses are dioptric, only very long focal length objectives utilize mirrors in order to be less cumbersome. A word should be said of these folded lenses before exclusively dedicating ourselves to dioptric lenses later in this chapter.

Catadioptric lenses often make use of Cassegrain-type beam folding schemes with mirrors. The entrance lens of the objective is annular, its center being occupied by a mirror said to be secondary. The beam undergoes two successive reflections on the primary mirror, also annular, and then on the secondary mirror, circular. These mirrors may be plane or spherical. The beam then comes out in the central part of the annular mirror to reach the sensor. The size of the objective is approximately divided by three along the axis, at the cost of a substantial increase in the diameter. By design, such an assembly often has a fixed aperture and only the exposure time has to be adjusted (as well as the sensitivity of the film). The impulse response of such a system differs significantly from the Airy disk of a regular objective lens (equation [2.9]). Such objectives are reserved for very special purposes: astronomy, sports, animal pictures, etc.

From now on, we will only refer to dioptric systems[1].

The starting equation of any optical formula is the relation between the focal length f of a spherical lens (both its sides are spherical caps) to the curvature radii R_1 and R_2 of the sides (these values are algebraic to

1 This section has to give full credit to Pierre Toscani's excellent site (www.pierretoscani.com), which presents a very complete and remarkably illustrated analysis of the main lenses on the market, specifically detailing the role of each optical element and by tracing the paths of light rays.

distinguish object sides and image as well as convex or concave lenses), to its thickness e on the optical axis as well as to the index n of glass:

$$\frac{1}{f} = (n-1)\left[\frac{1}{R_1} + \frac{1}{R_2} + \frac{e(n-1)}{nR_1R_2}\right] \qquad [2.14]$$

the variable $\frac{1}{f}$ being called the vergence of the lens.

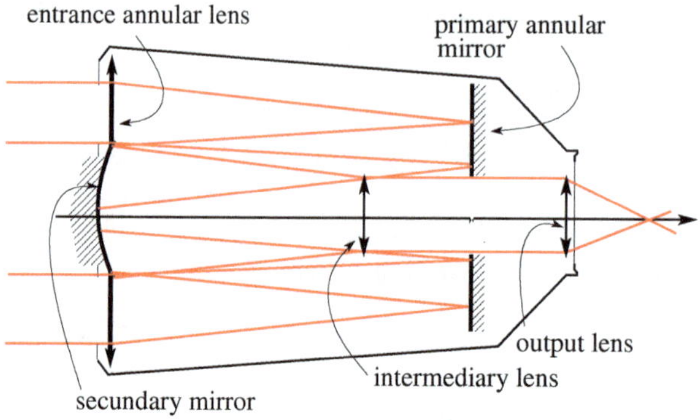

Figure 2.9. *Example of a catadioptric objective: Cassegrain-type assembly allowing the length of a lens to be reduced by a factor of the order of 3 by utilizing an annular lens at the entrance and two mirrors reflecting the beam, one of which is also annular*

2.4.1. *Of the importance of glasses in lenses*

The material constituting lenses (which defines the value of n) is, at least as much as their geometry, the important element of their optical properties. A high value of n can avoid overly significant curvatures, which cause aberration. Recently, new materials have emerged and are being used to manufacture lenses, fluoropolymers for liquid lenses[2] and polycarbonates or

2 Liquid lenses consist of a spherical meniscus (or interface) consisting of a liquid placed on a flat surface (usually a fluoropolymer). The wetting properties of the liquid drive the variation of the curvature of the interface under the effect of electrostatic, piezoelectric or acoustic forces [YEH 11]. By their ability to vary in shape without mechanical displacement, these lenses are particularly attractive for applications in mobile phones, for example.

polymers (known commercially as organic glasses)[3] for objective lenses of consumer systems. But, while these materials present numerous advantages, they still offer insufficient optical qualities to compete with mineral lenses on the market of quality objectives.

It is therefore of mineral glass that most of the photographic lenses of the market are made, with the notable exception of some very special elements made of fluorite. The designation mineral comes from silicon oxide (silica) which forms the basic component of glass, the additives being heavy metal salts. Glass is an amorphous compound, that is to say that the atomic structure is not inserted into a regular crystallographic mesh; it originates from the liquid phase where disorder is maintained below the fusion temperature, in a state precisely qualified as glassy. In contrast, fluorite is a cubic mesh crystal of calcium fluoride. The glassy state of the silica matrix allows manufacturers to add, in a very flexible manner, a large number of materials that give glass its optical qualities. The mixing is carried out at high temperatures, and is followed by a highly controlled cooling stage which allows both the material forming and the control of its surface states.

Additional components are numerous and varied: metal oxides (lead, barium, chromium and titanium), phosphate, chalcogenide or halides, etc. and each glass can use dozens of them. The result is a wide variety of lenses, all with good transparency qualities in the visible range, but with particularities that are essentially measured by two numbers:

– the refraction index, denoted by n, which is the ratio of the speed of light in vacuum to its velocity in the material;

– the constringence (or Abbe number, denoted by ν) which expresses the variations of the refraction index depending on the wavelength.

Several attempts to specify the reference wavelengths have not been fully successful and several wavelengths can be considered as candidates for

3 Organic glass is a plastic material with a very high transparency, used in optical sectors as well as in instrumentation in situations where the weight is a relevant factor (its relative density is less than 1.4, while mineral glass has a relative density comprised between 2.5 and 4). They have very good mechanical properties and can, therefore, easily constitute aspherical profiles. They can be coated on the surface as with glass and doped to provide, for example, index gradient profiles. Although highly resistant to shocks, they are, however, much softer than glass and are thus sensitive to scratches. They offer a low index latitude (from 1.5 to 1.75 only) and a strong chromatic dispersion (constringence between 30 and 60).

standardization. By chance, they lead to slightly different values. These are:

– in blue, the hydrogen line at 486.1 nm (denoted by F);

– in yellow, the helium line at 587.6 nm (denoted by d), or the doublet of sodium at 589.3 nm (denoted by D);

– in red, the hydrogen line at 656.3 nm (denoted by C).

The main refraction index n is fixed by the line d (respectively line D) and then designated by n_d (respectively n_D), generally two very close numbers; and the constringence is optionally expressed by one of the two equations:

$$\nu_d = \frac{n_d - 1}{n_F - n_C} \qquad \text{or} \qquad \nu_D = \frac{n_D - 1}{n_F - n_C} \qquad [2.15]$$

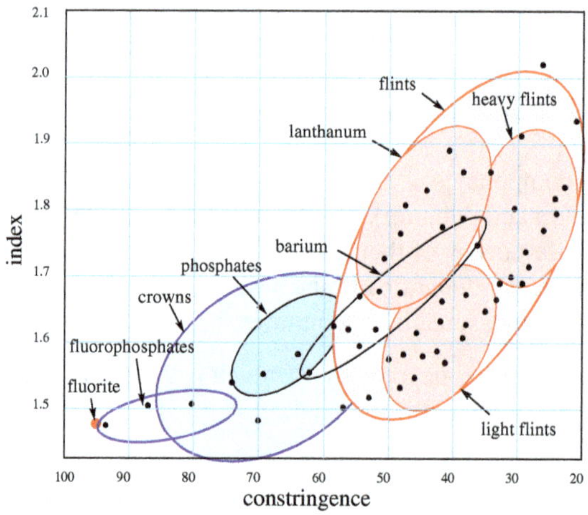

Figure 2.10. *Optical glasses distribution in the diagram (constringence and refraction index). Subdomains have been represented, where the influence of the components can be seen: lanthanum, phosphate or barium to modify the parameters for a given glass. Furthermore, light flints can be distinguished from heavy flints. Fluorite was represented at the end of the domain of the crowns*

A glass is, therefore, referenced on a two-dimensional diagram, depending on its main refraction index (glass indices range from 1.4 to 2) and on its constringence (it varies between 20 and 90) (see Figure 2.10). Strongly dispersive glasses have a low constringence, less than 50. They are called *flints*. These are heavy glasses, generally with a strong refraction, constituted of potassium and lead silicates, titanium and lanthanum oxides, barium. The others are *crowns*. Being lighter glasses, they thus have a constringence greater than 50, which can culminate above 80. Highly dispersive, they have a low index that made them rather unsuitable for use as optical components. They will be used again in centered systems in association with flints to correct their dispersion. They are obtained using potassium and sodium silicates.

We must mention fluorite again, which is presented in a singular manner in the diagram (constringence and index). With a very low index ($n_d = 1.43$) but a very high constringence ($\nu_d = 95$), this natural crystal appears like a "super-crown" which has played an important role in high-quality optics. Unfortunately, it is very expensive because, while its production in laboratory is possible, it remains very time-consuming (several weeks of chemical treatment).

Constringence and the main refraction index are complemented by the variation curve of the index n depending on the wavelength λ. This curve $n(\lambda)$ is measured experimentally. It has received more or less developed analytical forms [FOW 90]:

– Cauchy's formula: $n(\lambda) = A + \frac{B}{\lambda^2}$

– Sellmeyer's formula:

$$n(\lambda)^2 = 1 + \frac{B_1\lambda^2}{\lambda^2 - C_1} + \frac{B_2\lambda^2}{\lambda^2 - C_2} + \frac{B_3\lambda^2}{\lambda^2 - C_3} \qquad [2.16]$$

where the B_i and C_i are provided by manufacturers.

As shown in Figure 2.10, all the constringences are positive, therefore, all the lenses produce a higher deviation from the smaller wavelengths (the blue is more deviated than the red). If the objective is to compensate for this chromatic deviation, it is not possible to just vary the indices, it is also important to adjust the geometry of the interfaces in order to change the vergences (equation [2.14]).

2.4.2. *Chromatic corrections*

If the aim is to correct a lens such that, for two wavelengths (one blue at one end of the spectrum and one red at the other end of the spectrum), the system presents the same vergence, the result is a small dispersion over the whole spectrum and provided that [PER 94]:

$$f_1 \nu_{D_2} = -f_2 \nu_{D_1} \qquad [2.17]$$

Two lenses of different geometries must, therefore, be combined, one in flint and the other in crown, according to an arrangement called a doublet. The lens of the highest vergence (in absolute value) dictates the vergence of the whole assembly. It is desirable that this lens has the strongest constringence in order to be barely dispersive. As a result, a convergent achromatic doublet is preferentially constituted by a convergent in crown, followed by a divergent in flint (Figure 2.11). Nonetheless, the stigmatism is only correct for the two wavelengths used in the calculation. Some dispersion still remains for the other wavelengths which is reflected by a spreading along the optical axis of the focalizations of the various wavelengths. This is what is called the secondary "spectrum" of the doublet, responsible for small defects of chromaticism. A more complex arrangement using a triplet of lenses to even further reduce the secondary spectrum by canceling the chromatic dispersion for a third wavelength (these triplets are called apochromatic).

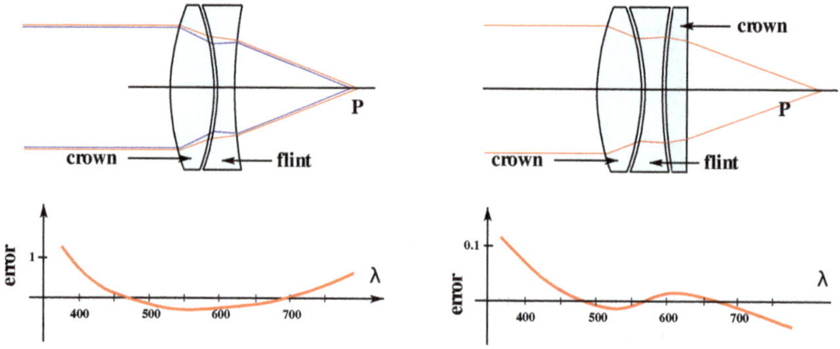

Figure 2.11. *On the left, convergent achromatic doublet "crown + flint" and the dispersion curve of the image focus relatively to the wavelength. The deviation between the point P and the convergence point of a wavelength λ constitutes the secondary spectrum of the doublet, responsible for a longitudinal chromatic residual aberration (the error is zero only in the case of the two chosen wavelengths). On the right, the same happens for an apochromatic triplet, the secondary spectrum is almost zero (it should be noted that the scales are different on both diagrams)*

	Curvature radius of the face R	Distance from the apex to the following apex d	Constringence ν_D	Refraction index of the medium found n_D
1	340.6588	4.2	64.14	1.516330
2	-340.6588	3	0	1
3	66.1639	1.8	26.3	1.784696
4	45.8341	8.8	81.61	1.486999
5	∞	6.3	0	1
6	...			
	...			

Table 2.1. *Table of the optical formula of a spherical central objective: it contains the list of all the interfaces found from the source to the sensor. The first and the last elements are usually air. They do not always appear. The apex is the entry point of the optical axis on the interface. The curvatures are positive if the centers of the spheres are in the image space. The diameters of the optics are not carried, nor surface coatings. A symmetrical biconvex lens can be recognized in the triplet given in the example, of the crown type, 3 mm apart from a convergent achromatic doublet (crown + flint) with a flat exit face, at a distance of 6.3 mm from the rest of the lens. If a part is mobile, the distance is replaced by a variable whose domain of variation is specified after the table*

2.4.3. *The choice of an optical system*

A spherical central system is characterized by the properties of each lens that it comprises. It is thus described by a succession of quadruplets, according to Table 2.1.

Software programs are able to calculate all the light rays, for any wavelength, from the knowledge of these elements. However, the cardinal elements of the objective can be derived analytically [SMI 90, FOW 90], often by a matrix approach:

– its focal distances, object and image;

– its main planes: conjugated planes with a transverse magnification of 1 whose knowledge enables the tracing of foci (the object focus is at distance f upstream of the main object plane and the image focus at f downstream the main image plane);

– its nodal points: totaling a number of 2 (one object and one image), they are equivalent to the optical center.

The formation of an image by a centered system verifies the same conjugation equations (Newton and Descartes formulas, equations [1.2]) as the thin lenses, but in practice, the cardinal elements being often unknown (with the exception of the focal length that is displayed on the adjustment ring), they can be used to focus by using these formulas. Focusing is, therefore, carried out by observing the image on a screen, or the system makes the adjustments from measurements from its rangefinder.

How does this focusing affect the disposition of the lenses?

We will look first at lenses with fixed focal length. The conjugate equation $(1/p + 1/p' = 1/f)$ tells us that if p varies, we can maintain the conjugate by acting on p', on f or on both at the same time.

In most objectives, the first solution is the one adopted: the whole optical block is shifted and as a result the flange distance varies. This solution is mechanically easy to implement, especially when the focal length is short because then the movements that must be done are minimal. By moving the entire optical block, the geometry of the nearly parallel rays that form the image when the object is very distant can be kept constant. Thus, the optimization which was achieved from the correction of the aberrations is maintained (Figure 2.12).

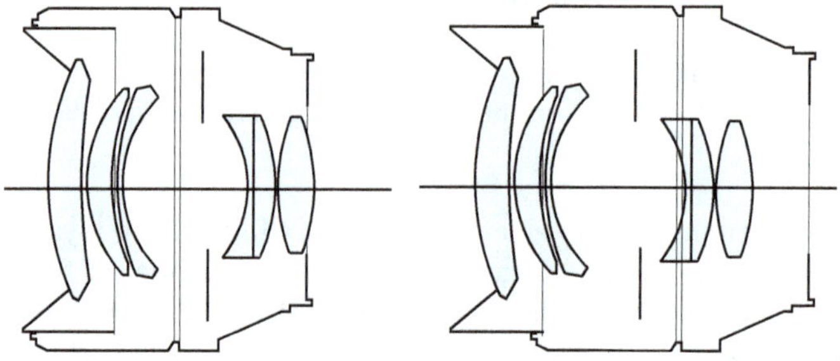

Figure 2.12. *Focus adjustment of a fixed 50 mm focal length lens. The focus between an object at infinity (on the left) and a very close object (at a distance of around 50 cm, on the right) is done by moving the whole optical block, here a displacement of about 10 mm*

With regard to the nearest objects and taking into account the aperture of the lens, a movement of the block can cause a variation of the angles of incidence. A compensation lens can then be used within the lens train in order to adjust to closer objects. The movement of this lens is paired to that of the objective but it is nonetheless different from it (as a result, it is sometimes called floating lens).

The solution consisting of transporting the lens overall is, however, not always the one adopted with long focal length objectives (telephoto lenses). For these objectives, the longitudinal magnification is very small, that is to say that a large displacement of the image plane is necessary to accompany a modest shift of the object, especially when it comes significantly closer. Moreover, the size of the lenses is often important which makes their displacement in groups a problem. It is then preferable to modify the focal length f to ensure the conjugation between object and image. In this case, even an objective known to have a fixed focal length can present different focal lengths according to the focus distance.

Telephoto lenses are often built on the model of telescopes, with a train of converging lenses at the front of the objective and a divergent train near the photodetector. The easiest solution to adjust the focal length consists of moving the frontal block. This nevertheless presents several defects. The first is that the size of the objective varies, and possibly in large proportions if the aim is to accommodate objects relatively close. A second defect arises from the weight and the size of the front lenses that, therefore, require powerful engines. For this reason, the second solution that consists of varying the distance between the front block and rear block is preferable, or a more complex but more effective solution in terms of image quality that consists of moving some intermediate lenses only (Figure 2.13).

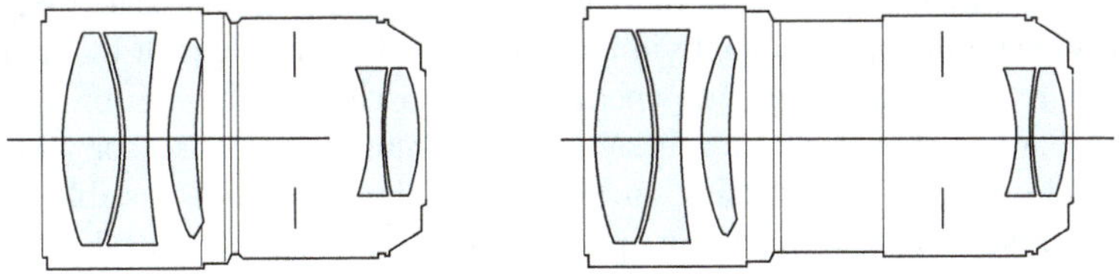

Figure 2.13. *Focusing of a long focal length objective (200 mm) by the sole movement of the frontal block (converging), the rear divergent block is fixed. In forward position, the focus is done at about 2 m. It corresponds to a displacement of the front lens of 25 mm, and an effective variation of the focal length from 200 to 175 mm*

In the case of zoom lenses, the problem that arises is different. As a matter of fact, it is desirable that focusing could change independently of the focal length and vice versa. We will examine this case a bit later on.

2.4.4. *Diaphragms and apertures*

2.4.4.1. *Diaphragm of a centered system*

In a central system, each lens introduces its own diaphragm. It is shown [PER 94] that a cascading combination of these diaphragms defines two important limitations for a lens:

1) an *aperture diaphragm* that controls the energy entering the unit. This is the one that is available to the user through the intermediary of a ring, or which is controlled by the system of the camera;

2) a *field iris diaphragm* (sometimes they may be several) which limits the extension out of the axis of the point objects that will be imaged. This field diaphragm is in practice defined such that all the sensor points are, therefore, images of points of the field being seen.

The aperture diaphragm is defined for an object point located on the optical axis. It ensures that all the rays it lets pass through will effectively reach the image plane. However, for an off-axis point object, not all the rays allowed by the field diaphragm will reach the screen: some rays will be stopped by internal diaphragms. An effect named *optical vignetting* can be observed: image areas appearing darker than object points with the same brightness located on the axis. Vignetting appears mainly in the corners and at the edges of the image. It is more significant at greater apertures. It is usually negligible for quality optics but can show up when these lenses are mounted on a sensor with an angle too wide, or when using an inadequate lens hood. For variable focal length lenses (zoom), vignetting appears usually at very short focal lengths, sometimes at very long focal lengths. In practice, vignetting is reduced by shutting the aperture diaphragm. Vignetting is the subject of section 2.8.4.

The image of the aperture diaphragm in the object space is called the entrance pupil. In the image space, its image is the exit pupil of the optical system[4].

4 The images of the field diaphragm are also called entrance and exit windows.

2.4.4.2. *Apertures*

We have briefly discussed the concept of aperture in section 1.3.4. We will recall this point to clarify some vocabulary elements and bring forward a few equations which will be useful later.

The aperture D of a lens is defined as the diameter of its entry pupil.

The f-number N is a characteristic of an optical system, defined as the ratio of the focal length and diameter of the entry pupil: $N = f/D$. It is a very important number in photography as we have seen, because associated with the exposure time, it controls the energy that the formed image will receive.

The numerical aperture N_n attached to a point P is another concept, used especially in microscopy, to express the angular resolution of a lens. It is defined by the formula: $N_n = n\sin(\theta)$ where n is the index of the propagation medium ($n = 1$ for photography) and $\sin(\theta)$ is the half angle under which point P sees the diaphragm (see Figure 2.14). The angle of resolution of the optical system is linked to the numerical aperture and the wavelength by a relation in $\lambda/2N_n$. The numerical aperture usually depends on the position of point P.

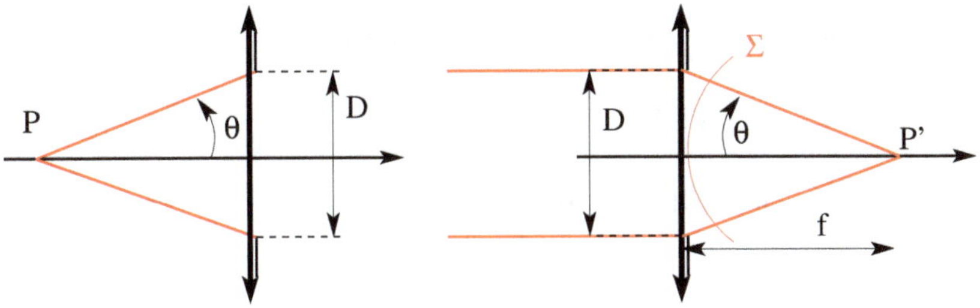

Figure 2.14. *On the left, definition of the numerical aperture of a lens seen from a point P :* $N_n = n\sin(\theta)$. *On the right, formation of the image of a point at infinity*

In the case of an object at infinity, $\tan(\theta) = D/2f$ and:

$$N_n = \sin(\theta) = \sin\left[\arctan(D/2f)\right] = \sqrt{\frac{D^2}{4f^2 + D^2}} \qquad [2.18]$$

and for small values of D/f:

$$N_n \sim \frac{D}{2f} = \frac{1}{2N} \qquad [2.19]$$

Equation [2.19] is approximated for a simple lens but accurate for a complex system whose spherical and coma aberrations have been corrected. It then yields the equation:

$$N_n = \frac{1}{2N} \qquad [2.20]$$

When the object is not at infinity, the definition of θ is no longer coming from the relationship $\tan(\theta) = D/2f$ but from the relationship $\tan(\theta) = D/2p'$, where p' is the distance from the optical center to the image: $p' = f+s$, and equation [2.19] becomes:

$$N_n \sim \frac{D}{2p} = \frac{1-G}{2N} \qquad [2.21]$$

where G is the transversal magnification of the optical system (equation [1.3]).

An "effective f-number" can then be defined by:

$$N_{eff} = N/(1-G) \qquad [2.22]$$

such that equation [2.20] can be rewritten in the form:

$$N_n = \frac{1}{2N_{eff}} \qquad [2.23]$$

G is negative for a camera (the image is inverted). Subsequently, N_{eff} is always smaller than N. G is usually very small, therefore, often negligible, recovering equation [2.19], except in close-up photography situations where it must imperatively be taken into account. This definition of the effective aperture f-number is often misinterpreted. It is sometimes included in the exposure energy balance where it has no role (N is the variable that steps in). It is only involved in the measurement of the resolution, such as the numerical aperture N_n.

Since N is indicated on the lens ring and f is sometimes relatively unknown for variable focal length systems, the term "effective focal distance" is sometimes found in the literature: f_{eff} to assume the default inseparable from the aperture angle in equation [2.22]:

$$N_{eff} = \frac{f_{eff}}{D} \qquad [2.24]$$

2.4.5. *Zoom*

As we said, a zoom function can be obtained by enabling some lenses to move within a central system.

The simplest way to design a zoom consists of inserting an afocal system, with variable magnification within an existing objective. By making use of a triplet (a diverging lens between two converging lenses), such a correctly and chromatically corrected afocal can be constructed. It maintains exactly the focus of the camera while varying the magnification for three specific dispositions of the lenses and, between these particular positions, and without being too greedy about the magnification excursion, a very reasonable focus can be obtained, and therefore a focus defect that can easily be compensated by a very small lens-to-subject distance. A lens of the triplet remains fixed during magnification excursion, while the other two follow relatively complex movements (the convergent moves continuously closer, while the front convergent moves away first then comes closer). However, this movement is easily translated by using guidance from the magnification ring (in the case of a manual setting) or by a suitable automatic control in the case of an automatic adjustment.

If more lenses could be allocated to zooming features, a greater quality could be still ensured. Today, quality zooms can have several dozens of lenses.

The quality of zooms has significantly progressed in recent years. They are about to become the principal lenses on the consumer market and cover very well the whole range of objectives, with the exception of fisheyes and very long focal lengths. They still show some difficulties in providing very good apertures and remain inferior to fixed objective lenses of comparable quality whenever the compensation for defects become very demanding.

A zoom lens is characterized by its magnification range which is the ratio of its maximal focal to its minimal focal length: for example, a 50–200 mm zoom has a $4\times$ range. Some zooms have today magnification ranges greater than $20\times$.

It is rare that a zoom offers the same aperture at every focal length because one of them (often that at the longest focal length) determines the physical diameter of the body. In the same manner, a fixed objective lens is described by a "focal length + aperture" pair, a zoom is described by two ranges: one of focal lengths, the other of f-numbers. As a result, a zoom lens will be described by {70–200 mm; f/4-5, 6}; the lowest aperture (1/5.6) is often associated with the longest focal length (200 mm).

Zooms are traditionally arranged in different categories according to the range of their focal lengths around the 50 mm "standard":

– wide-angle zoom lenses have focal lengths starting at about 10 mm and increasing up to 50 mm;

– trans-standard zooms generally start around 20 or 30 mm and cover up to 100 mm, "overlapping" thus the reference focal length;

– telezooms often begin at about 80 or 120 mm and can reach 300 mm or beyond.

However, the profusion of very varied sensor sizes makes these classifications slightly useless, a same zoom, mounted on a camera doted with a significant conversion factor, could be practically classified in the ordinary uses of another category.

2.4.6. *Zoom and magnification*

Zooms are often considered to produce a homothetic image enlargement. This is true if the object is plane and in a frontal position, but what happens for a complex scene consisting of more or less distant objects?

Relation [1.3] defines the magnification G in the focus plane once the focal length f is determined. Given p the distance of the object with size X to the optical center, the image of X will have a size: $x = fX/(p+f)$. The magnification of off-focal-plane objects can also be determined (see Figure 2.15), for example in front of ($q < p$) or behind ($q > p$) this plane, provided that they are still in the focus area. Displacing[5], therefore, the object

5 While ensuring that the conditions are correctly verified in order for the image to remain rather sharp (equation [1.4]).

at position Y at the distance Δ from X such that $q = p + \Delta$. The ratio of the two image sizes will be equal to:

$$r = \frac{y}{x} = \frac{p}{p + \Delta} \qquad [2.25]$$

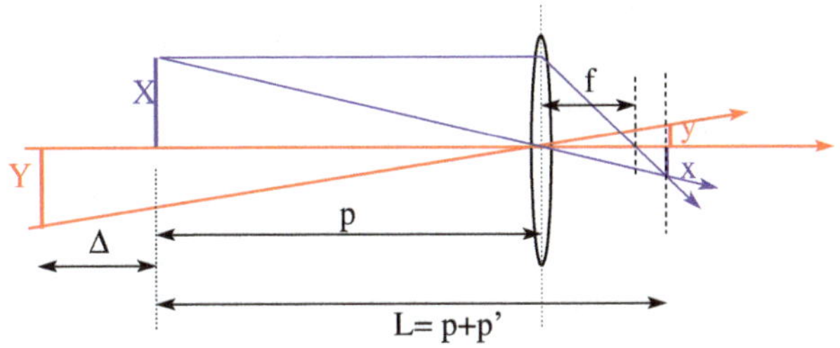

Figure 2.15. *On the condition that the depth of field conditions be verified, an (almost) sharp image of an object Y (of the same size as X) can be obtained after focusing on the object X. The image y of Y appears then (in this case where the object has moved backward) smaller than the image of X. For a color version of this figure, see www.iste.co.uk/maitre/pixel.zip*

If the movement is carried out in front of the object X ($\Delta < 0$), the object Y will appear larger than X, otherwise it will appear smaller, in accordance with the experience that we all have. But, the issue that we now raise here is different. If we maintain the scene fixed but vary the focal length of the zoom, how will the ratio r evolve? If it remains constant, this will indicate that the geometry of the scene remains identical, up to a homothety; the zoom has the mere effect of enlarging or diminishing the whole of the observed field. If this is not the case, there will be a geometric transformation of the whole of the image that will carry the specific information of the focal length being utilized.

The calculation is made as follows: after determining the observation point and the scene (the objects X and Y by the relations ($L = p + p'$) and the value of Δ), a focal length $\bar{f} = f + \phi$ is chosen whose magnification $\bar{r}$ is once again calculated (using equation [2.25]). The ratio $\eta = \bar{r}/r$ is then examined, in which the variables $\bar{p}$ are eliminated.

This result takes the form:

$$\eta = \frac{(L+\sqrt{A-4L\phi})(L+\sqrt{A}+2\Delta)}{(L+\sqrt{A})(L+\sqrt{A-4L\phi}+2\Delta)}$$
$$= 1 + \frac{2\Delta(\sqrt{A-4L\phi}-\sqrt{A})}{(L+\sqrt{A})(L+\sqrt{A-4L\phi}+2\Delta)} \quad [2.26]$$

with $A = L(L-4f)$.

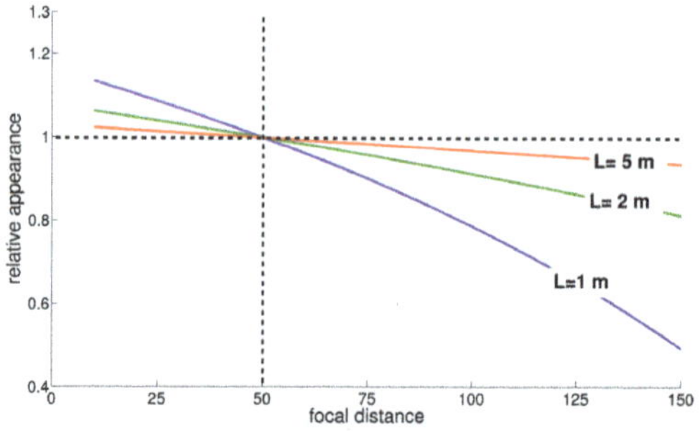

Figure 2.16. *Impact of the focal length on the relative apparent size of two objects with the same size, one placed at a distance L, the other placed at a distance 3L from the photographer, for an objective whose focal length varies from 10 to 150 mm (see Figure 2.15). The relative size seen by a 50 mm lens is taken as the reference. The three curves correspond to values of L of 1, 2 and 5 m. For a color version of this figure, see www.iste.co.uk/maitre/pixel.zip*

This result shows that if Δ is positive (therefore, with our convention, if the object is placed behind the focal plane), then η will be a decreasing function of ϕ, that is if we increase the focal length, the backgrounds will appear relatively smaller, whereas the foreground objects will be relatively larger than with a shorter focal length. Therefore, the change in focal length does not just enlarge the scene, but it also modifies the appearance by modifying the size ratio of the objects, which is generally called the perspective (see Figure 2.16). This property is well known by photographers. It will be used in computer vision to find important information such as the position of the shot or to calibrate the cameras.

2.5. Fisheye systems

Very wide-angle systems, referred to as fisheyes, form a class apart in photographic objectives, mainly due to the bias which drives their design, of not referring to the projection geometry of ordinary objectives. As a matter of fact, we already have previously met in the text the linear relationship existing between the location of the intersection in the image plane of a principal incident ray and the tangent of the angle it forms with the optical axis. Considering the growth of the tangent function, this relationship cannot be maintained with finite-dimension sensors when it approaches the limit angle of 90°. If we want to fit a camera with a field of vision so broad, it is important to modify the geometry of the image such that to shorten the angles on the periphery. As a result, very distorted images can be built that stand out at first glance from ordinary photographs. It is the *fisheye* effect. This effect can sometimes be sought for its artistic qualities. The image then generated will thus be retained as such. Nevertheless, it can also be sought for specific applications in scientific imaging, surveillance, robotic vision or to guide autonomous vehicles. These deformations are then often corrected or compensated for in order to print the image in a more common reference system. This is the reason why the deformations created by fisheyes and inverse transformations are simultaneously being studied.

2.5.1. *Projection functions*

There are multiple ways of distorting the image [MIY 64]. The study of the optical equations is very complex and leads to very singular forms of distortion. It is desirable to assimilate them to particular distortions due to their properties, defined analytically. The main constraint is to define a monotonic application of the largest possible interval from $[0, \pi/2]$ to $[0, 1]$, reasonably linear in the neighborhood of 0 such that the image will not be too distorted in its center.

In the pinhole camera system that has been used as geometric model, an incident ray issued from P forming an angle θ with the optical axis Oz passes through the optical center O (the opening of the pinhole) and creates an image P' of abscissa x in the image plane located at distance f from O, with:

$$x = f \tan(\theta) \qquad [2.27]$$

We will describe this scheme in Figure 2.17, taken from [HUG 10] where in the same half-space, the incident ray and image plane are represented, and we will use the [MIY 64] nomenclature. This diagram is that of the inversed pinhole which is convenient for not inverting the signs of the variables. We will use it again later (Figure 2.29).

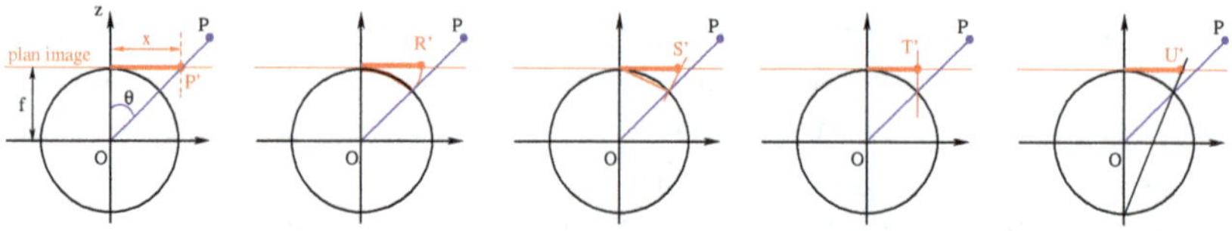

Figure 2.17. *Five schematic diagrams of image formation. On the left, the configuration of the pinhole camera and most of common objective lenses: the point image P' is located in the direction of the object point P. The other four configurations are designed to reduce the distances to the edge of the angle in order to accept wider viewing angles. From left to right: R' is the point of angular equidistance, S' corresponds to a view called equisolid, T' is an orthographic projection and U' is a stereographic projection (according to [HUG 10])*

In the case of the angular equidistant projection, the length of the arc of the intercepted circle is mapped to the image plane. The direct $x_R = \phi(x)$ and inverse transformation $x = \phi^{-1}(x_R)$ are then obtained:

$$x_R = f\theta \qquad x_R = f \text{Arc tg}(x/f) \qquad x = f \tan(x_R/f) \qquad [2.28]$$

In the case of the equal solid angle projection, the length of the intercepted cord is carried forward for x_S:

$$x_S = 2f \sin(\theta/2) \quad x_S = 2f \sin\left[\frac{\text{Arc tg}(x/f)}{2}\right]$$
$$x = f \tan\left[2\text{Arc}\sin(x_S/2f)\right] \qquad [2.29]$$

The orthographic projection projects the intersection of the circle of radius f on the map. It strongly compresses the wide angles and then becomes very difficult to invert. It thus cannot be often used for fisheyes:

$$x_T = f \sin(\theta) \quad x_T = f \sin\left[\text{Arc tg}(x/f)\right] = \frac{fx}{\sqrt{f^2 + x^2}}$$
$$x = \frac{f x_T}{\sqrt{f^2 - x_T^2}} \qquad [2.30]$$

The stereographic projection is obtained by reprojection from the point of the circle opposite to the center of the image. It provides a point U' such that:

$$x_U = 2f\tan(\theta/2) \qquad x_U = 2f\tan\left[\frac{\text{Arc tg}(x/f)}{2}\right] \tag{2.31}$$

$$x = f\tan\left[2\text{Arc tg}(\frac{x_U}{2f})\right] = \frac{f^2 x_U}{f^2 - \frac{x_U^2}{4}} \tag{2.32}$$

These direct and inverse functions are plotted in Figures 2.18 and 2.19, where we have denoted by "pinhole" the projection of an ordinary lens performing according to the paraxial approximation.

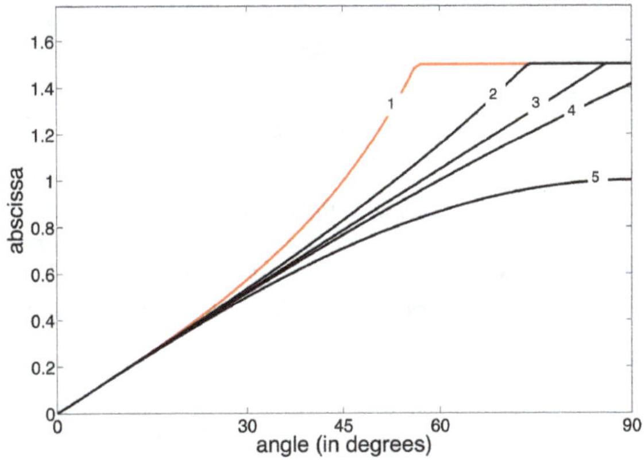

Figure 2.18. *The direct transformation of the projection of fisheyes. Angular equidistant projection (curve 3) is the reference. Two fisheyes are represented: equal solid angle projection (4) and orthographic projection (5), along with pinhole camera (1), and stereographic camera (2). The photodetector has a size 1.5 times the focal distance and the position of the image point (in ordinate) is expressed relatively to the focal distance. Object points that are projected beyond 1.5 are, therefore, no longer seen*

2.5.2. *Circular and diagonal fisheyes*

The purpose of fisheye lenses is to image the points on the periphery of the visual field. These points are located on a cone forming an angle close to $\pi/2$ with the viewing axis. They should, therefore, be inserted in a circle on the sensor of the image plane (Figure 2.20 on the left). This configuration is naturally called circular fisheye. It does not favor any direction but does not make a good use of the sensor, which does not make use of a large number of pixels.

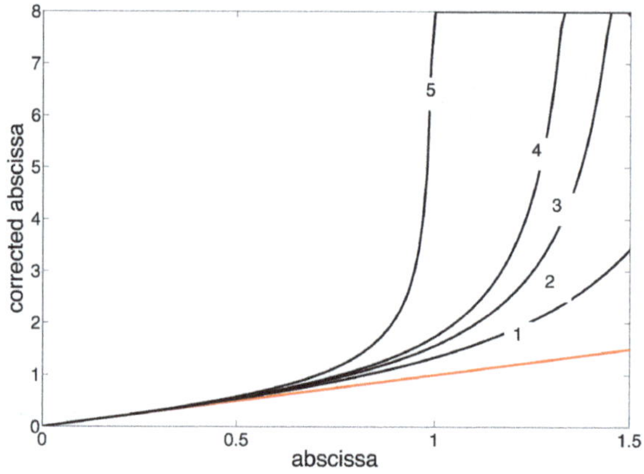

Figure 2.19. *The inverse transformation of the projection of fisheyes, under the same conditions as in Figure 2.18, which make it possible to recover an image identical to that a pinhole with the same angle would give. The ordinates have been limited here to 8 times the focal length*

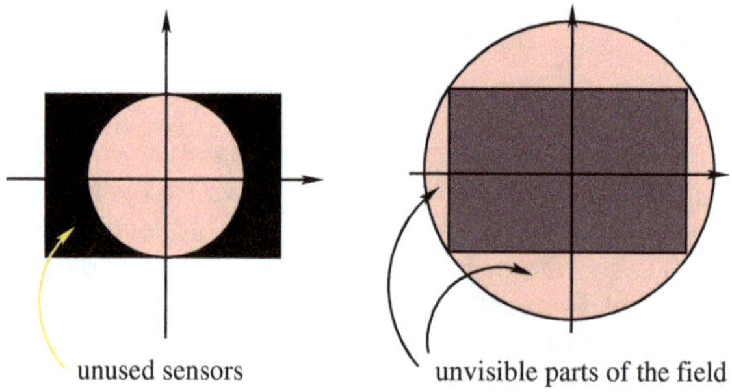

Figure 2.20. *Circular fisheye on the left, and diagonal on the right. These denominations come from the manner which the angle being seen is covered by the sensor. In a circular configuration, the entire angle is imaged but the sensor is not fully used; in a diagonal configuration, the whole sensor is used, but many points on the periphery of the angle are not seen. With a $24\,mm \times 36\,mm$ sensor, it is necessary to divide the focal length by 1.8 to change from the circular configuration to the diagonal configuration*

Another configuration has been popularized (in particular for scientific applications and industrial vision), which consists of adapting the beam such that all of the sensors are utilized. To this end, the diagonal of the sensor (its longest dimension) is adapted to the angle of the cone. This configuration is named diagonal fisheye. In this case, the whole angle of the photodetector is covered, but some areas of the visual field are not imaged.

The large variety of sizes of photodetectors leads today to many different configurations, a compromise between the circular situation and diagonal situation.

2.5.3. *Fisheyes in practice*

The choice of fisheye objectives on the market is not very wide because the general public seldom use them. The design of such an objective is also subject to numerous contradictory constraints. It is not surprising then that they are quite expensive.

The available models are usually developed for SLRs and top-of-the-range hybrids. They will then be able to make the use of relatively large sensors, a necessary condition to extend the viewing angle. The focal lengths of fisheyes must also be very short as we have seen. But, it is known that SLR systems do not allow the focal distance to be significantly reduced due to the presence of the mobile mirror that must be fitted between the lens and the image plane. Having designed a very short focal length system, it will be necessary to push its image plane away from the last lens. This is possible by placing a diverging lens upstream of the objective, in the exact inversed situation of a telephoto lens (this configuration is by the way named as "inverted telephoto lens"). By wisely adapting the position and vergence of the diverging lens to the parameters of the objective, the subject-lens distance can be increased (therefore, the space between the last lens and sensor) without significantly modifying the focal length. The last point remains then to deal with aberrations which certainly arise in optical paths so distant from the axis (see Figure 2.21). For this purpose, the process consists, as with all lenses, of replacing each lens by doublets of well-chosen glasses.

The size and weight of the best-performing fisheyes are often considerable because the entrance lenses are very broad and highly curved. Their price also increases in proportion.

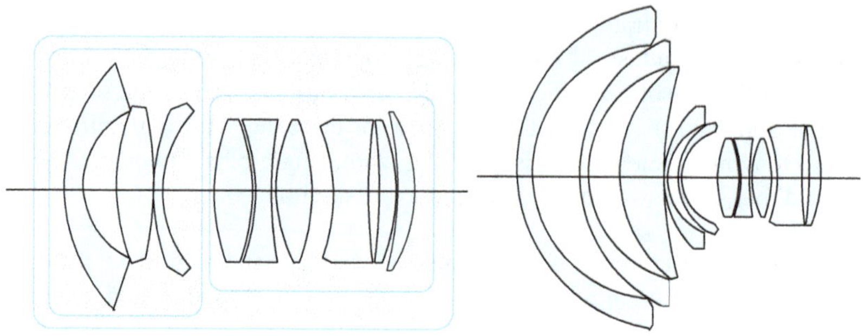

Figure 2.21. *On the left, very wide-angle objective lens. The three lenses of the train located on the left constitute a divergent lens; combined with the right-hand side (which is almost a short focal length lens), they make it possible to increase the subject-lens distance in order to allow the use of the reflex viewing. This is an inverted telescope configuration. On the right, layout diagram of a fisheye: the front lens is very broad and strongly protruding. It uses many convex-concave lenses with high curvature, nested, with variable index*

Most commercial fisheyes have focal lengths ranging from 5 to 18 mm. Very rare systems offer wider angles than the half-space (some propose a half-angle of 110°). Most of the systems available aim at a half-angle of 90° in diagonal configuration with focal lengths from 10 to 15 mm, adapted for the 135 format.

As previously seen, the images they produce present strong distortions, particularly sensitive on the periphery of the angle (see Figure 2.22). Lines, if they are not radial, are curved and all the more when they are on the periphery. In this respect, the orientation of the camera is an important factor in how the objects in the scene will appear (in contrast to an ordinary lens that makes an almost identical image of an object regardless of the framing). An object close to the edge of the angle usually takes on a considerable significance that it would not have at the center of the image. The position of the center of the viewfinder relatively to the horizon line, in landscapes, also makes it possible to move continuously from a closed space, even confined, to an open space. In this respect, fisheyes are an excellent medium for artistic creativity. The sometimes low quality of the images due to a low stigmatism, to chromatic aberrations and inevitable vignetting problems, appears often as secondary in this perspective.

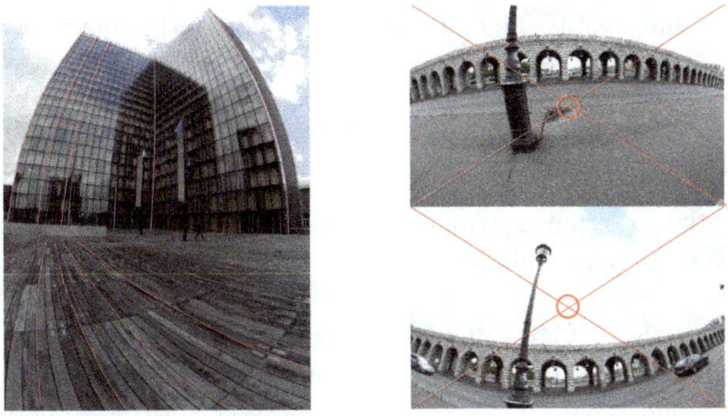

Figure 2.22. *Images from a fisheye lens. On the left, rectitude and parallelism of lines are lost. On the right, the same scene provides very different images depending on the direction of sight (represented here at the diagonal crossing)*

2.6. Diffraction and incoherent light

The results we have established so far make use of the formalism of geometric optics which consists of a simple modeling of physical reality. This modeling is quite satisfactory for most of the results presented, but does not allow the description of what happens at scales close to the wavelength, that is at the thousandth of a millimeter. But, this scale is roughly that of photosites that make up the sensor and we must, therefore, complement the previous findings by expressing more precisely what is happening at these scales. This will thus result in the correction of Figures 2.2, in the neighborhood of the critical values where they pass through a minimum.

2.6.1. *Coherence: incoherence*

First, we should recall a few elements of physics concerning image formation and the coherence of optical sources. An image is formed by the distribution on the image plane of the photons emitted by the various source points constituting the object. Each photon follows the Maxwell electromagnetism laws that in free space are simplified into the Fresnel and Fraunhoffer diffraction equations [PER 94]. The image is formed by the summation of the contributions of the various emitted photons. By definition, coherence is the quality that allows these photons to interfere, that is to combine their complex amplitudes. It consists of two elements:

1) a temporal coherence which ensures maintaining in time the phase relations of the sources during the time of the measurement;

2) a spatial coherence which ensures that the various source points maintain the same phase relation.

In photography, under natural light as well as under artificial light, the lighting is generally highly incoherent whatever the sources are because those sources are neither spatially (the photons are very short) nor temporally (the various sources are independent) coherent. The exceptions concern:

1) laser lighting (or by laser diodes) which provides, in the visible field, very monochromatic-stimulated waves and wavefront planes perfectly in phase;

2) microscopy, because studies are concerned with very small phenomena that may remain in the coherence length of the lighting being used and these lightings are often punctual and unique;

3) imaging out of the visible range: radar, ultrasounds and seismic, because in these areas of physics, only very coherent sources are achieved with coherence lengths of several hundred meters.

Here, we will neither examine the coherent sources (radar laser), nor the partially coherent microscopy sources.

In the case of the incoherent sources of ordinary photography, images are carried by photons in all the wavelengths of the visible domain. As has been said, these photons usually have a very short lifespan compared to the exposure time. They are emitted in a spontaneous way by the sources (the sun, a lamp, etc.), without any phase relation between successive and neighboring photons. If we are interested in a narrow range of wavelengths (range in which the photons could be combined), then it is shown that the interactions of photons without phase relations are on average non-existent and that only the interactions of each photon with itself are involved in the energy balance [BLA 55].

We will thus focus on the interaction phenomena of natural light with itself, known as incoherent diffraction.

The simplest case to consider is that of a photograph of an object at infinity, whose image is formed at the focal plane of the lens, a case that can be extended without difficulty to the general situation of image formation between conjugated planes, as long as the object is "far" from the lens[6].

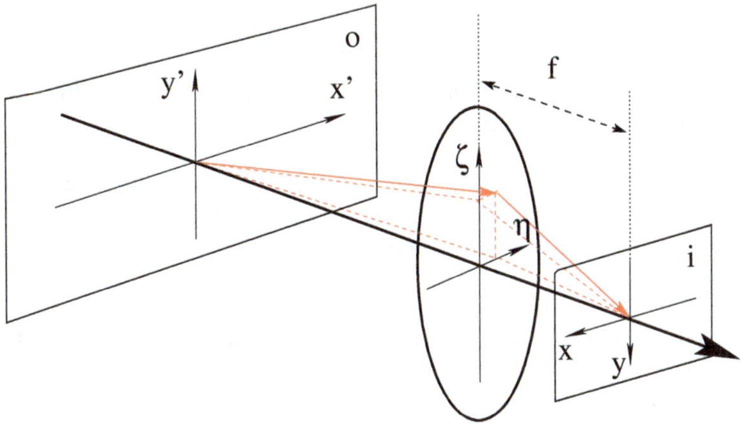

Figure 2.23. *Assembly used to calculate the influence of diffraction. The object is at infinity, its image in the focal plane. As a matter of fact, we will perform all the calculations in the image space* (x, y) *in order to not carry the magnification terms, infinitely small here*

2.6.2. *Definitions and notations*

We are facing the conditions represented in Figure 2.23. We reduce the objective to a single lens and the aperture to its diaphragm. We consider a plane object described by its brightness o (square of the modulus of the electromagnetic field) and a perfect focus, in absence of aberrations or any other defect. In order to simplify the notations, we will make all the

6 In the case of an object at distance p large compared to f (and not at infinity), it is preferable to take into account phase terms to return to the case at infinity and the result is essentially equivalent. In the case in which the object is very close, as in microscopy, the Fresnel diffraction (at infinity) can no longer be employed. It is necessary to model the propagation by the Fraunhoffer diffraction [PER 94], the situation is then much more complex and needs to be associated with the fact that the source is frequently partially coherent. Equations [2.35] and [2.36] that we are going to see can no longer be applied. They should be replaced by very complex formulations ([BOR 70] p. 418).

calculations in the image plane (described by its coordinates (x, y)). To this end, we will virtually transpose the light intensity of the object (therefore, without diffraction) into this plane. The object, in its plane, will be described by $o(x', y')$. Considering the image plane for our calculation, we will use $o(x, y)$, the variables (x, y) being derived from (x', y') by means of the transversal magnification G (relation [1.3], G is negative, image and object are inversed): $x = Gx'$, $y = Gy'$. We will extensively use the representation of the optical filtering by Fourier transform introduced by Duffieux and Dumontet [DUF 45, DUM 55].

In the lens plane, we use the space variables (η, ζ). The diffraction will be equivalently expressed between the object and lens (diffraction at infinity) or between the lens and image (diffraction at infinity reduced to the focal distance f by the lens).

Imaging would thus be perfect if the diaphragm were not diffracting. Note $K(\eta, \zeta)$ this diaphragm (this will often be the circle function of diameter D).

We are now going to break down the image by focusing on each wavelength λ. The theory of image formation under incoherent light can then be applied to the component o_λ ([BOR 70] p. 484).

For the wavelength λ, it is shown that image formation under incoherent lighting is governed by a linear relation (convolution relation) between the brightnesses in the object plane and the image plane. This convolution implements two functions, depending on the space dealt with:

1) the point spread function or impulse response $h_\lambda(x, y)$ which describes the image of a point object infinitely thin, denoted by PSF;

2) the modulation transfer function $H_\lambda(u, v)$ which expresses the attenuation of the spatial frequency (u, v). It is denoted by MTF.

2.6.3. *For a single wavelength*

The Fourier transform $O_\lambda(u, v)$ of the object is expressed according to the spatial frequencies (u, v), by:

$$\begin{aligned} O_\lambda(u, v) &= \mathcal{TF}[o_\lambda(x, y)] \\ &= \iint_{object} o_\lambda(x, y) \exp\{-2j\pi(ux + vy)\}\, dxdy \end{aligned} \qquad [2.33]$$

or by placing the coordinates (η, ζ) on the entrance pupil which are naturally expressed according to the spatial frequencies (u, v): $\eta = u/\lambda f$ and $\zeta = v/\lambda f,$:

$$O_\lambda(\eta, \zeta) = \iint_{object} o_\lambda(x, y) \exp\left\{\frac{-2j\pi(\eta x + \zeta y)}{\lambda f}\right\} dxdy \qquad [2.34]$$

By incoherent filtering by the diaphragm, the image brightness is obtained as well as the convolution of o_λ by the incoherent impulse response h_λ of the diaphragm:

$$\begin{aligned} i_\lambda(x, y) &= o_\lambda(x, y) * h_\lambda(x, y) \\ &= \iint_{object} o_\lambda(x_o, y_o) h_\lambda(x - x_o, y - y_o) dx_o dy_o \end{aligned} \qquad [2.35]$$

where the incoherent impulse response, h_λ, is the square of the modulus of the coherent impulse response [BOR 70]:

$$h_\lambda(x, y) = |k_\lambda(x, y)|^2 \qquad [2.36]$$

$k_\lambda(x, y)$ is the inverse Fourier transform of the pupil:

$$\begin{aligned} k_\lambda(x, y) &= \mathcal{TF}^{-1}\left(K(u, v)\right) \\ &= \iint_{lens} K(\eta, \zeta) \exp\left\{\frac{2j\pi(\eta x + \zeta y)}{\lambda f}\right\} d\eta d\zeta \end{aligned} \qquad [2.37]$$

Equation [2.35] can also be expressed in the Fourier domain:

$$I_\lambda(u, v) = O_\lambda(u, v).H_\lambda(u, v) \qquad [2.38]$$

as well as equation [2.36] according to the spatial frequencies:

$$H_\lambda(u, v) = K(u, v) * K^*(-u, -v) \qquad [2.39]$$

or according to the spatial variables in the entrance pupil:

$$H_\lambda(\eta, \zeta) = K(\eta, \zeta) * K^*(-\eta, -\zeta) \qquad [2.40]$$

We will have the opportunity to use the more general equations of incoherent diffraction when we will discuss the principle of coded apertures, at the end of this book (section 10.3.2).

The MTF is an important characteristic of the optical system: it measures how a pure spatial frequency (u, v) is attenuated by the system:

$$H_\lambda(u, v) = \frac{I_\lambda(u, v)}{O_\lambda(u, v)} \qquad [2.41]$$

It can be measured by photographing variable step (therefore, pure (u, v) frequencies) and fixed contrast sine test patterns and measuring the contrast in the image.

2.6.4. *Circular diaphragm*

Consider the most common practical case in photography, that of a circular diaphragm. In this case, it is rather advantageous to switch to polar coordinates since the optical system has revolution symmetry; given: $\rho^2 = \eta^2 + \zeta^2$, $w^2 = u^2 + v^2$ and $r^2 = x^2 + y^2$:

$$\begin{array}{ccc} K(\eta, \zeta) & \longrightarrow & K(\rho) = \text{circle}(2\rho/D) \\ h_\lambda(x, y) & \longrightarrow & h_\lambda(r) \end{array}$$

Starting from equation [2.36] and knowing the FT of the circle function, it is found that:

$$h_\lambda(x, y) = \left|\mathcal{TF}^{-1}(\text{circle}(2\rho/D))\right|^2 \longrightarrow h_\lambda(r) = \left[\frac{2J_1(\pi r D/\lambda f)}{\pi r D/\lambda f}\right]^2 \qquad [2.42]$$

where J_1 is the Bessel function of the first kind and first order.

The impulse response h_λ is called Airy disk (see Figure 2.24). It cancels out for $r_o D/\lambda f = 1.22$, therefore:

$$r_o = 1.22\frac{\lambda f}{D} = 1.22\lambda N \qquad [2.43]$$

where N is the f-number of the objective lens. The calculation of the transfer function $H_\lambda(u, v)$ is carried out directly from equation [2.39] and using Figure 2.25 on the left. It yields:

$$\begin{aligned} H_\lambda(\rho) &= \tfrac{D^2}{4}\left[2\text{Arc}\cos(\tfrac{\rho}{D}) - \sin(2\text{Arc}\cos(\tfrac{\rho}{D}))\right] \\ &= \tfrac{D^2}{4}\left[2\text{Arc}\cos(\tfrac{\rho}{D}) - \tfrac{\rho}{D}\sqrt{1-\tfrac{\rho^2}{D^2}}\right] \end{aligned} \quad [2.44]$$

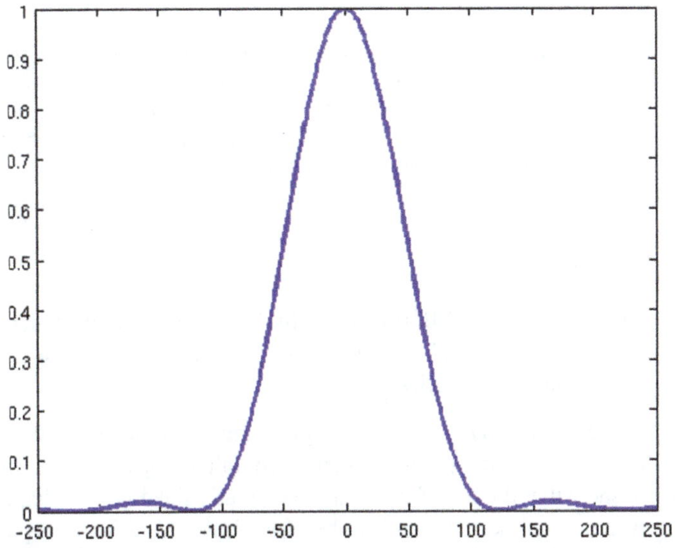

Figure 2.24. *Monochromatic incoherent diffraction by a circular diaphragm, or Airy disk. The zero of this function is situated at the position $x' = 1.22\lambda f/D = 1.22\,N\lambda$, λ being the wavelength ($\lambda = 0.5\,\mu m$ for the green), f the focal length of the lens, D the aperture of the diaphragm and N the f-number. For a perfect 50 mm objective open at $f/2$, the diffraction figure has, therefore, a diameter of the order of 1.2 μm*

It is a revolution symmetry function, regularly decreasing according to an almost linear distribution and which cancels out for $\rho = \sqrt{(\eta^2 + \zeta^2)} = D$ (see Figure 2.25 on the right).

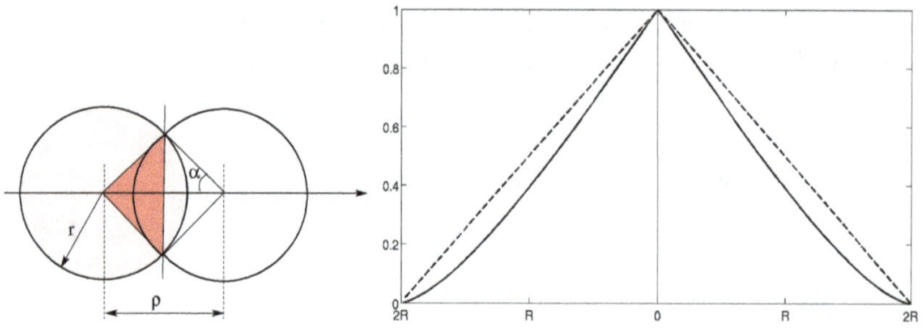

Figure 2.25. *On the left: calculation of the incoherent transfer function of the circular diaphragm. The value of H is equal to the area of the two lunules between the two circles. On the right, this transfer function, (the linear decrease in a cone of diameter $4R$ has been drawn in dashed line). In coherent light, this transfer function would have been equal to 1 between $-R$ and R*

2.6.5. *Discussion*

Equations [2.36] and [2.39], illustrated in the case of a circular diaphragm by equations [2.42] and [2.44], deserve to be discussed because in their simplicity, they conceal subtleties:

– we have expressed $H_\lambda(\eta, \zeta)$, in a slightly incorrect manner, with respect to the spatial variables (η, ζ) directly measured on the objective lens, in order to correctly show that the transfer function is directly linked to the diaphragm, since it is its autocorrelation function. Therefore, it would suffice to replace the circle of Figure 2.25 to adapt it exactly to a 9-rounded-blade diaphragm, for example, if desirable to obtain a better suited MTF for a given camera;

– strictly speaking, H_λ is actually a function of the spatial frequencies (u, v) that describe the object $o(x, y)$ or the image $i(x, y)$. It is by means of the spatial frequencies that the wavelength λ is involved. λ does not appear in equation [2.44], for example, but it can nevertheless be brought forward by rewriting it in the form:

$$H_\lambda(\lambda f w) = \frac{D^2}{4}\left[2\text{Arc}\cos(\frac{\lambda f w}{D}) - \sin(2\text{Arc}\cos(\frac{\lambda f w}{D}))\right]; \quad [2.45]$$

– the spatial frequencies are not an extravagant display of mathematical calculations introduced by the Fourier transform, but physical variables. Directly related to the propagation of the electromagnetic wave during image formation, these variables are closely related to the inclination of the optical rays interacting with the diaphragm.

– it is thus explained that if the autocorrelation function of the diaphragm $K(\eta, \zeta)$ is independent of the wavelength, the MTF that is originated therefrom when light is traveling through the lens H_λ, for its part depends thereupon, and consequently the impulse response h_λ. The larger λ is, the larger the impulse response is (and therefore the blurry image);

– it should be observed that expression [2.35] is symmetric to that which would have been obtained under coherent lighting [PER 94], but where this time the variables o and i would be the complex amplitudes of the wave. In this coherent case $h(x, y) = k(x, y)$, the calculations are affected in complex amplitude, and we revert back to intensity only at the time of the detection: the square of the modulo of $i(x, y) : |i(x, y)|^2$ allows the image to be formed from the complex wave.

It should be noted that the transfer function of the incoherent optical system being the autocorrelation of the coherent transfer function, it is twice as spread out as that of coherent optics. Can we infer that it will be richer in details? Yes and no. Yes, because its cutoff frequency will be twice as large and the inconsistent image will, therefore, have high frequencies that will not be contained in the coherent picture. No, because its module will be generally lower than that of the coherent MTF. The different frequencies of the continuum will thus be generally more attenuated in the incoherent case[7].

2.6.6. *Case of a wide spectrum*

Recovering here the construction of an image in white light, that is with a wide wavelength spectrum.

Each component X (X = R, G or B), contributing to the three channels of the image, is itself composed of a set of wavelengths that will all undergo a slightly different filtering. Taking $\sigma(x, y, \lambda)$ the sensitivity of the sensor into account (for example, using the curves such as those presented in Figure 5.23), a component X of the image is obtained given by:

$$i_X(x, y) = \int_{\lambda_{min_X}}^{\lambda_{max_X}} [o_\lambda(x, y) * h_\lambda(x, y)]\, \sigma(x, y, \lambda) d\lambda \qquad [2.46]$$

7 It should also be highlighted that coherent imaging generally suffers from a more significant defect: it is affected by granularity noise (*speckle*) because of the interference of coherent waves due to the microirregularities of the surfaces. It is in practice very difficult to overcome this granularity that considerably affects the very high frequencies of the image (see [GOO 76]).

In Figure 2.26, on the left, the impulse response is presented with two limit wavelengths in the case of a circular diaphragm: $\lambda = 0.7\,\mu m$ (red) and $\lambda = 0.5\,\mu m$ (green), as well as the mean PSF of a spectrum that would be uniform between these two values. It should be observed that the average impulse response, in contrast to monochromatic impulse responses, does not cancel out any longer. Finally, in these same conditions in linear scale, global PSFs are presented at various apertures (Figure 2.26 on the right).

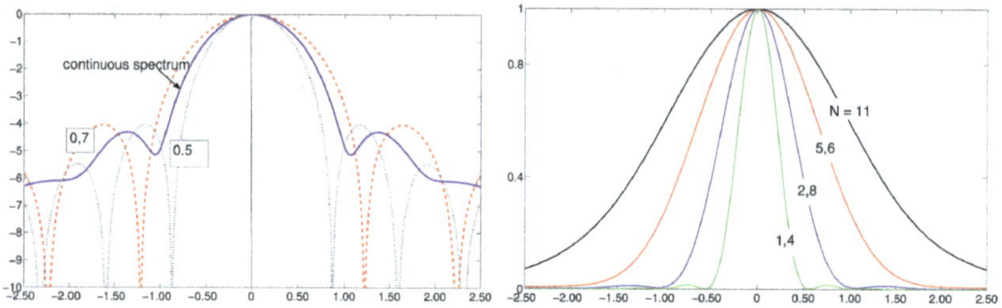

Figure 2.26. *Diffraction (on the left in logarithmic scale) of uniformly distributed light between red (0.7 μm) and green (0.5 μm). The circular objective is open to $f/2.8$. On the right, uniformly distributed light diffraction (in linear scale) for 4 f-numbers: 1.4, 2.8, 5.6 and 11. The abscissa scale is in micrometers*

Figure 2.27 presents the example of a white light PSF for a flat spectrum between 0.4 μm and 0.8 μm. On the left: the three color curves of the sensors; in the center, the white light PSF; on the right, the PSF as it would be if all wavelengths had the same PSF (an average PSF at 540 nm). It can be observed on the right, a uniformly monochrome figure, while in the center, while the diffraction spot is mostly white, the red wavelengths which decrease more slowly are dominant at the edge of the PSF, resulting in a reddish hue in the contour of the spot. It should be noted that the ring of the second maximum is not visible with linear dynamics.

In order to conduct an accurate study of the effect of diffraction on image formation, it should finally be necessary to take into account the filtering by the filters of the Bayer matrix, thus eventually as by the infrared and ultraviolet filters.

There is, therefore, no true overall impulse response of a lens under polychromatic illumination, but a mixture of impulse responses, a mixture that changes with the spectral content of the imaged object.

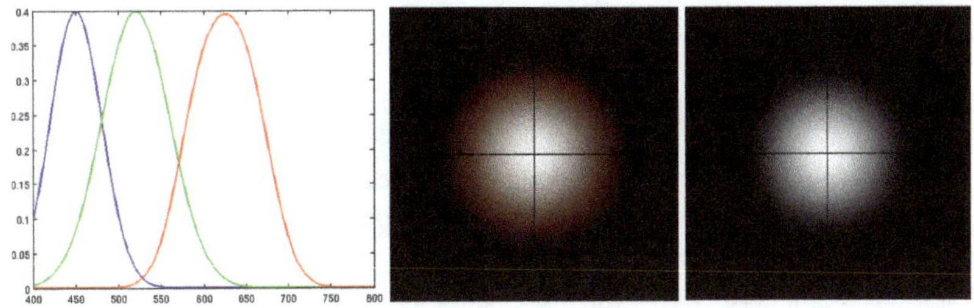

Figure 2.27. *PSF obtained with white light. The sensor has three sensitive CCDs according to the curves of the figure on the left. The light spectrum is uniform on the visible band from 0.4 to 0.8 μm. In the center, the PSF obtained by summation of the contributions of all wavelengths: the spot is dominated by the red wavelengths outside the axis. On the right, hypothesis of an identical average diffraction for all the wavelengths. The spot would be monochrome. For a color version of this figure, see www.iste.co.uk/maitre/pixel.zip*

Equation [2.42] used for the green wavelengths (λ in the order of $0.5\,\mu m$), however, gives a good average value of the impulse response due to the diaphragm (which is usually called "diaphragm diffraction").

2.6.7. *Separation power*

The separation or resolving power of an optical system is the minimum distance that must separate two image points so that they can be seen separately. In practice, the resolving power takes into account all the defects of the lens. But since it has the diaphragm diffraction as the theoretical limit, the separation power of a circular aperture has been the subject of a particular study.

By convention, the resolution limit of a circular lens is assimilated to the extension of the Airy disk. Two image points will be referred to as separated as soon as the position of one of them is beyond the first zero of the Airy disk of the other (this is the Rayleigh's criterion), (see Figure 2.28). The minimum is then at about 73 % of the maximum, when the two points are equal in energy.

2.7. Camera calibration

2.7.1. *Some geometry of image formation*

We have so far dealt with the geometric aspects of image formation by the camera in the camera coordinate system (pixel coordinates), but a very important problem in computer vision consists of connecting the coordinates of a point in the scene to the position of the image in the picture, according to specific parameters of the camera: its coordinates and orientation, as well as the focal length of the lens. This issue involves the determination of these parameters (usually unknown); this is called the calibration of the camera. Calibration has generated a very large amount of literature [FAU 93, DHO 03]. We will here only make use of a few particularly simple elements of these geometric properties.

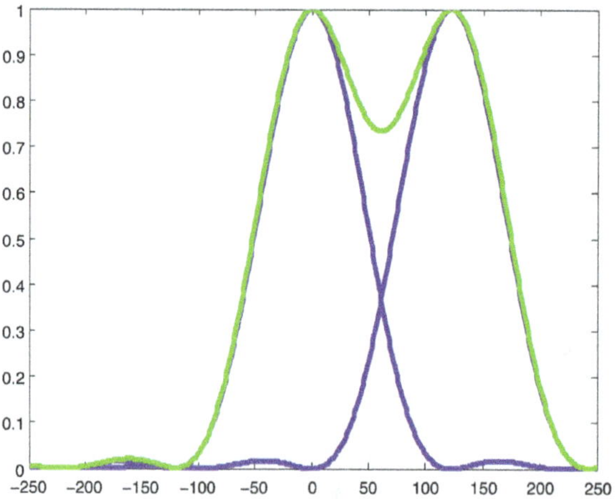

Figure 2.28. *Separation power of an optical system limited by a circular diaphragm. By definition, it is the limit distance that enables two objects to be distinguished as separate and by convention, when the PSF is an Airy disk, it is the distance to the first minimum of the Bessel function of Figure 2.26*

Similarly to the vast majority of authors on this subject, we believe a camera is perfect, that is without aberrations. We model its geometry with that of the pinhole (see section 1.3.3), and in order to avoid the manipulation of negative magnifications related to the inversion of the image, we choose the inverted pinhole model (Figure 2.29) which is fully justified when the observed objects are much further than the focal length. We will also use the system of homogeneous coordinates that makes it possible to manipulate in a same linear formulation, rotations and translations[8].

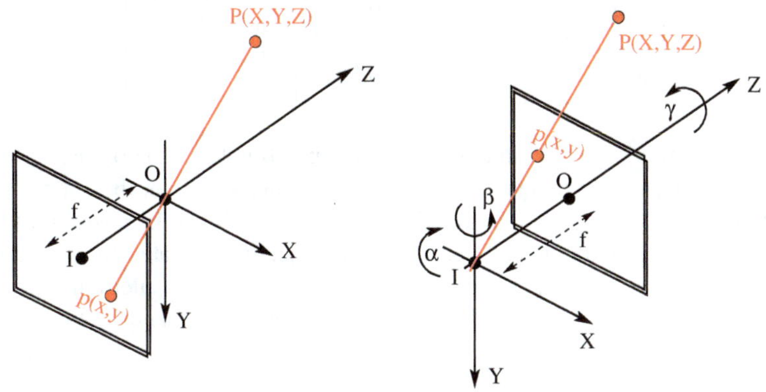

Figure 2.29. *Geometric camera model. On the left: the pinhole, on the right: the inversed pinhole which is adopted by most vision models. The two models are equivalent (with the difference that the image is inverted) if the observed object P is at a distance Z much larger than the focal length f*

In this representation, the coordinates (x, y, s) of the point p expressed in the image are linked to the coordinates $(X, Y, Z, 1)$ of the point P, in a universal representation related to the Earth, by the very general matrix equation:

$$\begin{bmatrix} sx \\ sy \\ s \end{bmatrix} = M_{3\times 4} \begin{bmatrix} X \\ Y \\ Z \\ 1 \end{bmatrix} \qquad [2.47]$$

8 The homogeneous representation maps a point of a space $\mathbb{R}^n$ in a space of dimension $n + 1$ by multiplying its n coordinates by a scalar s and by assigning s to the $(n + 1)^{th}$ coordinate. The point at infinity of $\mathbb{R}^n$ is then represented by $s = 0$. It is demonstrated in [COX 13] that homogeneous coordinates are particularly adapted to tackle projective geometry problems encountered in vision and image synthesis, and in particular changes in coordinate systems and image formation by means of matrix computations.

and the matrix M, called calibration matrix, is written in the form [DHO 03]:

$$M = \begin{bmatrix} 1/dx & 0 & x_0 \\ 0 & 1/dy & y_0 \\ 0 & 0 & 1 \end{bmatrix} \begin{bmatrix} f & 0 & 0 & 0 \\ 0 & f & 0 & 0 \\ 0 & 0 & 1 & 0 \end{bmatrix} \begin{bmatrix} R_{11} & R_{12} & R_{13} & O_x \\ R_{21} & R_{22} & R_{23} & O_y \\ R_{31} & R_{32} & R_{33} & O_z \\ 0 & 0 & 0 & 1 \end{bmatrix} \qquad [2.48]$$

where dx and dy are the dimensions of the pixel (photosite size), (x_0, y_0) are the coordinates in the image of the point I at the intersection of the optical axis and the image plane, f is the focal length, (O_x, O_y, O_z) is the position of the optical center of the objective lens O in the object coordinates and the terms R_{ij} of the 3×3 rotation matrix $\mathcal{R}$ are expressed using the Euler angles α, β and γ of the rotation with respect to the three axes Ox, Oy and Oz[9] by:

$$\begin{aligned} \mathcal{R} &= \begin{bmatrix} \cos\gamma & -\sin\gamma & 0 \\ \sin\gamma & \cos\gamma & 0 \\ 0 & 0 & 1 \end{bmatrix} \begin{bmatrix} \cos\beta & 0 & \sin\beta \\ 0 & 1 & 0 \\ -\sin\beta & 0 & \cos\beta \end{bmatrix} \begin{bmatrix} 1 & 0 & 0 \\ 0 & \cos\alpha & -\sin\alpha \\ 0 & \sin\alpha & \cos\alpha \end{bmatrix} \\ &= \begin{bmatrix} \cos\gamma\cos\beta & \cos\gamma\sin\beta\sin\alpha - \sin\gamma\cos\alpha & \cos\gamma\sin\beta\cos\alpha + \sin\gamma\sin\alpha \\ \sin\gamma\cos\beta & \cos\gamma\cos\alpha + \sin\gamma\sin\beta\sin\alpha & \sin\gamma\sin\beta\cos\alpha - \cos\gamma\sin\alpha \\ -\sin\beta & \cos\beta\sin\alpha & \cos\beta\cos\alpha \end{bmatrix} \end{aligned} \qquad [2.49]$$

Starting from equation [2.47], eliminating the intermediate variable s and expressing the position of point p in pixels (x', y'), rather than in distances (x, y), the collinearity equations of photogrammetry are obtained:

$$\begin{cases} x' = f\frac{R_{11}X+R_{12}Y+R_{13}Z+O_x}{R_{31}X+R_{32}Y+R_{33}Z+O_z} \\ \\ y' = f\frac{R_{21}X+R_{22}Y+R_{23}Z+O_y}{R_{31}X+R_{32}Y+R_{33}Z+O_z} \end{cases} \qquad [2.50]$$

These equations thus show a homographic relationship between the coordinates of P and those of p (in contrast to the homogeneous equation [2.47] which is linear).

The parameters $\{x_0, y_0, f, dx, dy\}$ are the intrinsic parameters of the camera.

The parameters $\{R_{ij}, (i, j = 1, \ldots, 3)\}$ (or α, β, γ) and O_x, O_y, O_z are the extrinsic parameters of the calibration.

---

9 Note that the order in which the transformation is applied is very important, since the rotation is not commutative.

There are 15 unknowns in the calibration problem, since the 16 terms are defined up to a coefficient. But, there are also constraints between these variables (the three rotations are unitary transformations each introducing one constraint) which must be taken into account. Therefore, it is necessary to know the coordinates of at least six points of the object space and their coordinates to calculate M up to the scaling factor s. In practice, the aim is to try to oversize the problem by including significantly more points in the solution.

A very large number of methods have been proposed to solve the problem of calibration, either in the form [2.47], or in the form [2.50], using artificial (test patterns) or natural targets (landmarks) [FAU 93, LAV 03, DER 09]. They are mainly distinguished by their robustness to take into account the inevitable small errors in the pointing of the images of these points as well as the distortions of the objectives that we have so far neglected.

2.7.2. Multi-image calibration: bundle adjustment

If ν images taken from different shooting angles are made available, but if there is no information about the points of the scene (no value (X, Y, Z)), we can try to solve the problem of the simultaneous determination of the calibration parameters and the unknown coordinates of points which will be used to position the images, respectively, to each other. This approach, conventional in photogrammetry, is called *bundle adjustment*. It works iteratively by refining the calibration parameters from a good initial estimate.

To this end, $3n$ unknown (X_i, Y_i, Z_i) are added to equations [2.50], corresponding to the unknown positions of points M_i whose ν projections m_i^j are identified for each one of them (one in each image j: (x_i^j, y_i^j)). It is assumed that the association of these points is known and reliable, either because they have been manually associated, or because they are the result of a robust detection algorithm.

Starting from good initial values of the intrinsic and extrinsic parameters, using equations [2.50], a measurement of the positioning error of the image of these points can be formed using the approximate settings. The sum of these errors forms a quality criterion of calibration. By linearizing equations [2.50] with respect to the calibration parameters, and proceeding to a gradient descent, the error was brought down. With $2\nu n$ equations and $9 + 6\nu$ unknowns, an invertible system is obtained when, for example, $n = 11$ distinct points are available viewed on $\nu = 3$ different images.

Naturally, the practical implementation of this approach depends considerably on the experimental conditions (availability of a reliable starting value, camera position, disposition of the points in the scene, etc.) [DER 09].

2.7.3. *Fisheye camera calibration*

Fisheye cameras, having very short focal lengths and very wide angles of view, work very differently from the paraxial approximation (Gaussian optics) of the thin lenses optics that we have used in this chapter. The pinhole camera model is quite inappropriate and our previous conclusions cannot be applied to it. We have described the fisheyes in section 2.5. Here we will discuss the aspects related to their calibration. We will only consider circular fisheyes which consist of spherical lenses and not diagonal fisheyes which do not cover the whole front half-space.

A very general approach, relying on this assumption of sphericity, exploits the fact that the images are distorted according to a radial law, which allows us to introduce aberration terms in our previous equations, terms that we have neglected until now.

By using equations [2.50] and defining $r = \sqrt{(x - x_0)^2 + (y - y_0)^2}$, we can express the position r' of an ideally incident point in r subject to aberrations that deviates it from the Abbe conditions:

$$r' = r(1 + a_1 r^2 + a_2 r^4 + a_3 r^6 + a_4 r^8 + \ldots) \qquad [2.51]$$

where (x_0, y_0) is the projection center in the image, r is the radial eccentricity of the pixel and the a_i expresses the magnitude of the aberration (see section 2.8). Only the paired terms of the expansions are retained, as factors of the term r, according to equation [2.53]. This adds, therefore, a few unknowns (four if constrained to the 8th order of the expansion) but mainly, a number of delicate problems regarding digital conditioning that appear because of the high values of the order of the collective terms, problems that must dealt with on a case-by-case basis [LAV 00, MIC 03].

Camera calibration is now well understood and constitutes an essential building block in many application domains: photogrammetry, simulation, virtual reality special effects, etc. Numerous products are available on the market. In addition, software programs are available free to access for several years to allow a mono- or multi-camera calibration such as "Camera Calibration Toolbox for Matlab"[10], "OCamCalib"[11] and "MultiCamera"[12].

2.8. Aberrations

As we have mentioned, the thin lens is only stigmatic (that is it forms an image point of an object point) under conditions of paraxial approximation (that is when rays make small angles with the optical axis). This is not the general case of shooting situations in photography in which wide angles are often desirable. An object point then has a spread out image in the photo. In addition, materials are not perfect and are also the cause of defects [SMI 90]. These two causes add up together and are reflected in practice by systematic defects of the image, called aberrations.

2.8.1. *Chromatic aberration*

In section 2.4.1, we have discussed the implications of the chromatic dispersion of glasses. We have indicated that it is essentially to limit the chromatism effects that complex optical formulas are being used. We have also stressed the quality of the solutions found which often reduced these defects. They still remain present and significantly affect images. They occur in two slightly different manners:

– the longitudinal chromatic aberration affects the quality of the focus anywhere in the angle of view. It translates into a focalization somewhat different from the various wavelengths which in the image plane is manifested by a lightly tinted enlargement of the contours. In practice, it is very difficult to bring this defect forward other than by an overall loss of sharpness and chromatic contrast of the image;

– the transverse chromatic aberration (see Figure 2.30) ensues from the same defect but is characterized by a transverse color shift expressed by an iridescence of strong achrome contrasts. A brutal transition from black to white then again to black will result in a black area followed by a blue line, then from

10 Camera Calibration Toolbox for Matlab from Jean-Yves Bouguet: *http://www.vision.caltech.edu/bouguetj/ calib_doc/*.

11 OCamCalib from Davide Scaramuzza: *https://sites.google.com/site/ scarabotix/ocamcalib – toolbox* relies on J.Y. Bouguet's libraries.

12 Multi-Camera Self-Calibration from Tomas Svoboda: *http : //cmp.felk.cvut.cz/ ∼ svoboda/SelfCal/*.

the white area followed by a red line (the order of colors depends on the optical combination to oppose this defect as shown in Figure 2.34).

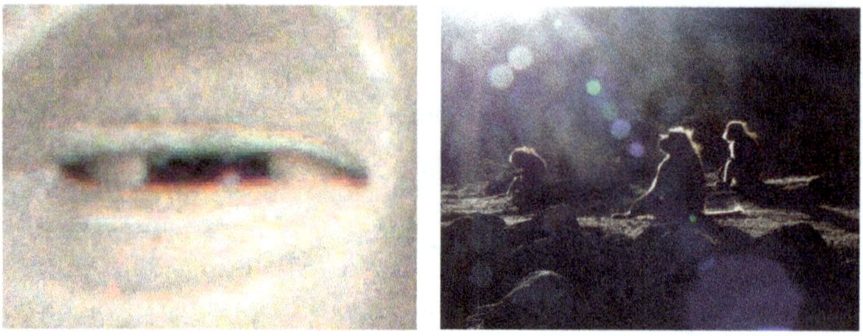

Figure 2.30. *On the left, chromatic aberrations. The red or blue edging that affects the contours above and below the eye can clearly be seen (the full image where this detail is taken from is in Figure 8.2). Right image: multiple reflections on the various lenses of the objective: the heptagonal shape of these reflections comes from the diaphragm composed of seven plates opening like a hand fan. The shape of the diaphragm is involved in these reflections, but also in the shape of the blurring of the objects that are focused. For a color version of this figure, see www.iste.co.uk/maitre/pixel.zip*

2.8.2. *Geometrical aberrations*

These aberrations come from the complexity of the optical paths in real situations that do not validate the Gaussian optics hypotheses. We will examine them from an analytical perspective in this section; they can be found with more physical descriptions in [GUE 90].

Since they originate from approximations made during the calculation of the image of a point by a flat lens in conditions that do not validate Gaussian optics, we will examine the two critical situations, when the rays are traveling far away from the optical center (high lens aperture) and when the object point is distant from the axis (wide angle). The deviation terms from the image of a point, according to their dependence on these two parameters (inclination of the radius and distance to the axis), are defects which are been classified by Seidel. Here, we will make use of Seidel's classification.

We should recall in the simplified approximation we have adopted until now (that of geometrical optics under the small-angle approximation) that: an image point P' is derived from a point object P by a simple magnification G of the source point. If a complex notation is adopted to describe P in the object plan and P' in the image plane: $X = x + iy$, $X' = x' + iy'$, the relations between object and image (emerging from equations [1.2] and [1.3]) can be written as:

$$X' = GX \qquad [2.52]$$

This equation is perfectly linear.

When we abandon the small-angle hypothesis, these equations are only approximate, there is no more linearity; point P' is extended to a more complex figure because the various rays originating from P do not converge in a single point. The obtained figure P' depends on the distance of P to the optical axis, and on the inclination (angles α and β) of the ray transmitted from the source point. Given: $\Omega = \alpha + i\beta$. Rewriting equation [2.52] using a limited expansion around the ideal position above, according to the variables X and Ω [PER 94], the general equation is obtained:

$$X' = \sum_{\mu\nu\tau\upsilon} C_{\mu\nu\tau\upsilon} X^{\mu} X^{*\nu} \Omega^{\tau} \Omega^{*\upsilon} \qquad [2.53]$$

To satisfy the hypothesis of revolution symmetry, only the odd terms of the expansion remain. The first-order term is that of a defect-free lens: equation [2.52]. By grouping the terms of third order according to their dependence on X and Ω[13], the various defects can be identified:

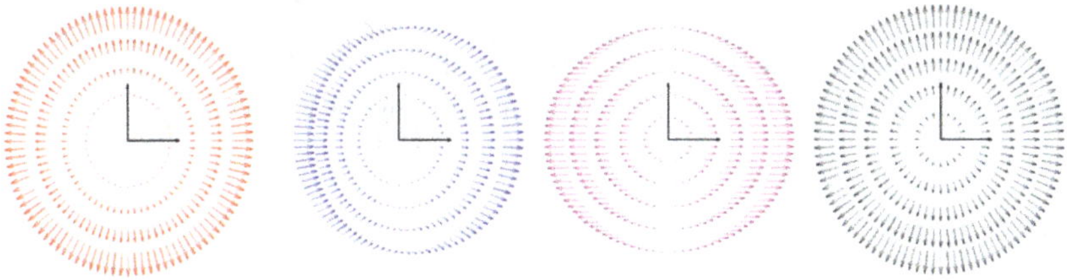

Figure 2.31. *Spherical, coma, astigmatism and field curvature aberration, according to [TAN 13]. The graphics represent the deformation of the image point according to its position in the field. For a color version of this figure, see www.iste.co.uk/maitre/pixel.zip*

13 In section 2.7.3, we will perform an expansion to higher orders to trace the rays of fisheye lenses.

1) The *spherical aberration* is due to the curvature of all the entry faces of the optical system and translates into a spreading of an image point along the optical axis, all the stronger as the lens opens wider; it gives a term in Ω^3; it preserves the revolution symmetry but focuses the sources points far away from the axis out of the image plane.

2) *Coma aberration* emerges from rays that make very large angles with the axis and that undergo a different magnification to those passing through the center of the lens, thus creating an asymmetrical spot, in the form of an egret (a "coma", hence its name); it gives terms in $X\Omega^2$.

3) *Astigmatism*, similarly to the field curvature below, originates from the particular transformation of the rays that are not in a sagittal plane (i.e. a plane containing the optical axis). It, therefore, affects a pencil traveling from a off-axis source: such a pencil, passing through the optical, is no longer a cone, but a complex surface presenting two focal lines (a sagittal and a tangential) instead of converging in a single point; similarly to the field curvature, it gives terms in $X^2\Omega$.

4) The *field curvature*, which reflects the fact that the focus should rather be on a paraboloid than on an image plane (coma defect of the system) (terms in $X^2\Omega$).

5) The *distortion* is only dependent of the distance from the point source to the optical axis. If the image point P' is shifted outward the image, it is called pincushion distortion (a pincushion would be the form that a square would take); if it moves inward, it is called barrel distortion. It gives terms in X^3.

A very technical approach can be found in [MAT 93] dedicated to the calculation of the aberrations in view of optimizing lenses in cameras.

2.8.3. *Internal reflections*

These are not aberrations, but quality defects in image formation. Composed of multiple optical elements, photographic objectives are subject to reflections at all their interfaces, reflections which are normally minimized by the use of anti-reflective coatings but can, however, be significant if the scene includes powerful sources (sun and lamp). These reflections yield bright images that reproduce so frequently the geometry of the internal diaphragms of the objective (see Figure 2.30). According to the materials and the incidence angles, the reflections on a dioptric system are of a few percent and may reach 20%. After an antireflection coating (by thin layering), these reflections can be reduced to 0.3%.

2.8.4. *Vignetting*

This generic term groups together all the effects of light decay between the center and the edge of the image (see Figure 2.32 on the left) due to the difference in the optical paths of the rays. These effects are of various origins but generally add up. They are particularly sensitive at the corners of the image. They are not always unwanted: on the one hand, they introduce a somewhat vintage effect that may be sought for style purposes, but on the other hand, they may bring forward the objects in the center of the picture[14].

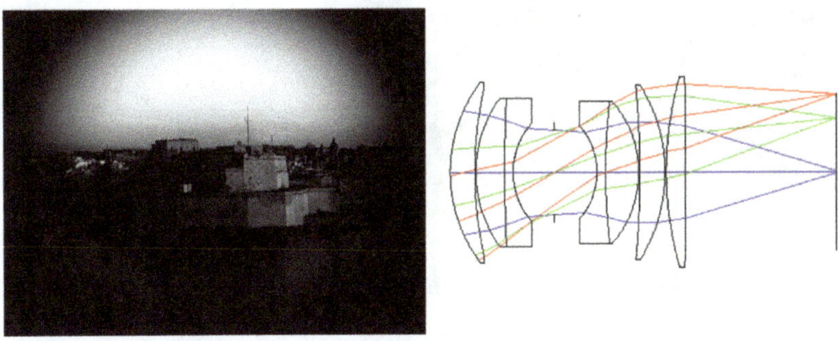

Figure 2.32. *On the left: vignetting effect on a wide-angle image. On the right: tracing of the optical paths in an objective. The red radius corresponds to an image point of the edge of the field. Only the rays traveling from the lower half of the objective will cross the diaphragm in the center of the objective (©Radiant Zemax). For a color version of this figure, see www.iste.co.uk/maitre/pixel.zip*

We will note four possible sources of vignetting, which result in very different mathematical expressions. These sources add up, so their full evaluation is very difficult to model, if not effected through a comprehensive review of the optical transitions from the source to the photo-sensor, which is only achieved by the designers of the objectives.

2.8.4.1. *Vignetting by the objective, or optical vignetting*

We have seen in section 2.4.4 that a modern lens is composed of multiple lenses, each diaphragmming down the optical beam in its path. When the ray

14 The term "vignetting" comes from the word "vignette". A vignette, in the field of printing, designates the ornaments which decorate the printings (see Diderot's Encyclopedia) and in particular that surround titles. It was often vine-stems, hence their name (from *vigne* in French).

makes an angle with the optical axis, the useful extent of the beam is reduced (see Figure 2.32 on the right) and therefore the brightness of the corresponding points is lower. The evolution of the aperture (high or low) is shown in Figure 2.33. As explained in the legend of the figure, an exact calculation of this vignetting cannot be done on simple geometrical considerations based on the image of the diaphragms, it is necessary to take into account the exact path of the rays inside the lens. A calculation outline is available in ([SMI 90] p. 135).

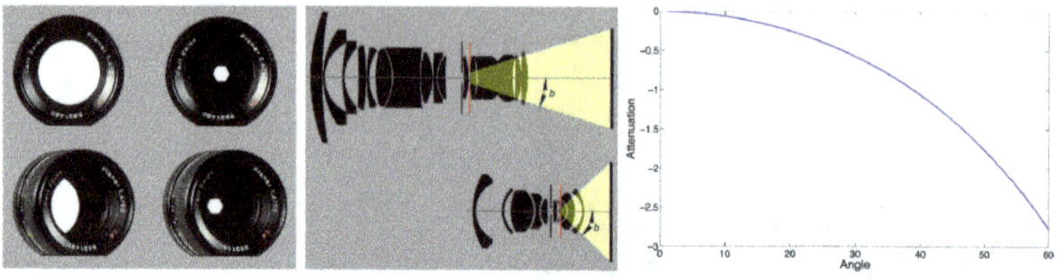

Figure 2.33. *On the left: optical vignetting – the observer is situated here at an object point, above on the optical axis, below off the optical axis. The apparent aperture left free by the objective appears clearly. It can be seen that if the diaphragm is closed, there is no vignetting. Attention, this view does not allow the effective attenuation of light to be determined by the decay of the central light since this figure does not represent the propagation of rays inside the lens (some rays will enter in the dark part of the entrance face, will be deflected by the lenses and will emerge until the sensor). This, however, clearly explains the role of the diaphragm and the importance of the length of the objective lens. In the middle, vignetting by the camera. The two objectives (on top, a Distagon, at the bottom a Biogon from Carl Zeiss) create the same image, but with very different geometries. The value of the angle b controls the vignetting which will be stronger in the Biogon ©Paul van Valree. On the right: the vignetting effect (expressed in diaphragm apertures) depending on the angle β of the beam (in degrees)*

Lens vignetting is corrected by reducing the aperture of the objective lens.

Body vignetting: what occurs here is the energy balance received by the sensor, originating from the exit lens. This balance is summarized in the literature in the law of dependence in $\cos^4 \beta$, (β being the angle, respectively, to the optical axis of the incident ray on the sensor) which is rather blindly attributed to all the manifestations of vignetting. Its original cause should be explained from Figure 2.33 at the center, where rays incident on the sensor with an angle β are represented.

First, the sensor sees an apparent aperture of the lens with an area proportional to $cos^2\beta$, on the other hand, the sensor also forming an angle β with the ray only receives at each pixel an amount of energy here yet proportional to $cos^2\beta$, respectively, to a pixel that would be perpendicular to the incident rays. It is, therefore, a term in $cos^4\beta$ that affects the luminous flux (see Figure 2.33 on the right). This term affects all the objectives, with the exception of objectives especially designed, such that the rays at the edge of the field see a larger exit pupil.

Vignetting by the body is independent of the optical aperture.

Sensor vignetting: this vignetting, also called pixel vignetting, is highly specific to the technology for the production of photosites. It may be possible that some sites are less accessible to the rays that are not perpendicular to the component, either because of elements located upstream of the sensor (microlenses and chromatic filters), or due to the geometry of the sensors which are hidden. The signal ensuing from these photodetectors will, therefore, be lower for the same incident lighting.

These effects are generally compensated in part during the construction of the sensor by a different amplification of edge signals according to a typical lens.

Screen or mechanical vignetting: this should in fact not be included here, however, it is well known of photographers (mostly amateur) who are forced to place elements upstream of the lens: sun visors, additional lenses, filters, etc. The frames of these elements can act as a screen between the lens and scene being observed. This is thus reflected by an extinction of the image on the periphery of the field. This extinction is all the quicker and more brutal if the optical aperture is small (because the transition is in fact the convolution of the screen by the aperture).

Such a vignetting is thus logically corrected by replacing the parasite accessory by another more suitable.

2.8.4.2. *Vignetting correction*

Gradient filters, brighter at the edges of the field, have been proposed to compensate for vignetting. They are in fact mainly used in photo printing in laboratories.

For digital cameras, programs are available to achieve a digital compensation of the effects of vignetting, after learning from defects by placing the couple objective/camera on a test bench. Other software programs rely on the tables available for the objectives and the most popular cameras. These corrections have the merit of taking into account all the sources of vignetting (with the exception of screen vignetting which generally does not affect professional images).

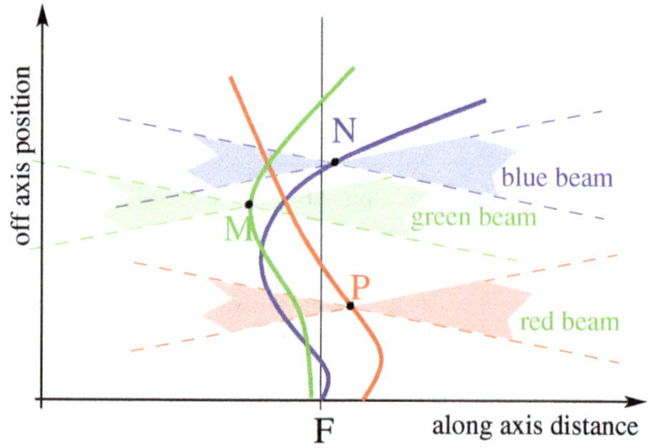

Figure 2.34. *Example of the imperfect result of a correction of aberrations. Complex compromises are made without, however, completely canceling the defect. Here, the blue is too corrected at the edge of the field, while the red is not sufficiently corrected on the axis. Three beams red, green and blue focus on three points M, N and P on the corrected focal nappes. It is very difficult to determine the real focal plane F of such a system. For a color version of this figure, see www.iste.co.uk/maitre/pixel.zip*

2.8.5. *The correction of the aberrations*

We can see in Figure 2.34, a typical result of the optimization of an objective, expressed for three wavelengths. Focalization places (the focalization place is considered the point in space where the dispersion of rays traveling from various points of the pupil is minimal) of these three monochromatic beams taken for reference are complex nappes that cannot be found in a unique fronto-parallel plane. It is, therefore, not possible to define focusing otherwise than as a compromise between different errors. It is also likely that the compromise obtained for a picture will be different from that of another image with a very different spectral content. By chance, there is, therefore, again plenty of room for the photographer's talent.

Is it possible to *a posteriori* correct this image with a computer? The answer here is not clear.

Yes, it is possible to considerably improve the image if precise information about the manner it was built is available. To this end, it is necessary to accurately have access to the parameters of the objective (Table 2.1) and to the exposure parameters: aperture, subject-lens distance and zoom. Nevertheless, more accurate information is probably needed on the association of the objective, sensor and all the elements encountered on the light path (chromatic filters in particular). These pieces of information are only reliable if they are measured on the exposure system itself, by a fine calibration of the various distortions and aberrations. When all these elements are known, then the implemented corrections can be highly effective, to the point that some manufacturers consider that it is now more important to design very powerful correction software programs rather than to optimize optical components at the expense of costly compromises.

Nonetheless, this approach has limitations. We have seen that the spectral composition of light induced specific defects to each wavelength. Furthermore, the spectral composition disappears as soon as the photon-electron conversion is achieved. The user has only three components left to take account of a much wider diversity. There is an intrinsic limitation that prohibits the exact correction of chromatic aberrations or diffraction, for example. It will be necessary to comply with compromises that can be very good but do not have much chance of being perfect.

3

The Digital Sensor

The sensor is naturally the key element of the digital photographic camera. Though the first solid detectors were metal-oxide semiconductor (MOS) photodiodes, at the end of the 1960s, it is the charge-coupled devices (CCD) technology, discovered in 1969[1], which has for many years constituted most of the sensors of photo cameras[2]. Since the year 2000, the complementary metal-oxide semiconductor (CMOS) transistor technology, developed in the 1980s, has concurrently spread for reasons we will see later; it is rapidly becoming the standard. Today, other materials are being developed to improve the performance of the sensor, especially quantum dots in graphene nanostructures [JOH 15, LIU 14] or colloidal crystals [CLI 08], which could eventually deliver exceptional performance regarding both the efficiency of photon/electron conversions (greater than 1) and detection speed. We will quickly address these sensors in the chapter on color, in section 5.4.1.2.

The efforts to develop sensors have taken several directions that all result in a complexification of the manufacturing processes:

– a densification of the photosites[3];

– an increased performance: sensitivity, linearity, immunity to noise;

---

1 The discovery of the CCD by W. Boyle and G.E. Smith in 1969 earned them the Nobel Prize in 2009.

2 A history of solid sensors and a review of their various commercialized forms can be found in [BOU 09].

3 The photosite is the elementary structure where the optical signal is transformed into electronic current. It is therefore the place where pixels are measured, the irreducible elements of the image. These terms are often used interchangeably.

– an integration of processes at the sensor to improve the signal and unburden the central processor of systematic operations.

It is according to these three axes that our presentation will be developed.

3.1. Sensor size

The race for miniaturization has been driven with two objectives:

1) to reduce the size of the sensor, in particular for mobile phones (today, the aim is a sensor with a side of a few millimeters, maintaining at the same time a high number of pixels) (see Figure 3.1);

2) to increase the resolution for a fixed image field size (the format $24\times$ 36 mm being used as commercial reference).

Consequently, we should examine the various aspects of the physical size of the sensor, a very important point in image quality as for the whole optics and electronics architecture of the camera.

3.1.1. *Sensor aspect ratio*

The aspect ratio is the ratio between the large and the small side of the image. Two different trends are found in competition in the early days of digital photography:

– the ratio 3/2 = 1.5 which is the most popular format in film photography (the 24×36 of the 135 film);

– the 4/3 = 1.33 report that corresponds to super video graphics array (*SVGA*) video screens and allowed a direct full-field display.

These two aspects have been chosen today for the majority of commercialized sensors. Cameras still offer images with a broad variety of aspect ratios (of which, for example, the 16/9, originating from cinema, which predominates today in video). Nonetheless, when the aspect ratio of the pictures is not identical to that of the sensor, the image is obtained by trimming the original image and discarding pixels when storing, which can be achieved *a posteriori* in a similar fashion.

It should be noted that the square format which has a certain success in photography circles, is not found in digital sensors, the medium formats that make extensive use of them have nowadays opted for rectangular sensors. Yet, it is the most appropriate regular format to exploit the properties of an objective rotationally symmetric by nature. However, the square format is poorly adapted to a reflex assembly because the mirror requires a large amount of space to move. The 4/3 format, closer to the ideal square, is a step toward the optimization of the couple (optics vs sensor) that reveals its full meaning in non-SLR assemblies (compact and hybrid).

Finally, some cameras now offer a stereoscopic capture (through a single lens) and to this end reserve the right-hand half of the sensors for the right picture and the left-hand half for the left image (separated by their paths in the objective). The format of the resulting image is then half that of the sensor.

3.1.2. *Sensor dimensions*

The integration of sensors with microelectronics technologies has met, in the early days of the digital camera, the limitations of masking techniques that hardly covered wide angles with a single exposure. This has very much promoted the development of small-size sensors. Gradually, by overcoming these constraints, sensors with larger dimensions have emerged (see Figure 3.1).

We have highlighted that the *Full Frame* format (24 × 36 mm) is the commercial reference. Originating from film photography, the 24 mm × 36 mm format is also called 35 mm (due to the width of the film) or designated by the code 135. It was not achievable in the early days of photography and manufacturers have used conventional but smaller photographic formats, such as the advanced photo system (APS) or new original formats (the CX, the Micro 4/3). Others, especially in the area of compacts and of telephones, have sought inspiration from the world of electronics for denominating the sensors developed by following the practices being used for video tubes: a sensor is qualified by the inch fraction occupied by 1.5× the diagonal of the sensor (the factor 1.5 comes from the optical mount that surrounds a video tube but has no relationship with photo editing itself).

Thus, a 7.5 mm × 5.6 mm sensor will be referred to as 1/1.7″ because its diagonal is about 10 mm and $10 \times 1.7 \times 1.5 \sim 25.4$ mm. This denomination is only approximated and allows the definition of classes of sensors (see Figure 3.1), rather than a fine identification.

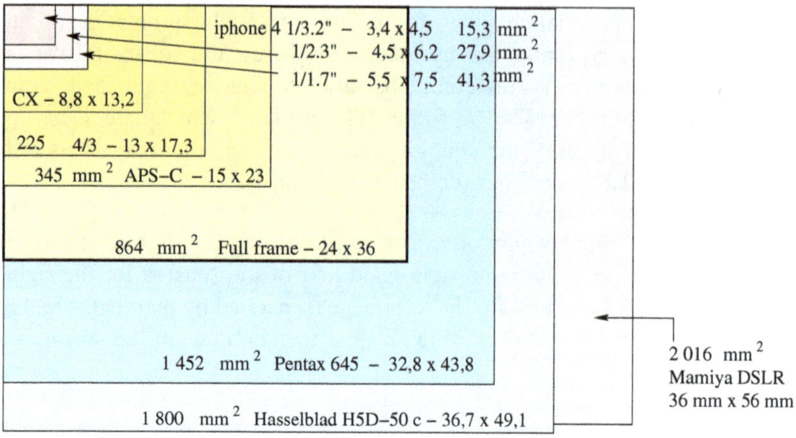

Figure 3.1. *A few sensor formats available in 2016. The reference in photography is the 24 × 36 mm format, referred to as full frame in this diagram. A very large number of sensors are smaller today than the sensor of the iPhone4 (3.39 mm × 4.54 mm), in particular on mobile phones. They are not represented here. Intermediate sizes, shown here along with their technical denomination, either designed by photography (APS, 4/3, CX), or sometimes video manufacturers/companies (in inches fractions), do not always follow a standard size (for example, the APS-C can vary a few percent depending on the manufacturer). Medium format sensors, larger than full frames, (Pentax, Mamiya or Hasselblad here) are rare and expensive*

Despite the success of smaller sizes, the 24 mm × 36 mm format has remained the holy grail of photographers because of the long history that is associated with it and the range of objectives and accessories available.

"Full-frame" cameras are now fairly widespread among high-range SLRs. They are optically compatible with the objectives designed for analog cameras (same covered angle, aberrations also compensated), but the features of these objectives are often unsuited for modern developments (stabilization, focus, zoom, etc.). They are mainly in demand because of the quality of their image, this being in theory, directly dependent on the size of the sensor as it will be discussed in Chapter 7.

Full-frame sensors have also recently emerged in hybrid cameras that are yet sought for their compactness and their lightness.

Most quality cameras that are not full frame have either 34 mm diagonal APS[4] or 22 mm diagonal Micro 4/3 formats. Compacts have formats ranging from the CX to the 1/2.3″, while mobile phones are reduced down to millimeter sizes.

However, manufacturers of "medium format" cameras (see Chapter 1) have gradually developed sensors with larger surface areas, first by juxtaposing smaller sensors, and then by gradually developing sensitive surfaces that can exceed 20 cm^2, often motivated by scientific research (astronomy and remote sensing).

3.1.3. *Pixel size*

This is a highly variable parameter depending on the camera. The race for huge numbers of pixels has contributed to reducing the size of sites, but at the same time, the increase in surfaces obtained by photolithography has allowed maintaining significant pixel dimensions whenever aiming at very high image quality. Today (in 2016), the very-large-scale integration (VLSI) technologies being implemented make use of 65 nm microelectronics integration processes and even of 45 nm for the most advanced prototypes. This allows us to design pixels smaller than the micron; for example, [WAN 13] refers to 0.9×0.9 μm pixels. Renowned sensors, like the Nokia Lumia 1020, despite being primarily a phone, display 41 million pixels at a step of 1.1 μm on a 1/1.5″ sensor. However, several cameras propose images of 20–25 million pixels on full-frame sensors (thus, 24 mm $\times$ 36 mm), which were recorded by 7.5 μm photosites that is to say a surface 50 times bigger.

The aspect ratio of the pixels is often equal to 1 (the pixels are square).

3.2. The photodetector

3.2.1. *Image detection materials*

The detection by the photodetector is achieved by photoelectric effect [NEA 11].

If the photon has an energy ($h\nu = \frac{hc}{\lambda}$) larger than the bandgap of the material, there is transformation of a photon in an electron–hole pair. For a

4 There are several slightly different APSs.

photon in the visible spectrum, with $\lambda < 0.8.10^{-6}$ m, $h = 6.626.10^{-34}$ J.s and $c \sim 3.10^8$ m.s^{-1}, it is therefore necessary that the energy of the gap band E_g be smaller than $\sim$ 2.5.10^{-19} J, that is about 1.55 eV.

It should be remembered that the energy of a photon of wavelength λ (in micrometer) can be expressed simply in electronvolt by the formula:

$$E\,(\text{in}\,\text{eV}) = \frac{1.24}{\lambda\,(\text{in}\,10^{-6}\text{ m})} \qquad [3.1]$$

In silicon, the difference in energy between the valence band and the conduction band is 1.12 eV. The silicon will be sensitive to all radiation whose wavelength is less than 1.1μm (thus those of the visible spectrum): 0.4–0.8 μm. It can also be predicted that it will be desirable to filter the wavelengths comprised between 0.8 μm and 1.1 μm to ensure a good signal-to-noise ratio.

Another important variable is the penetration depth of the photon into the material. It is defined as the distance for which the flux is reduced by a factor of $e = 2.718$ relatively to the incident flux (only 37% of the photons are remaining). It increases as the square root of the wavelength. It is equal to about 100 at 0.4 μm (in the blue) and to 1,000 at 0.8 μm (in the red).

Silicon-based sensors are therefore at the heart of virtually all cameras because their gap is well suited to the range of the visible wavelengths. III–V component-based sensors (AsGa and its alloys, HgCdTe, etc.) also exist, whose gap is much smaller, allowing detections in the infrared. They are very expensive and their use is reserved for very specific applications (military or scientific applications, ultra-fast cameras, etc.). We also reported at the beginning of this chapter a few new and very promising materials that are emerging on the market.

3.2.2. *CCDs*

CCD technology has been predominant due to the very regular and simplified structure of its architecture. Multiple slightly competing schemes exist depending on whether the photosensitive area covers the entire surface of the sensor (this configuration is also sometimes known as *full-frame CCD* even though it has no relation with the size of the sensor that we have thus also denominated above), for example, for applications where very high sensitivity is desired, or that blind areas be reserved in order to help support the movement of charges. In photo cameras, these latter patterns are prevailing.

Two different technologies have been used for CCD sensors: the P-N junction on the one hand, and the p-MOS capacity on the other hand [MAR 10]. The latter having gradually become a standard, and with its scheme also being used in CMOS, we will not describe any other and we refer the reader interested in the functioning of the P-N junction to more general texts ([NEA 11] pp. 137–190).

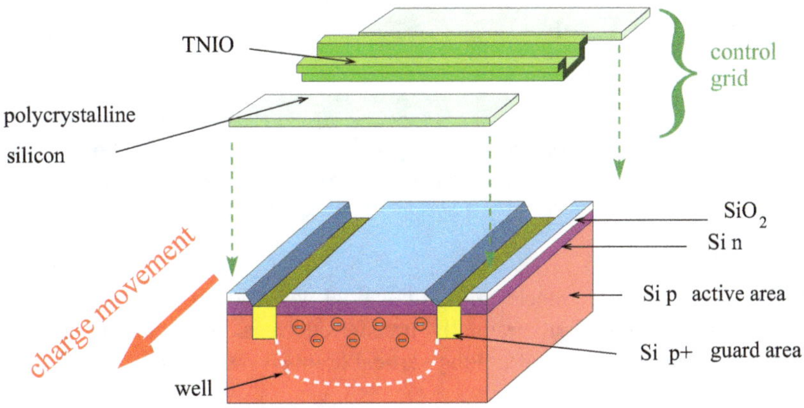

Figure 3.2. *CCD: detail of a photosite. The silicon dioxide SiO_2 layer functions as an insulator between the active area and the control grids. The silicon substrate is usually doped p with boron. The doped layer n called* buried channel *is doped with phosphorus. Its role is to keep away the charges of the interface Si/Si O_2 where they could be trapped. The potential well extends under the sensitive area and is limited by the doped guard regions p. The grids controlling the movement of charges (metal grids) placed on the photosensitive structure in contact with the insulating layer SiO_2, have been displaced upwards in the picture for better readability. Indium tin oxide (called ITO) is chosen in modern configurations because it presents a good transparency in the visible range*

Photodiodes, consisting of p-MOS capacitors ([NEA 11], p. 404), pave a portion of the sensor according to a regular matrix. They are separated by narrow guard regions (doped *p*). The charge transfer is secured from site to site by applying a sequence of potentials, in accordance with Figure 3.3. It is the silicon base which is used as a shift register. Moreover, transfer columns are created that vertically drain the charges up to a row of registers. The latter in turn collects each column horizontally until the analog-to-digital conversion stage.

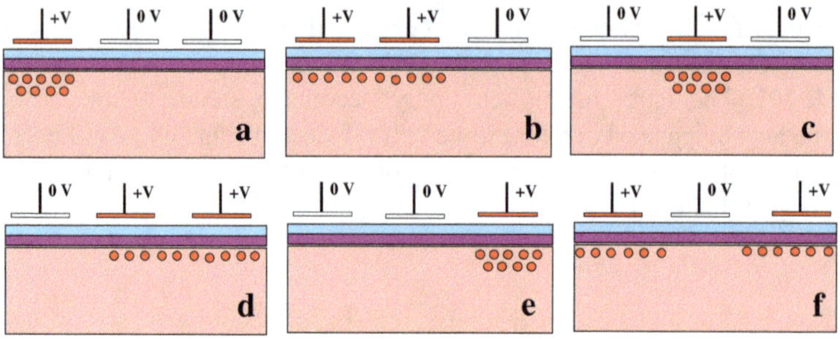

Figure 3.3. *CCD: movement of electrons. The view is a cross-section made perpendicular to that of Figure 3.2. The sequence of the potentials applied to the various sites allows that charges can be displaced from a site every two clock peaks*

The electrons are collected in the potential well (doped p) maintained at each photosite, and then gradually transferred by a cyclic alternation of potentials applied to the grids, along vertical drains (*vertical bus*).

At the end of the bus, an identical scheme evacuates the charges along a single horizontal bus that thus collects all photosites columns successively. The converter located at the end of the horizontal drain is therefore common to all pixels. The charge-to-voltage conversion is usually carried out using a voltage-follower amplifier. An analog-to-digital converter transforms the voltage into grayscale. These steps being common to all pixels, the CCD architecture ensures the high homogeneity of the image that it collects. The grayscale will have a strictly identical gain and will be affected by the same electronic noises.

An additional advantage of the CCD architecture comes from the very small space occupied by the pixel readout "circuitry" which reduces the blind area. These blind regions consist of the grids on the one hand (but the choice of transparent electrodes allows the effect to be reduced), as well as of the weak areas insulating the pixels, which makes it possible to avoid charge spillover particularly in case of blooming. The fill-factor of the the sensor is defined by the ratio of the active area and the total area of the sensor. CCDs offer a very good fill-factor, close to 100% for *full frame CCDs*.

One final advantage of the CCD ensues from the simplicity of the base motif that lends itself well to a very strong reduction of the site size, and therefore to a very strong integration compatible with 20 megapixel sensors or more, or to very small sensors (mobile phones for example, or embedded systems). Photosites with sides smaller than 2 μm have been common for several years in CCD technology.

Nevertheless, the CCD manufacturing technology is specific to this component and is not compatible with the processes used for the electronic features essential to the development and to the processing of the acquired image. The CCD is thus developed separately and scarcely benefits from the considerable effort achieved to improve CMOS technology. This explains why the photographic industry has progressively turned toward the CMOS which today holds the top end of the market among the photography sensors.

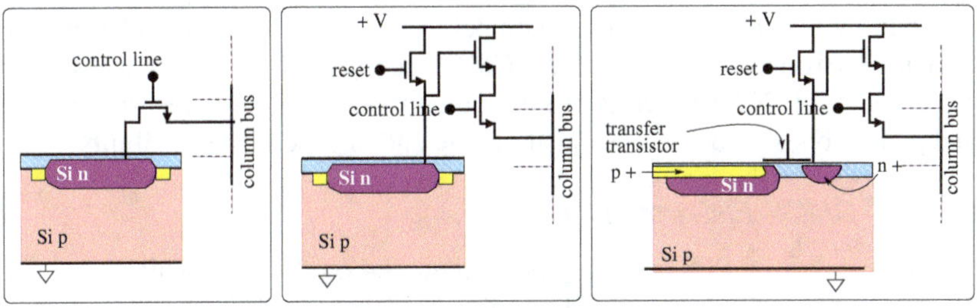

Figure 3.4. *On the left, passive CMOS assembly (PPS) around a pixel: this is a reverse polarized P-N junction, mounted with a transistor to measure charge loss at the end of the exposure and to transmit it on the readout line. At the center, in the case of an active APS-type assembly, each pixel has a recharging transistor* (reset) *(which restores the junction potential after a measurement such that the next measurement is performed in good conditions) and an online measurement transistor. On the right, a PPD assembly (APS with PIN diode) (according to [THE 08])*

3.2.3. *CMOSs*

The CMOS technology used until the year 1990 [ELG 05] was the passive pixel sensor (*PPS*) technology for reasons of integration capacity (see Figure 3.4 on the left). This technology however offered poor performance compared to CCDs (very noisy signal and poor resolution). As a result, it was therefore reserved for cheap equipment or professional applications requiring that the sensor be integrated to other electronic functions (for example, in guidance systems).

The CMOS APS technology uses the PPS photodiode scheme, but integrates an amplifying function at each pixel (see Figure 3.4 at the center); it has allowed the performance of the CMOS sensor and in particular its signal-to-noise ratio to be greatly improved. These advances can be complemented by the integration of an analog-to-digital converter to each pixel, allowing the parallelization of the acquisition of the image and thus the shooting rate to improve. The quality of the photodetection can also be improved by using a more complex PIN diode-type junction P + /N/P (P *intrinsic* N) instead of the PN junction.

Nonetheless, these improvements are only possible at the cost of a complexification of the CMOS circuit requiring very large integration capabilities that resulted in truly competitive sensors for high-end cameras only in 2010s. Nowadays (in 2015), CMOS sensors show performances very close to that of CCDs in terms of image quality and of integration dimension, and greater ones in terms of acquisition rate and of processing flexibility.

3.2.3.1. *CMOS PPS operation*

In the case of a PPS assembly (see Figure 3.4 on the left), which contains a single transistor and three link lines, the acquisition cycle is as follows:

– at the beginning of the exposure, the photo diode is reverse polarized by a positive voltage (typically 3.3 V);

– during exposure, the incident photons cause the decrease of this polarization voltage by photoelectric effect;

– at the end of the exposure, the voltage is measured. The voltage decrease expresses the photon flux during the exposure. This voltage drop (analog) is just inserted at the end of the list in the column register;

– a new polarization allows the cycle to be resumed.

The PPS assembly does not provide a good measurement of the electron flow because the electric capacitance of the junction is low compared to that of the readout bus. Furthermore, as in the CCD, it does not give a digital signal but an analog signal. The analog–digital conversion will thus be carried out column-wise, at the end of the bus.

An important variable to characterize a sensor is the full-well capacity of a photosite. For a higher number of photons, the photosite will be saturated. We say that this is dazzling. The full-well capacity of a good quality detector is of several tens of thousands of electron–hole pairs. It not only depends on the material of course, but also on the surface of the site and on its depth.

3.2.3.2. *CMOS improvements: CMOS APS*

Based on this simple mechanism, more complex architectures have been developed to correct the defects of the CMOS-PPS [ELG 05, THE 08, THE 10].

In the case of an APS assembly (see Figure 3.4 at the center), the acquisition cycle is identical to that of the PPS, nevertheless, the photo diode is first recharged by the reset transistor before being polarized, such that each measurement takes place under optimum conditions. As a result, the signal is of much better quality since it is only dependent on the absorption of the incident photons and no longer on the previous signals more or less compensated during the polarization of the diode.

At the end of the exposure, the charge is converted (into gray levels) by the follower-amplifier attached to each pixel. Such assembly is designated by 3T-APS, indicating that each pixel has three transistors available.

However, this assembly occupies more space on the component and thus causes a drop in the luminous flux sensitivity. Moreover, the reseting of the potential V of the region n is accompanied by a new noise, the reset noise (see section 7.4).

3.2.3.3. *From the CMOS to the ICS*

The progressive complexity of the image sensor and the integration of processing functions have resulted in the emergence of new terminology: CMOS imaging sensor (*CIS*) to designate these architectures that lead naturally to *smart sensors* where the processing function becomes decisive alongside that of imaging.

Important progress has been achieved around 1995, with the introduction of a fourth transistor into the APS scheme giving rise to an architecture designated by 4T-pinned-photodiode device to *4T-PPD* (see Figure 3.4 on the right). In this architecture, a PIN diode is incorporated that makes it possible to decouple the photoreceptor of the output bus and to increase the sensitivity while reducing the thermal noise. This architecture is now widespread on all sensors (and in addition, it also works on CCDs). It presents the great advantage of allowing a double measurement (before and after exposure), which by subtracting from the value obtained after measuring the charge available just after the reset, makes it possible to determine more precisely what is due to the photon flux (by subtraction of the recharge noise, of the thermal noise and the reset noise of the follower amplifier). This differential sampling operation is called correlated double sampling (CDS) [MAR 10]. However, it costs an additional transfer transistor and the fill-factor is therefore accordingly decreased. Furthermore, by reserving the surface of the CCD to incorporate pixel control, it leads to somewhat complex pixel shapes (typically in "L") (see Figure 3.5, on the right), which affect the isotropy of the signal and the MTF (see for example the exact calculations of the impulse response of these complex cells in [SHC 06]), and hence the quality of the image.

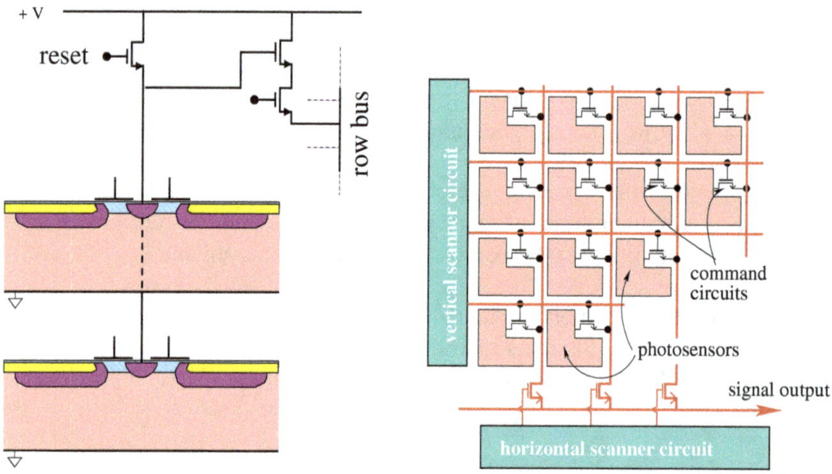

Figure 3.5. *On the left, an assembly sharing the same output circuitry for a block of 2×2 photosites in a 4T-PPD configuration (based on [THE 08]). Such an architecture now presents eight connections and seven transistors, therefore 1.75 transistor per pixel. On the right, general architecture of a CMOS circuit*

In order to maintain a good fill-factor, architectures have been therefore proposed where adjacent pixels share certain transistors.

This results in 1.75T (1.75 transistor per pixel) configurations by sharing the register of a 4T-PPD assembly between four neighbors (see Figure 3.5 on the left). The price to pay is a slightly more complicated readout cycle since it is then necessary to temporally multiplex the four pixels affected by the same reset transistor, the follower-transistor, the addressing transistor and the transfer node.

There are of course many other configurations on the market: 1.25T, 1.5T, 2.5T, 5T, etc., each manufacturer looking for a particular performance: high sensitivity, high measurement accuracy or addressing flexibility, according to its positioning on the market.

In addition to complexifying the architecture of the electronic path of each photosite and reducing the surface area of the photodetector dedicated to receive the photons, the integration of increasingly more advanced features very close to the photosensitive area also presents the disadvantage of complicating the current feeding circuitry and that of instructions for the control of the LEDs. This circuitry is laid out during the construction of the VLSI as a metal grill, between the chromatic filters and the active area, but in several layers to ensure the various tensions. It therefore constitutes a chicane for light, able to divert photons or by reflection to send back parasite photons normally destined to the neighboring sites (see Figure 3.6, on the left). It is therefore very important to pay the greatest attention to the manner these control grids are laid out. The BSI assembly, as we will see later, is a good response to this problem.

Moreover, in recent years, proposals have been made to dynamically reconfigure the matrices of photodetectors in order to achieve specific performance according to the shooting conditions (see section 5.4.2.3). The Fuji-EXR sensor is an example [FUJ 11], which allows coupled measurements to be programmed between neighboring pixels, either to increase the sensitivity (the signals traveling from the two photosites are then added up), or to increase the dynamics (a pixel is assigned a higher gain such that to determine the least-significant bits), or to optimize the resolution (the pixels are then independent). Such operations naturally require specific electronics that we will not describe here.

3.2.4. *Back-side illuminated arrangement (BSI), stacked arrangement*

Notable progress has been made by introducing a slightly different assemblage than that traditionally adopted. It consists of transferring the control grids of the charges movement to the back of the photodetector. In practice, the photodetector is flipped over after the silicon slab (*wafer*) has been thinned, in such manner that the photons reach the doped region n after crossing the doped region p, without passing through the metallic connections (see Figure 3.6, on the right). Consequently, the sensitivity of the sensor and its overall performance are significantly increased. This solution, called backside illuminated or BSI assembly, was proposed in 2007 by Omnivision and despite proving difficult to implement for manufacturing reasons, appeared in 2009 in Sony cameras. It has gradually been adopted by all manufacturers. It allows a gain in sensitivity of the order of 10–20% and a slight gain with respect to the signal-noise ratio [SON 12b]. The essential difficulty of BSI configurations lies in the obligation that a sufficiently thin silicon base is necessary not to absorb the photons in excess. The fact that wafer thinning can lead to a decrease of the available full-well capacity should also be taken into account. Whereas traditional arrangements have a base thickness in the order of a millimeter, BSI assemblies require a base reduced to a few microns. It is this delicate stage which has delayed the marketing of BSI-CMOS.

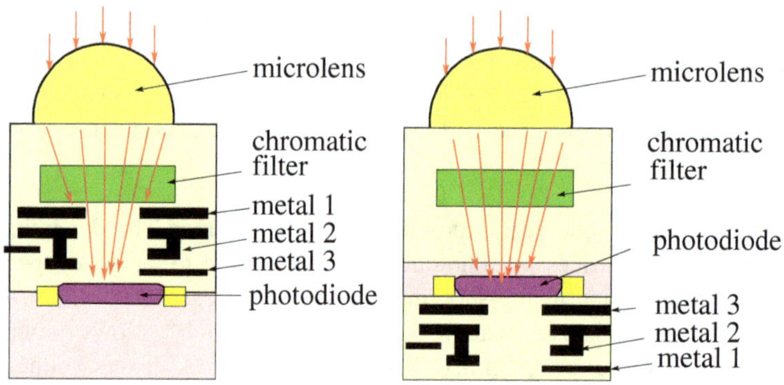

Figure 3.6. *On the left: diagram of a cross-section of a conventional CMOS photodetector, highlighting the control grids located upstream of the sensor which create a chicane for the light path (giving particularly rise to the "pixel vignetting" defect that affects the pixels bordering the edge of the field, see section 2.8.4). On the right, backside illumination arrangement. The silicon substrate is highly thinned and the photons traverse through before reaching the sensitive area*

3.2.5. *Stacked arrangements*

Sony proposed in 2012 (and marketed in 2013 under the name of Exmor-RS) solutions based on BSI where the pixel control electronics, and no longer simply the photodiode gates, instead of being placed next to the sensitive area (which reduces fill-factor), is rejected behind the photosite and no longer overlaps the sensitive area [SON 12a] as described in Figure 3.7 on the right. These arrangements are called *stacked* to indicate that the capture and processing functions are now stacked in two superposed VLSI layers and not any longer juxtaposed on the sensitive surface. The connection problems between the layers that appear then benefit from the progress achieved in the development of multilayer memory that are faced with the same problem.

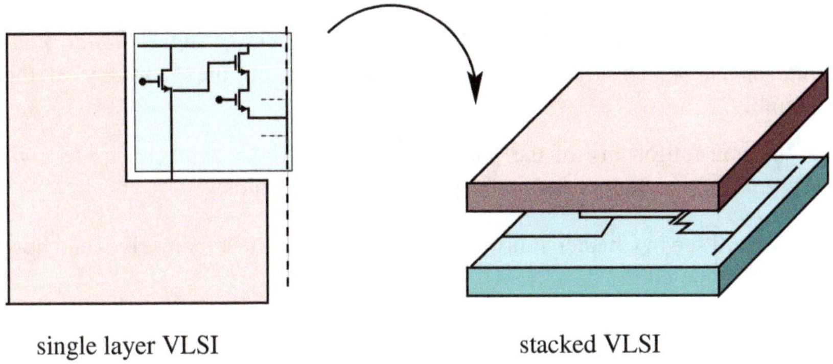

Figure 3.7. *Stacked schema. On the left, diagram of a conventional CMOS: a portion of the sensor is reserved for the electronics VLSI and thus does not contribute to the measurement of the optical flow. On the right: stacked sensor. The whole sensor is available for the optical path (on top), the electronic path is delegated to another circuit (at the bottom) which is placed on the rear face of the optical path*

The emergence of stacked technology is in effect at least as important for future cameras as back-side illumination, but for very different reasons and still not fully exploited. The juxtaposition of an optical part and an electronic part in a same VLSI drives manufacturers to compromise in order to prevent processing, desirable for the optical path (as for example, the implementation of an optical guide avoiding the loss of photons during their way from the microlenses), from deteriorating the electronic path. These compromises are reflected by fixes to defects introduced or by partial solutions to known problems. By separating the layers, separated processing of the two paths will be possible which should result in better performance.

3.2.6. *Influence of the choice of technology on noise*

We have seen that the image of a CCD sensor benefits from a great processing homogeneity. The main cause of image inhomogeneity derives from the effects of geometry which penalize the peripheral sensors [MAR 10]: the pixel vignetting that we have seen in section 2.8.4.1.

However, in the case of CMOS, counting can be affected by a precision up to a pixel; if it is flawed, the pixel is erroneous. This can have two different causes:

– a non-uniformity of the dark signal (DSNU = *dark signal non-uniformity*), which depends on the material and on the geometry of the assembly;

– a non-uniformity of the photo-reponse (PRNU = *photo response non-uniformity*), which depends on the processing electronics.

We will see in Chapter 7, how these problems manifest themselves and how they are modeled.

3.2.7. *Conclusion*

The considerable progress accomplished with the CMOS sensor makes it today the best photodetector for photography. They fully benefit from the continuous progress of microelectronics while CCDs can only rely on investments specific to this sector. A very large amount of flexibility has just been allowed by the development of stacked technology. Furthermore, it is very likely that we will see in the forthcoming years an increase in the performance of these sensors, by fully taking advantage of the benefits offered by the new silicon surfaces made available.

3.3. Integrated filters in the sensor

We have just examined photoreceptors and their associated electronics. However, it is not possible to disregard the case of optical elements which are built jointly to the photodetector. In addition to anti-reflective layers which are essential to limit losses at the interface level, two optical accessories can be added to the sensors: chromatic filters and lenses that enable the convergence of the selected channels toward the photosite. In some cases, an item can be added not by integrating it with the photodetector but by attaching it on top: the anti-aliasing filter (see section 3.3.2).

3.3.1. *Microlenses*

As shown in Figure 3.8, the microlenses are practically placed on the photodetector and are as numerous as the photosites (of the order of 10–20 million per sensor) and of the same size (less than 10 μm). Their objective is not to form an image, but to group the rays in order to ensure that the largest number easily reaches the sensor. This significantly loosens the constraints on their optical characteristics.

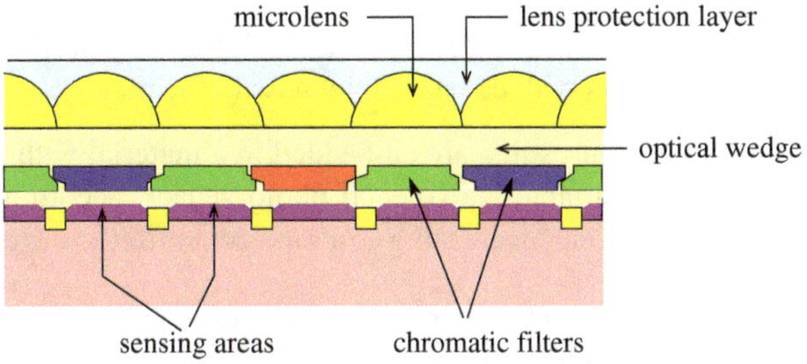

Figure 3.8. *Cross-sectional view of a CMOS circuit comprising, in addition to the layers of the photodetector, the chromatic selection elements (Bayer mosaic) and the microlenses. The dimension of such a device is typically in the order of 5 μm between adjacent lenses*

Most microlens arrays are made of a unique material (usually a high index polymer, but very transparent), whose profile is adapted: it is often a half-sphere built on a base (for manufacturing and rigidity reasons) whose other face is flat. In certain cases, we encounter aspherical profiles. It is also possible that plates with lenses on both faces be found [VOE 12].

These lenses are usually manufactured by molding from a liquid polymer completed by electrolytic machining. Electro-optical processing then takes place to treat and harden the material thus formed (this is the technique known as wafer level optics *WLO*). The positioning of the optical layers on the photodetector wafers is a very delicate stage, controlled today with an accuracy of 0.25 μm. The cost of the fine positioning of the lenses is higher than that of the lens itself [VOE 12].

During the design of the sandwich (sensor plus microlenses) in high-end systems, pixel vignetting phenomena at the edge of the field can be taken into account (see section 2.8.4). Either the microlens array then presents a step slightly different from that of photosensors, or the lenses offer an adapted profile on the edges of the field, or finally the lenses index varies regularly from the center to the edge. It should be noted that, when the dimension of the photosites becomes very small (in practice less than 1.5 μm), the diffraction effects in the material itself which constitutes the sensor as well as the protection surfaces make the lenses very ineffective. It is imperative to use thinner or back-side illuminated sensors (Figure 3.6) [HUO 10].

Two variants can be seen to emerge and that can substitute for spherical microlenses arrays:

– micro-Fresnel lenses that replace the half-sphere by portions of spheres, or even by simple rings with constant thickness (zone plates);

– gradient-index lenses that are embedded in a material with parallel faces but whose index is variable above each photo sensor, according to a radial symmetry distribution law decreasing from the center to the edge.

The role of microlenses is very important, especially when using lenses with short focal lengths (*fisheye*-type) and very open objective lenses ([BOU 09], page 100).

3.3.2. *Anti-aliasing filters*

The elements that have an influence on the resolution of a digital image are:

– the sensor resolution, that is the size of the elementary cell that collects the energy;

– the focusing, which takes into account as we have seen in section 1.3.4, the relative position of the object point with respect to the plane of focus as well as the aperture of the lens;

– the aberrations that degrade this focusing (see section 2.8);

– the diaphragm diffraction (see section 1.3.3).

In some configurations, these filters are insufficient to prevent spectrum-folding phenomena (or *aliasing*) which show in the image unwanted frequencies on periodic textures. In order to reduce this effect, many devices have an optical anti-aliasing filter which, paradoxically, will physically degrade the impulse response of the camera to improve the image [ZHA 06]. Others carry out a strongly nonlinear *a posteriori* filtering of the image (we will examine these methods during the demosaicing stage in section 5.5). Moreover, it is the trend in modern cameras.

We will examine how an optical anti-aliasing filter works.

It is mounted on a glass layer placed just in front of the sensor (see Figure 3.9, on the left), generally in front of infrared (warm filter) and ultraviolet (cold filter) filters when they exist and in front the chromatic filters which carry out the matrix conversion of the colors. The filtering function is ensured by a birefringent layer (for example, a lithium niobate crystal). The image is carried by an electromagnetic wave that can be polarized (if such a filter is employed in front of the lens) or not. During the traversal of the glass, a uni-axis medium, the wave is split into two orthogonally polarized rays (see Figure 3.9, on the right): one ordinary, the other extraordinary. This second ray is deflected by an angle ε in accordance with the mechanisms of birefringence [PER 94]. It exits the glass parallel to the incident beam, but laterally shifted to $\delta = e\tan(\varepsilon) \approx e\varepsilon$. It is essential that e be equal to the interpixel distance, such that two neighbouring cells receive the same signal, thus dividing by two the resolution and therefore the bandwidth of the image and thus reducing aliasing in this particular direction. In order to also reduce the aliasing in the second direction, a second glass layer of $LiNb0_3$ is placed crossed with the first. Four image points are thus obtained instead of just one. The resolution has been degraded by two in each direction to reduce (without removing it) the aliasing. In non-polarized light, the four images carry the same energy (if using a polarizing filter upstream from the objective, for example, to eliminate reflections on water, the image intensity will depend on the orientations of the axis of the crystal and of the polarizer).

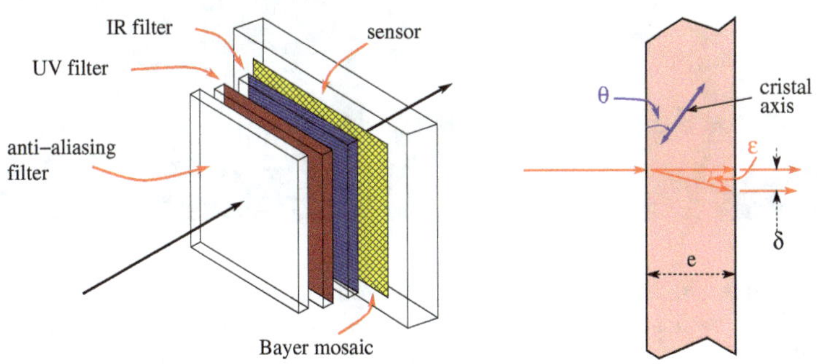

Figure 3.9. *Anti-aliasing filter: on the left, positioned in front of the sensor; on the right, beam traversing the lithium niobate crystal, giving rise to two offset beams*

How should the axis of the crystal be positioned in order to form a good quality image? Note that the image beam is not a parallel beam when it arrives on the sensor, but it is open with an angle ω fixed by the diaphragm. It is necessary that both ordinary and extraordinary beams remain converging on the sensor. It should be recalled that the extraordinary index n_θ seen by a wave forming an angle θ with the crystal axis verifies:

$$\frac{1}{n_\theta^2} = \frac{\cos^2\theta}{n_o^2} + \frac{\sin^2\theta}{n_e^2} \qquad [3.2]$$

For a beam composed of rays ranging between $\theta - \omega$ and $\theta + \omega$, it is necessary to choose θ in the vicinity of an extremum of equation [1.3] such that the excursion of the index $n_{\theta\pm\omega}$ be minimal. This is achieved if the optical axis makes an angle of 45° with the face of the layer and corresponds to an extraordinary index of 2.24 (for lithium niobate, $n_o = 2.29$ and $n_e = 2.20$), and a deflection angle of 45 milliradians between the ordinary and the extraordinary beam. A 100 microns thick layer therefore results in an offset $\delta = 45.10^{-3}.100 = 4.5\,\mu m$ compatible with the interpixel step.

Figure 3.10, on the one hand, presents the effects of aliasing on periodic structures, on the other hand, the effect of the anti-aliasing filter on the image.

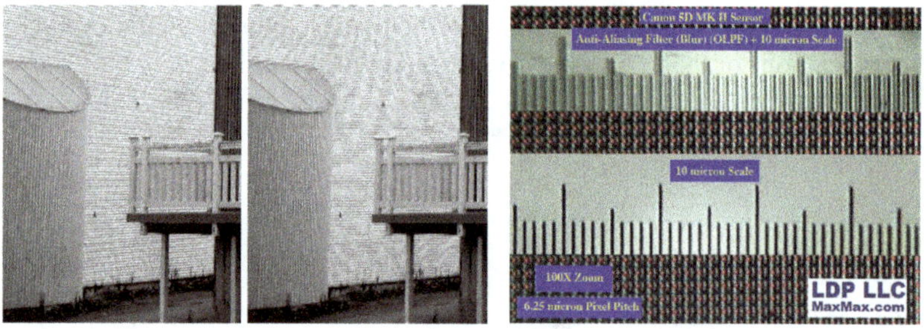

Figure 3.10. *On the left: two images presenting spectrum folding defects on periodic structures of corrugated iron: moderate on the left and very accentuated at the center. On the right: effect of the anti-aliasing filter. The image is the macrophotography of an object containing, on the one hand, a Bayer mosaic superimposed on a CMOS sensor, on the other hand, a small scale whose gradations are 10 micrometers apart. The upper part is obtained with a layer of anti-aliasing filter: the image is then doubled at a very close distance of the photodetectors step. At the bottom, the image is taken without anti-aliasing filter and therefore presents a much sharper image, in this image without any risk of aliasing (©MaxMax : www.maxmax.com/)*

In 2014, the first switchable anti-aliasing filters made their appearance. Instead of using a naturally birefringent glass layer, they use a material with induced birefringence by application of an electrical field (electro-optic effect), either by the Kerr effect or Pockels effect. They allow switching quickly from an anti-aliasing filtering position to a position without filtering.

However, a tendency to delegate the processing of spectrum folding to software programs during the demosaicing process, without any involvement of an anti-aliasing filter is also very sensitive. This fact suggests that ultimately the solution with a polarizing filter could be reserved for systems only equipped of very tiny photosites and without enough computing power available.

3.3.3. *Chromatic selection filters*

These are very important elements of the camera since it is through them that color images are created from mainly panchromatic sensors. We will examine in a very detailed manner later in this book, on the one hand, the geometry of filters (in particular of the Bayer array which constitutes the bulk of sensors on the market) and its influence on image quality (section 5.4), and on the other hand, their colorimetry and the inversion process of subsampling, a process called demosaicing (see section 5.5).

We provide a few elements about the physical production of these filters [MIL 99], a stage that for a long time was one of the most extensive of the manufacture of sensors, causing manufacturers to design them separately from photoreceptors and then to assemble them. Those days are over and chromatic filters are now designed during the same process.

They are laid down by photolithography during the final stages of sensor manufacture and therefore on the sensor surface, and will only be covered by the microlenses. Infrared and ultraviolet filters (*hot* and *cold*) will be added at a later stage and mechanically fixed.

They are usually constituted of organic dyes or pigments in a photoresist array. The dyes are distinguished from pigments by a strong chemical interaction between the substrate and the absorbing molecule. Pigments offer a better resistance to temperature and to light exposure necessary for their layering; as a consequence, they have been favoured over dyes in recent years.

The layers are very thin, of the order of the micrometer. Photoresists and dyes are successively layered down and combined to achieve good selections of wavelengths in order to produce positive filters (RGB) or negative (CMY) (see Chapter 5). Complex chemical and thermal processes are brought forward in order to guarantee the success of the processes of, on the one hand, lithography and on the other hand substrate coloration. They often have contradictory effects, which explains the slowness of the generation processes which initially ran for several weeks but are now reduced to a few hours.

4

Radiometry and Photometry

During the recording of a photograph, the light wave emitted by the scene is transformed into an image on film or on a photodetector. The image is the result of the transformation of the energy of the photons into electrical or chemical energy. The role of radiometry and photometry is to describe the parameters involved in this flux of photons.

Radiometry and photometry are two closely related sciences which have the energy effects of light radiation as a common subject of study. Therefore, they describe two complementary aspects:

– radiometry is only concerned with the objective and physical aspects of radiation;

– photometry considers the subjective, perceptual aspects of radiation and therefore the energy balances involved, such as the measurements of a reference observer, i.e. the human eye, along with its metrics.

Nevertheless, according to several authors, the term "photometry" is also associated with the physical aspects of radiation in the visual spectrum (radiometry being extended to areas not perceptible to humans). As a precaution, energy or objective photometry will be distinguished from visual or subjective photometry throughout this chapter [KOW 72, LEG 68, SMI 90].

In the field of photography, one is driven to maintain a close connection between these two aspects since a physical sensor is available. This is precisely related to the first aspect, but most often it is sought to assimilate it as closely as possible to criteria ensuing from the second.

4.1. Radiometry: physical parameters

4.1.1. *Definitions*

To photograph is to achieve the image of a point M on a sensor M'. In this chapter, we will consider the transfer of radiation energy between these points. We will first ignore all the optical elements which are used to form the image, and will consider that the photodetector is directly exposed to the radiation originating from the object M, as shown in Figure 4.1. Then, we will examine the role of the optical system. We will subsequently designate the point M as "source" since the light received by the sensors comes from there, but it may be a secondary source which merely reflects light from the primary light source (the Sun, a lamp, etc.). This aspect will also be considered.

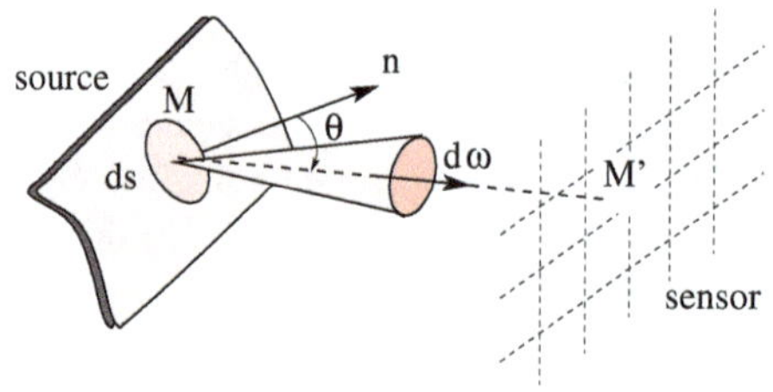

Figure 4.1. *Geometry in the neighborhood of source point M of the surface imaged in a point M' of the sensor*

We assume the scene is fixed during the short exposure time δt. The energy E emitted by the source M in the direction of the detector is proportional to the exposure time. The spectral energy $E(\lambda)$ is the share of this energy emitted in the small interval of wavelength $\partial\lambda$[1].

4.1.1.1. *Luminous flux*

The luminous flux represented by Φ, expressed in watts, is defined as the power emitted by this source: $\Phi = E/\delta t$, and the spectral luminous flux

1 Radiometry and photometry have seen the emergence of a very large number of parameters of which only a small fraction is reflected in this text. The diversity in the terminology being employed is often justified to express the complexity of the situations where the measurements depend on a very large number of parameters. We have chosen to assume the notations and conventions adopted by the ISO standard [ISO 96] and the AFNOR [DUP 94], but we have also presented a few terms of common use vocabulary when it seemed that they were needed in the literature.

(expressed in watts per meter (SI unit), or more commonly in watt by micrometer[2]) is denoted by $\Phi(\lambda)$.

4.1.1.2. *Radiant intensity*

The radiant intensity of the source is by definition the flux per solid angle surrounding an observation direction θ. It is expressed in watt/steradian and, anticipating on notations, it is, therefore:

$$\mathcal{I} = \frac{\partial \Phi}{\partial \omega} = \iint \mathcal{L} \cos \theta ds \qquad [4.1]$$

We will later precise the integration domain. The spectral intensity is defined similarly:

$$\mathcal{I}(\lambda) = \frac{\partial \Phi(\lambda)}{\partial \omega} = \iint \mathcal{L}(\lambda) \cos \theta ds \qquad [4.2]$$

In the calculation of intensity, the source is considered as a whole. The distribution of the emission on its surface is not distinguished.

4.1.1.3. *Irradiance*

The irradiance[3] of an object is defined as the intensity per area unit in the direction of observation. It is, therefore, the luminous flux emitted by the elementary surface ∂s of the source in a small solid angle $\partial \omega$ around the direction θ (Figure 4.1):

$$\mathcal{L}(\theta) = \frac{\partial^2 \Phi(\theta)}{\partial (s \cos \theta) \, \partial \omega} \qquad [4.3]$$

The dependence of irradiance with respect to the direction of observation θ is a property of the material. If $\mathcal{L}(\theta)$ is not depending on θ, it is said that the source is Lambertian; Lambertian emitters are good approximations for a large number of matte or rough surfaces in nature photography. Conversely to the Lambertian surfaces, there are monodirectional sources such as lasers and specular surfaces that only reflect in a single direction.

---

2 The micrometer is sometimes referred to as micron, in the community of opticians, but the micron is not a unit of the SI system [DUP 94].

3 The terms "brightness" or "luminosity" will sometimes be used instead of irradiance.

Similarly, the spectral irradiance can be defined from the spectral intensity $\Phi(\lambda, \theta) = \frac{\partial \Phi(\theta)}{\partial \lambda}$:

$$\mathcal{L}(\lambda, \theta) = \frac{\partial^2 \Phi(\lambda, \theta)}{\partial (s \cos \theta)\, \partial \omega} \qquad [4.4]$$

Irradiance is expressed in watts per square meter, and steradian ($W.m^{-2}.st^{-1}$) and the spectral irradiance in $W.m^{-3}.st^{-1}$.

When M is a primary source, $\mathcal{L}(\theta)$ is known by its radiation pattern (also called far-field pattern). When M is a secondary source, it is defined using its reflectance (see section 4.1.4).

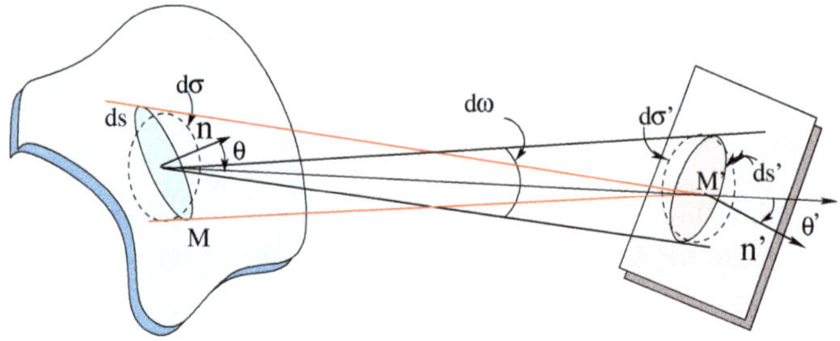

Figure 4.2. *Geometry between the point M of the imaged object and the point M' of the sensor for the calculation of the illumination. Surfaces ds and ds' are carried on the one hand by the object, on the other hand by the sensor. The surfaces $d\sigma$ and $d\sigma'$ are the straight sections of the two elementary beams, one such as the sensor sees the object, the other such as the object sees the camera*

4.1.1.4. *Radiance or exitance*

Exitance[4] $\mathcal{M}$ (in watts$/m^2$) characterizes the source, in each of its points, and expresses the luminous flux emitted per unit area:

$$\mathcal{M} = \frac{\partial \Phi}{\partial s} = \int \mathcal{L} \cos \theta d\omega \qquad [4.5]$$

4 Radiance, radiosity, emittance and excitance for the photographer designate identical quantities (see [BUK 12, ISO 96]). In thermodynamics, radiosity covers both the light emitted and the light reflected from the surface, while the excitance (term used in thermodynamics) and the emittance (term used in optics) rather refer to the light emitted by the source. We prefer the term "radiance" which has a long history in the field of photometry.

and the spectral radiance:

$$\mathcal{M}(\lambda) = \frac{\partial \Phi(\lambda)}{\partial s} = \int \mathcal{L}(\lambda) \cos \theta d\omega \tag{4.6}$$

For a Lambertian source ($\mathcal{L}$ independent of θ), as $d\omega = 2\pi \sin \theta d\theta$:

$$\mathcal{M} = \pi \mathcal{L} \tag{4.7}$$

4.1.1.5. *Irradiance*

Irradiance is the incident flux per unit area received by the sensor originating from the source. It depends both on the luminous intensity emitted by the source in the direction of the camera and on the orientation of the sensor, given $\partial s'$ the surface on the sensor around the point M', described by the beam traveling from M and with a solid angle $\partial \omega$. The surface $\partial s'$, including the normal, makes an angle θ' with MM' and receives the flux $\partial^2 \Phi$ emitted by the source. We have:

$$\partial \omega = \frac{\partial s' \cos \theta'}{r^2} \tag{4.8}$$

if r is the source sensor distance $r = d(MM')$.

By definition, the irradiance fraction of $\partial s'$ provided by the surface ∂s of the source on the sensor is given by:

$$\partial \mathcal{E} = \mathcal{L} \frac{\cos \theta \cos \theta'}{r^2} \partial s \tag{4.9}$$

4.1.1.6. *Luminous exposure*

It is the integral of the flux during the exposure time per area unit of the receiver. In practice, it is expressed, in each point of the sensor, by:

$$\mathcal{X} = \int \frac{\partial \Phi cos \theta'}{\partial s'} dt \tag{4.10}$$

and is measured in $joule/m^2$.

4.1.1.7. *Total irradiance*

M', expressed in watt/m^2, is given by the integral of all points of the object contributing to the pixel, that is to say the inverse image of the pixel:

$$\mathcal{E} = \iint \mathcal{L} \frac{\cos\theta \cos\theta'}{r^2} ds \qquad [4.11]$$

Similarly, the spectral irradiance is equal to:

$$\mathcal{E}(\lambda) = \iint \mathcal{L}(\lambda) \frac{\cos\theta \cos\theta'}{r^2} ds \qquad [4.12]$$

In photography, the observation distance of the various points of the object and the angle of the sensor are almost constant and thus, for small objects, we have:

$$\mathcal{E} \sim \frac{\cos\theta'}{r^2} \iint \mathcal{L} \cos\theta ds = \frac{\mathcal{I} \cos\theta'}{r^2} \qquad [4.13]$$

4.1.2. *Radiating objects: emissivity and source temperature*

We have used here the equations of radiometry between an object and the sensor based on the knowledge of the energy E or the spectral energy $E(\lambda)$, emitted by the object, assumed to be known. But how can these quantities be determined? We must consider two different situations: that where objects emit light and that where objects reflect it.

If most of the objects in a scene are passive and re-emit the light from a primary source, some radiate by themselves (flame, filament, etc.). These bodies whose behaviors are very varied prove generally difficult to characterize. This is achieved by reference to the black body defined in statistical thermodynamics as an ideal body which absorbs and re-emits any radiation to which it is exposed. The properties of the black body depend only on its temperature which thus dictates all its behavior. A black body at temperature T emits a spectral irradiance (expressed in $W/m^2/sr/\mu m$) that follows Planck's law [LEG 68, MAS 05]:

$$\mathcal{L}^o(\lambda, T) = \frac{2\pi hc^2}{\lambda^5} \frac{1}{e^{\frac{hc}{\lambda kT}} - 1} \qquad [4.14]$$

where:

– h is Planck's constant: $6.626 \times 10^{-34}\, m^2.kg/s$;

– k is Boltzmann's constant: $1.381 \times 10^{-23}\, m^2 kg/s^2/K$;

– c is the speed of light: $299,800,000\, m/s$.

Irradiance is calculated independent of the angle θ: the black body is Lambertian. In the SI system (see Figure 4.3), Planck's law is written as:

$$\mathcal{L}^o(\lambda, T) \sim 3,75\ .\ 10^{-16}\ \lambda^{-5}\ \frac{1}{e^{\frac{0.0144}{\lambda T}} - 1} \qquad [4.15]$$

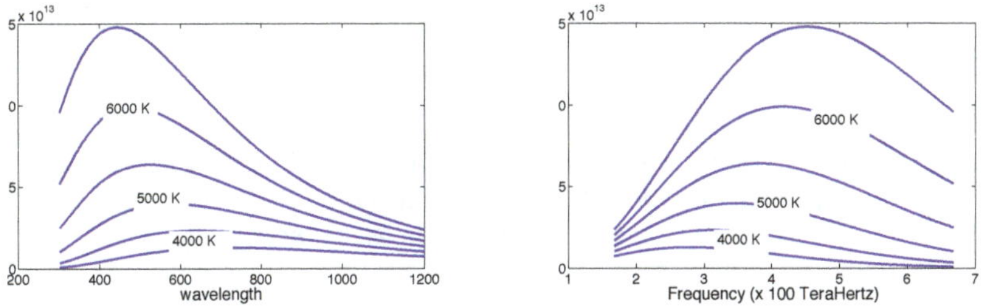

Figure 4.3. *Emission of the black body temperature (in steps of 500 K) for wavelengths ranging from the ultraviolet to near-infrared, on the left as a function of wavelengths, and on the right as a function of frequencies (the visible spectrum lies between 700 (purple) and 400 (red) THz)*

The total irradiance of the black body $\mathcal{L}_{cn}$ in the visible spectrum is, therefore:

$$\mathcal{L}_{cn}(T) = \int_{0.4\ 10^{-6}}^{0.8\ 10^{-6}} \mathcal{L}^o(\lambda, T) d\lambda \qquad [4.16]$$

In the visible spectrum and for a source of incandescence ($T \sim 3,000\ K$), the exponent of the exponential of equation [4.14] acquires values in a range between 5 and 10. Ignoring the term -1 in the denominator of Planck's law, thus obtaining Wien's law, is often sufficient to solve problems in photography (Figure 4.4):

$$\mathcal{L}^o(\lambda, T) \sim 2\pi h c^2 \lambda^{-5}\ e^{-\frac{hc}{\lambda k T}} \quad \sim \quad 3.75\ 10^{-16}\ \lambda^{-5}\ e^{-\frac{0.0144}{\lambda T}} \qquad [4.17]$$

But, this approximation is often not sufficient in near-infrared.

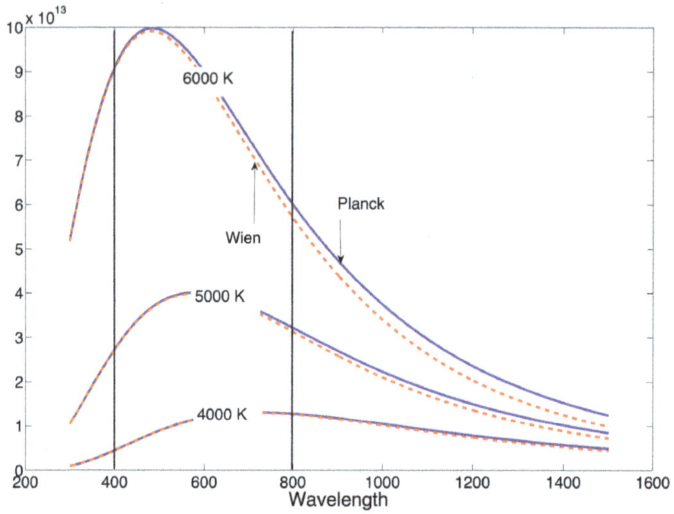

Figure 4.4. *Wien's (dashed line) and Planck's (solid line) laws for the black body at temperatures of 4,000, 5,000 and 6,000 K. Radiations are virtually indistinguishable in the visible spectrum*

The emittance of the black body is derived from Planck's law by integration in the whole of the half-space and all wavelengths:

$$\mathcal{M}^o(T) = \sigma T^4 \qquad [4.18]$$

This is Stefan's law, and σ, the Stefan constant, is $5.67\ 10^{-8}\ W.m^{-2}.K^{-4}$.

Another consequence of Planck's law, the abscissa λ_m of the maximum irradiance curves, according to the wavelength, varies with the inverse function of the absolute temperature of the black body (formula referred to as Wien's displacement law): $\lambda_m T = 2,898 \times 10^{-6}$ m.K.

4.1.2.1. *Color temperature*

For a given temperature T, Planck's law (equation [4.14]) gives the maximal emittance that a body emitting light can achieve. In nature, where the emittance of a material is always less than that of the black body, it can be expressed by:

$$\mathcal{L}(\lambda, T) = \varepsilon(\lambda)\, \mathcal{L}^o(\lambda, T) \qquad [4.19]$$

where $\varepsilon(\lambda)$ is the spectral emissivity of the material, between 0 and 1.

If $\varepsilon(\lambda)$ is not dependent on λ, it is said that the body is gray. It uniformly absorbs for all wavelengths a part of the radiation that the black body would reflect. A gray body is fully described by the temperature T^o of the corresponding black body and the value of $\varepsilon : T = T^o$.

If $\varepsilon(\lambda)$ is dependent on λ, a search is launched among the black body curves for the curve that has the same wavelength behavior of the material being studied. If this behavior is reasonably similar, it is then said that the object behaves as the black body at temperature T^o. It yields that $T \neq T^o$. T^o is called the color temperature of the object. How is the value of T^o found? Starting from equation [4.19], $\varepsilon(\lambda)$ is expressed on the basis of $1/\lambda$ for all temperatures T using equations [4.17] and [4.19] and allowing the tracing of: $\log(\varepsilon(\lambda)) = f(1/\lambda)$. The most linear curve designates the most suitable temperature T^o because it has a slope proportional to $1/T - 1/T^o$ [WYS 82].

We will return to the transition from a color temperature to another using Wien's formula in Chapter 9 where we describe the filters that can equip the objectives. This transition is based on filters which optical density varies inversely to the wavelength and proportionally to the difference between the inverses of the source temperatures (see section 9.7.5).

Nevertheless, this thermodynamic definition of color temperature is inconvenient because it requires a good knowledge of the sources, a condition hardly satisfied in practice. Therefore, in reality, empirical approaches are frequently used, strongly based on spectral distribution hypotheses in the scene. We will encounter these approaches when we discuss white balance (section 5.3).

4.1.2.2. *The tungsten case*

Tungsten is used a lot for filaments of light bulbs because it has the highest melting point of all the metals (3,422 K). This allows us to create high brightness sources and a relatively high color temperature. The color temperature of tungsten is approximately 50 K higher than its real temperature in its operating range in the vicinity of 3,000 K. It should, however, be noted that the color temperature of tungsten remains low compared to that of daylight. To simulate daylight, this color temperature is altered by inserting a colored filter in front of the source (or the sensor), as we will see in section 9.7.5.

The International Commission of Illumination (CIE) has standardized source A as the prototype of the incandescent sources with tungsten filaments [LEG 68]. It corresponds to a filament temperature of 2,855.6 K. Standard illuminants denoted by B and C are derived from this source in order to, on the one hand, express the daylight at midday, on the other hand, the average daylight. These standard illuminants are obtained by filtering the illuminant A by filters which are themselves standard. The standard illuminants A, B and C are still found but are used much less nowadays. The family of standard illuminants D, which will be discussed later, is preferred over these.

4.1.2.3. *Sunlight*

The Sun's emission is reasonably similar to that of a black body at a temperature of 5,780 K. Its maximum[5] is roughly the wavelength of 500 nm. Naturally, the atmosphere considerably filters this flux, but in the visible spectrum, the curves remain almost identical (although attenuated) between the top and bottom of the atmosphere; these curves are, however, very different in the infrared, where the absorption bands of water vapor stop the spectrum received at the level of the sea.

Systematic studies on the spectral density of daytime radiation at various moments of the day (related in [JUD 64]) have been conducted by Judd *et al.*, who have proposed an empirical law of this radiation, derived from a polynomial approximation of the curves thus obtained. Even if they are implicitly dependent of unknown variables, they are often used to determine the whole of the spectrum from a very small number of measurements (therefore, ignoring the hidden variables):

$$\mathcal{L}(\lambda,T) = \mathcal{L}_0(\lambda,T) + M_1(T)\mathcal{L}_1(\lambda,T) + M_2(T)\mathcal{L}_2(\lambda,T) \qquad [4.20]$$

Furthermore, daytime radiation is represented with very good accuracy as the combination of three terms:

1) a term (denoted by $\mathcal{L}_0$) obtained as average, at a fixed wavelength, of all measurements;

2) a term (denoted by $\mathcal{L}_1$) which is dependent of the presence of clouds and introduces a yellow-blue antagonism term;

3) a term (denoted by $\mathcal{L}_2$) which is dependent on the amount of water vapor in the atmosphere and translates into a red-green antagonism term.

---

5 The maximum is found at this wavelength only because of the choice of variables (here, as we will often do, the spectrum is expressed according to the wavelengths linearly spread). If the choice is to represent the emission curve based on the frequencies of the electromagnetic field, as shown in Figure 4.3, it is found that the spectral luminance maximum of the solar spectrum significantly moves toward the infrared, at about 1 μm ([LEG 68].

The functions $\mathcal{L}_0, \mathcal{L}_1$ and $\mathcal{L}_2$ have been tabulated for specific temperatures: 5,500 K, 6,500 K, etc.

These curves reflect the spectral evolution. How does the solar radiation evolve, in energy, during the year? The flux received by a point on the Earth also depends on the Earth–Sun distance and, therefore, regardless of the climatic conditions, solar irradiance follows an annual periodic law. If the Sun irradiance is referred to by $\mathcal{E}_{solar}(j)$ on day j, it can be derived from the irradiance $\mathcal{E}_{solar}(j = 0)$ of the 1st January 1950, taken as a reference, by the formula [HAG 14]:

$$\mathcal{E}_{sun}(j) = \mathcal{E}_{sun}(0) \left[1 - 0.01673 \cos \left(\frac{2\pi}{365.3}(j - j_0 - 2) \right) \right] \qquad [4.21]$$

where j is the date in days, j_0 is the date on 1/01/1950 and $\mathcal{E}_{sun}(0) = 1,367\ W/m^2$.

4.1.2.4. *The standardized sources*

The International Commission on Illumination has standardized light sources on several occasions according to technological needs. The ISO makes use of several of them as references for photography [ISO 02]. Their curves are normalized in order to have a value of 100 at the wavelength of 560 nm. These are mainly the ones derived from the standard illuminant A:

– the standard illuminant A, itself derived from the radiation of the tungsten filament at a temperature of 2,855.6 K, which is used as reference for many studio lightings;

– the illuminant B, corresponding to direct daylight with the sun at its zenith. Obtained from source A, filtered by a standardized filter B, it has been assigned a color temperature of 4,874 K;

– the illuminant C corresponds to an average day. In a similar manner to B, it also derives from a filtering from source A. It is assigned a color temperature of 6,774 K.

These are illuminants mathematically defined to match the outdoor and natural lighting. They have been obtained from Judd's studies that we have cited above. From these remarks, the family of illuminants D has been inferred, which is declined using two numbers which designate the particular temperature associated with a given moment: D50 at 5,000 K, D55 at 5,500 K which is the ISO reference of bright sun, D65 at 6,500 K, D75 at 7,500 K. The illuminants D are not on the Planckian locus (as defined in Section 5.2.1, but they are close to it (Figure 4.5).

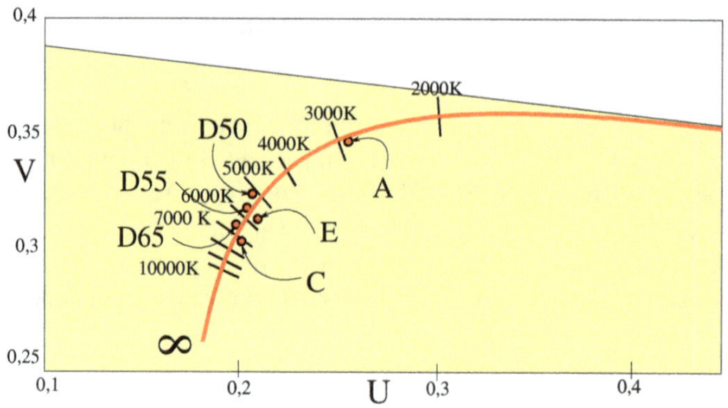

Figure 4.5. *Positioning of the various reference illuminants with regard to the Planckian locus. The chromatic representation space is here the LUV space (see section 5.2.4)*

Furthermore, illuminant E is used as reference (also referred to as W_0) which corresponds to a constant wavelength spectrum. It is also not a black body, but its color temperature is very close to the black body at 5,455 K.

Finally, illuminants F (from F1 to F12) have been defined corresponding to various fluorescent sources (without reference to a temperature). They cover various types of emission with more or less wide lines, more or less numerous, superimposed to a more or less significant continuous spectrum.

4.1.3. *Industrial lighting sources*

Industrial sources are far removed from the black body, particularly since they comprise narrow lines. Although it is still possible to define a color temperature for these sources, this temperature poorly expresses the way colors really look. In addition, the photographer knows that it is very difficult to change images made with these sources to make them look like they were taken with lighting such as the Sun.

A new quantity is, therefore, introduced that makes it possible to express the distance of a spectrum to that of a black body: the color rendering index (CRI), which expresses the mean deviation that an observer would perceive between reference objects illuminated on the one hand by a black body, and on the other hand by this source. The CIE has set the source and the reference objects to a total of 14. A perceptual distance is adopted that we will see in

Chapter 5 and the CRI is calculated as 100 minus the average of the distances of the reference objects. All black and white reference bodies have a CRI of 100, expressing that they actually comprise all colors. Below 70, a source is considered poorly suitable for comfortable lighting, this is evidentially the case of many public thoroughfare lightings, such as yellow sodium lighting (CRI $\sim$ 25), metal halide lamps (CRI $\sim$ 60) as well as certain fluorescent tubes (60 $\leq$ CRI $\leq$ 85).

4.1.3.1. *The case of light-emitting diodes*

Light-emitting diodes (LEDs) have become quite quickly accepted as white light sources. Nevertheless, because of these commercial developments, they are still neither stabilized in their technology nor *a fortiori* standardized.

White LEDs can be designed according to various principles which give them significantly different spectral properties:

– diodes formed from three monochromatic red, green and blue diodes exist but are seldom used. It is the proportion of the sources and the position of the lines that secures the quality of the white color obtained. Green LEDs are unfortunately less effective than red and blue;

– the sources consisting of a single blue diode (of yttrium aluminum garnet (YAG) or gallium nitride (GaN) surrounded by a phosphorus-based material whose luminescence is excited by the LED, provide a fine band (blue) accompanied by a broad band (the luminescence emission), placed around the yellow (complimentary color of blue);

– sources constituting a diode in the ultraviolet, exciting a fluorescence covering the entire visible spectrum, are therefore white.

LEDs are characterized by a CRI greater than 80, some exceeding 90.

4.1.4. *Reflecting objects: reflectance and radiosity*

4.1.4.1. *Reflectance*

In photography, most of the objects are secondary sources and the flux they emit toward the sensor originates from a primary source (the Sun and a lamp). How is the radiometry of these objects related to the sources? A new quantity is introduced: the reflectance of the object (also referred to as the reflection coefficient).

The reflectance of an object is defined as the ratio of the reflected flux Φ to the incident flux Φ_S (Figure 4.6):

$$\rho = \frac{\Phi}{\Phi_S} \qquad [4.22]$$

The reflectance is always less than 1. It is an intrinsic property of the material, therefore depending neither on the source nor on the sensor. It depends on the wavelength λ, on the direction of the incident light ξ and on the direction of observation θ.

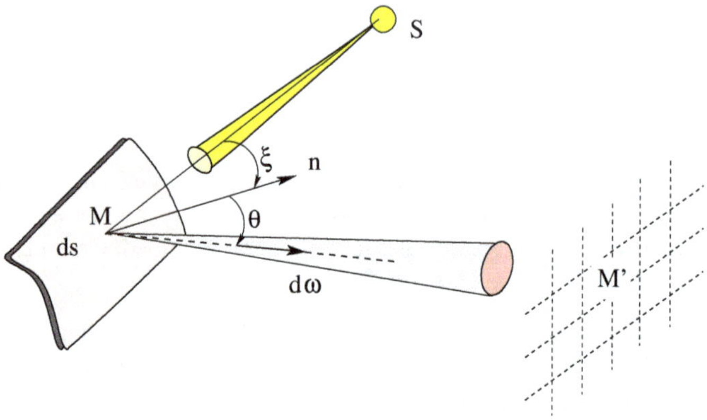

Figure 4.6. *Definition configuration of the object reflectance. S refers to the primary source. It lies in the direction ξ with respect to the normal n to the object. The sensor is in the direction θ*

Reflectance can also be written depending on the incident illumination $\mathcal{E}_S$ and on the luminance $\mathcal{L}$ of the object by the equation:

$$\rho = \frac{\pi \mathcal{L}}{\mathcal{E}_S \cos \xi} \qquad [4.23]$$

When the dependencies of the reflectance are expressed with respect to all its variables, a function of five variables is built (the wavelength, two angles to identify the source and two angles to identify the sensor), which is then referred to as bidirectional reflectance distribution function (BRDF), which is a tool widely used in image synthesis.

The albedo is defined as the ratio of the spectral irradiance attenuation in the direction θ due to the diffusion in other directions, and the attenuation due to the absorption in the direction θ and to the diffusion in all the other directions. It is especially used in remote sensing and planetology. It is a dimensionless quantity.

4.1.4.2. *The case of distributed sources*

The case of strongly dominant point sources allows the luminance of the objects in the scene to be determined, as we have just seen. A more complex, but very common, case is met when there is no dominant source in the scene, either because the sources are very numerous, or because they are diffuse. The first case often occurs in night photos, for example in urban areas. The second case is, for example, illustrated by outdoors photographs under overcast skies.

We are then faced with a complex situation which is dealt with using the formalism of radiosity.

In this approach, each point in the scene is considered as receiving a luminous flux that it partly re-emits according to the rules of its reflectance in the whole of the scene. The thermodynamic balance of the scene is then sought, exclusively limited to the wavelengths of the visible spectrum. Such a complex integral problem (where each source point interacts with its neighbors) can only be solved from a very large number of approximations, for example:

– by considering materials as Lambertians (see equation 4.7);

– by subsampling the diffuse sources;

– by limiting the number of reflections of the beams;

– by ignoring fine interactions: diffraction, dispersion, etc.

Radiosity is primarily a tool for image synthesis [COH 93, PHA 10, SIL 94] and is only rarely employed to interpret real scenes. Yet it is capable of explaining the presence of a signal in the areas of the image which do not see the primary sources. With the developments of cameras that give good-quality signals even with a very small number of photons, radiosity will assume an increasingly important role in the interpretation of images.

4.2. Subjective aspects: photometry

Photography seems *a priori* concerned with the physical effects of radiation: the number of photons leaving the object to strike the sensor, then the transformation of the photon into signal, chemical with film, photoelectric with phototransistors. It is, however, the aspects of subjective photometry that guide the works, since the beginning, to improve photographic cameras in order to assimilate them more closely to the human visual system. It is, therefore, not surprising that whole swathes of photographic literature, based on the instruments developed, have given major significance to the eye and to its performance and that most guides utilize subjective photometry units rather than those of objective photometry.

Thus, the Sun at its zenith emits approximately 2×10^6 W/sr giving an illuminance of 1.5×10^5 lux, while the Moon which emits approximately 6 W/sr, only giving an illuminance of 0.2 lux, proportionally 3 times lower.

It should also be noted that a significant proportion of the energy radiated does not really contribute toward image formation because it corresponds to frequency domains to which photographic equipment are not sensitive. The subjective quantities are, therefore, in direct relation to their effect on the image. In the case of an incandescent bulb, the 100 W it consumes; is essentially dissipated into heat; its power radiated in the visible spectrum is merely 2.2 W (approximately 1,380 lumens).

We will consider these visual photometry units.

We can review all of the properties that we have presented until now by examining how they act on the human visual system. Since the latter has a very wide variety of operating modes (in day or photopic vision, in night or scotopic vision or in intermediate vision, known as twilight, mesopic vision, with a wide or a narrow field), it is essential to properly define the working conditions.

There is also a wide diversity of visual capabilities in the population. However, the International Commission on Illumination has devised a standard model from statistical studies carried out on reference populations. This has allowed the luminous efficiency curves indispensable to subjective photometry to be established.

4.2.1. *Luminous efficiency curve*

The luminous efficiency curve expresses the relationship between the luminous flux received by the human eye and the electromagnetic radiation power received in specified experimental conditions [LEG 68].

These reference curves have been defined by the CIE from experimentation on perception equalization for different lengths, conducted under specific conditions (ambient luminous level, size and position of the stimuli, duration of the presentation, etc.). These experimentations rely on the CIE's photometric reference observer and lead to the curves (daytime or nighttime) in Figure 4.7. They are usually referred to as $V(\lambda)$; if necessary, this should include: $V_p(\lambda)$ in photopic vision (daytime) and $V_s(\lambda)$ in scotopic vision (night). We will only mention day vision from now on, the most important for photography. The calculations are identical in night vision.

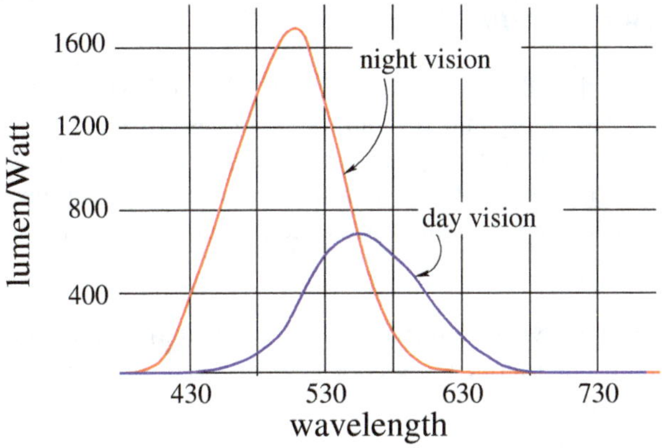

Figure 4.7. *Luminous efficiency under night vision conditions (left curve: the maximum is in the green, close to 507 nm where it is equal to 1,750 lm/W) and daytime conditions (right curve: the maximum is around 555 nm and is equal to 683 lm/W in the green-yellow). It is important to see that the two curves are just expressed in the same reference by convention and the fact that the scotopic curve has a maximum higher than that of the photocopic curve is just a consequence of this convention*

It should be noted, however, before we move on, that night vision confuses the visibility of the luminous spectra: two spectra, one red (at 610 nm) and the other green (at 510 nm), originally perceived at the same level, come across each other after one hour of dark adaptation, with a ratio of 63. If the concept of color has disappeared in favor of just the intensity modulation, the red becomes completely black, while the green is perceived as a light gray.

It should also be observed that the two day and night curves cannot be simultaneously defined in the same reference. As a matter of fact, they would require that an experiment be simultaneously conducted during the day and night since the curves are derived from the equalization of the sensations for an observer during the successive presentation of spectra. Absolute curve fitting is obtained by convention, by deciding whether the black body at the melting temperature of platinum (2,045 K) is similar to that of day and night [LEG 68]. This leads to the curves in Figure 4.7 which could falsely suggest a greater night visibility. To this end, it is often preferred to normalize the two curves by their maximum in order to show that these curves reflect mainly relative relations.

4.2.2. *Photometric quantities*

Based on the definitions established in section 4.1.1 (equations [4.3]–[4.12]), and taking into account the luminous efficiency curves $V(\lambda)$, it is possible to define new perceptual quantities. For each quantity, a new name and a new unit are designated to it which makes photometry become rather complex. We will keep the same variable to designate them, but with an index v for "visual".

For example, the visual luminous flux $\delta^2\Phi_v$ is inferred from the knowledge of the luminous flux (defined in section 4.1.1) for each wavelength by:

$$\delta^2\Phi_v = ds\; d\omega \int_{380nm}^{800nm} V(\lambda)\; \mathcal{L}(\lambda,\theta)\; \cos\theta \; d\lambda \qquad [4.24]$$

This visual luminous flux is measured in lumen. It should be observed that the terminals of the integral (here 380 and 800 nm) have little importance as long as they are located in the region where $V(\lambda)$ is zero.

The visual irradiance $\mathcal{L}_v$ is thus defined, the visual luminous intensity $\mathcal{I}_v$, the visual emittance $\mathcal{M}_v$ and the visual illuminance $\mathcal{E}_v$.

The quantities being manipulated are summarized in Table 4.1.

4.3. Real systems

We have the energy elements available to determine the complete photometry of the photographic camera. However, in order to take into account the lens between the source and detector (see Figure 4.8), an important property needs to be introduced: etendue.

Quantity	*Symbol*	*Physical photometry radiometry unit*	*Visual photometry unit*
Luminous flux	Φ	$watt$	$lumen$
Spectral intensity	$\Phi(\lambda)$	$watt.m^{-1}$	$lumen.m^{-1}$
Luminous energy	E	$joule$	$lumen.s^{-1}$
Irradiance	$\mathcal{L}$	$watt.m^{-2}.sr^{-1}$	$lumen.m^{-2}.sr^{-1}$
Spectral irradiance	$\mathcal{L}(\lambda)$	$watt.m^{-3}.sr^{-1}$	$lumen.m^{-3}.sr^{-1}$
Luminous intensity	$\mathcal{I}$	$watt.sr^{-1}$	$candela = lumen.sr^{-1}$
Emittance, radiosity and excitance	$\mathcal{M}$	$watt.m^{-2}$	$lux = lumen.m^{-2}$
Spectral emittance	$\mathcal{M}(\lambda)$	$watt.m^{-3}$	$lux = lumen.m^{-3}$
Illuminance	$\mathcal{E}$	$watt.m^{-2}$	$lux = lumen.m^{-2}$
Exposure	$\mathcal{X}$	$joule.m^{-2}$	$lux.second$
Reflectance	R	Dimensionless	Dimensionless
Spectral reflectance	$R(\lambda)$	Dimensionless	Dimensionless

Table 4.1. *Energy and visual photometric properties and associated units in the SI system. Photography is a priori just concerned with radiometric metrics. However, a long tradition has promoted visual photometric units, easier to interpret for an observer*

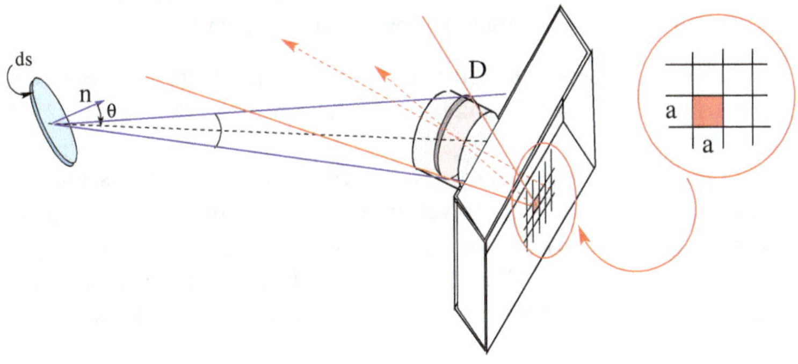

Figure 4.8. *Inside a camera, the beam originating from the object is captured by the objective through its entrance diaphragm and then imaged on the photodetector. The photodetector step is a, the focal length is f and D is the diameter of the diaphragm*

4.3.1. *Etendue*

The etendue (or geometric etendue) U characterizes the dispersion of the light beam when it reaches the receiver. It is the coefficient by which the irradiance of the source must be multiplied to obtain the luminous flux. The elementary etendue, according to equation [4.3] and assuming the source locally flat, is defined by:

$$\partial^2 U = \frac{\partial^2 \Phi}{\mathcal{L}} = \partial s \cos\theta \; \partial\omega \qquad [4.25]$$

Using equation [4.8], it is also written in a symmetric form between source and sensor:

$$\partial^2 U = \frac{\cos\theta' \cos\theta \partial s \partial s'}{r^2} \qquad [4.26]$$

which is also expressed according to the elementary surfaces $d\sigma$ and $d\sigma'$ (see Figure 4.2):

$$\partial^2 U = \frac{\partial\sigma\partial\sigma'}{r^2} \qquad [4.27]$$

It should be emphasized that:

– seen from the source, the etendue is the product between the source surface and the solid angle subtended by the photodetector;

– seen from the photo detector, the etendue is the product between the detector surface and the angle under which the sensor sees the source.

The importance of the etendue comes from the fact that this quantity is maintained when light rays are traveling through the optical systems (it is the expression of the conservation of energy within the optical system, performing according to Abbe's hypotheses [BUK 12]). This is used to determine the incidence angles knowing the surfaces of the sensors or vice versa.

4.3.2. *Camera photometry*

It should be recalled that we have defined the field diaphragm in section 2.4.4 as an important component of a photographic objective lens. This element limits the etendue and therefore allows us to establish the energy balance. We are faced with the case of a thin lens whose diaphragm D limits the rays viewed from the source as viewed from the detector. We will mention a couple of ideas about thick systems. Let f be the focal length of the lens. The objective is, therefore, open at $N = f/D$ (N = aperture f-number (see section 1.3.4)).

We will especially examine the frequent situation in photography where $r \gg f$. In this case, the transversal magnification is expressed by $G \sim f/r \ll 1$, otherwise it is expressed in the general case by $G = p'/p$.

Let dt be the exposure time and τ be the transmission rate of the lens. Finally, we will consider a sensor consisting of photodetectors with an $a \times a$ side.

In the case of an object at infinity and of small angles on the sensor (no vignetting), considering the luminous flux traveling from the object and passing through the diaphragm D and building of the object on the photosite, the irradiance received by the photosite is written as:

$$\mathcal{E} = \mathcal{L}\frac{r^2}{f^2}\frac{\pi\tau D^2}{4r^2} = \frac{\pi\tau\mathcal{L}}{4N^2} \qquad [4.28]$$

and the energy received during the exposure time dt:

$$e = \mathcal{E}a^2 dt = \frac{\pi\tau\mathcal{L}a^2 dt}{4N^2} \qquad \text{or} \qquad e = \frac{\pi\tau\mathcal{L}_v a^2 dt}{4N^2 V(\lambda)} \qquad [4.29]$$

depending on whether $\mathcal{L}$ is expressed in $Wm^{-2}sr^{-1}$ or that $\mathcal{L}_v$ is expressed in visual photometry units, $lm^{-2}sr^{-1}$, with $V(\lambda)$ the luminous efficiency (defined from Figure 4.7).

For a Lambertian object, it is possible to take advantage of $\mathcal{M} = \pi\mathcal{L}$ in the previous equations to express the energy based on a measurement of the radiance.

4.3.2.1. *In the general case*

Equation [4.28], taking into account an object which is not at infinity and therefore would not form its image at distance $p' = f$ but at distance $p' = f + s' = f - fG$ (G the transversal magnification being negative in photography where the image is inverted on the sensor) and taking into account the vignetting effect at the edge of the field (angle θ'), a slightly more complex but more general formula is obtained that can be found in numerous formulations [TIS 08] (especially in the standard that establishes the sensitivity of sensors [ISO 06a]):

$$\mathcal{E} = \mathcal{L}\,\frac{\pi\cos\theta'^4\tau}{4N^2(1-G)^2} \qquad [4.30]$$

Assuming that all the photons have the same wavelength λ, starting again from the simplified equation [4.29], the number of photons incident on the photodetector can be calculated (see section 4.1.2):

$$n(\lambda, \delta t) = \frac{\pi \eta \lambda \tau \mathcal{L} a^2 dt}{4hcN^2} \qquad [4.31]$$

where h is Planck's constant, c the speed of light and η the external quantum efficiency of the sensor (see section 7.1.3).

Figure 4.9 shows the number of photons reaching the photosite for several luminance values according to the size of the photosite and for defined experimental conditions: a wavelength $\lambda = 0.5\,\mu m$, an f-number $N = f/4$, an exposure time $dt = 1/100$ s, a quantum efficiency $\eta = 0.9$ and an optical transmission $\tau = 0, 9$, leading to the very simple relation:

$$n = 10^3\ \mathcal{L} a^2 = 0.6\ \mathcal{L}_v a^2 \qquad [4.32]$$

with a in micrometers and $\mathcal{L}$ in $W sr^{-1} m^{-2}$ or $\mathcal{L}_v$ in $lm \times sr^{-1} m^{-2}$ (because V_p(500 nm) = 1,720 lm/W).

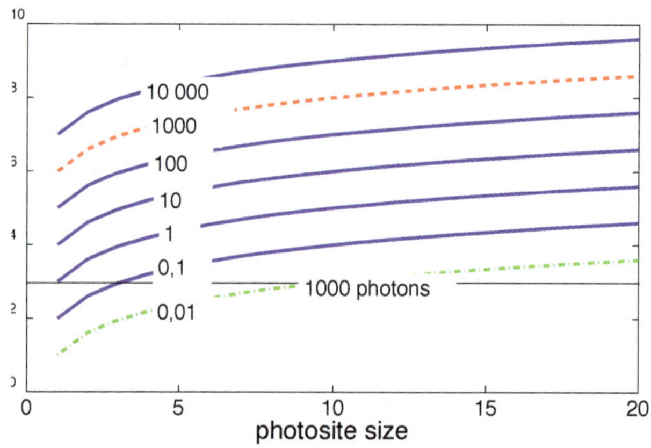

Figure 4.9. *Number of photons collected by a square photosite depending on the value of the side (from 1 to 20 μm) for incident flux values of 10,000 at 0.01 W/m^2. The dashed curve corresponds to a full sun light (1,000 W/m^2), while the dotted curve corresponds to a full moon lighting ($0.01 W/m^2$). The line of the 1,000 photons corresponds to the value above which photon noise is no longer visible in an image (see section 6.1.1)*

4.3.2.2. *Discussion*

From these equations, we can draw a few elements of photography practice:

– as expected, for a given scene, the received energy depends on the ratio dt/N^2. In order to increase the energy, the choice will have to be made between increasing the exposure time or increasing the diameter of the diaphragms;

– when magnification G is small, that is under ordinary photography conditions, the distance to the object has a very small role in the expression of energy. In effect, the magnification G of equation [4.30] is then significantly smaller than 1 and will only be involved in the energy balance in a very marginal manner which is almost independent of the distance to the object if it is large enough. The energy received from a small surface ds through the lens decreases in inverse ratio of the square of the distance, but on a given photosite the image of a greater portion of the object $\int ds$ is formed, collecting more photons and virtually compensating this decrease;

– this observation does not hold for micro- and macrophotography where G is no longer negligible compared to the unit and must imperatively be taken into account. Recalling that G is negative, the energy balance is always penalized by a stronger magnification;

– from these same equations we can see that, if we maintain an f-number N constant, then the focal length f has no influence on the exposure time. With N constant, it is therefore possible to zoom without changing the exposure time;

– it is nevertheless difficult to keep N constant for large focal lengths because $N = f/D$. If the diaphragm D is limiting the lens aperture, it can be seen that f can only be increased at the expense of a longer exposure time dt. We can then expect to be very quickly limited in the choice of large focal lengths by motion defects due to long exposure times associated with large magnification. This is the dilemma of sports and animal photography.

4.3.2.3. *More accurate equations for thick systems*

We have conducted our calculations with a thin lens and small angles. The lens objectives utilized are more complex and allow the small angles hypothesis to be abandoned. They lead to more accurate results at the price of greater complexity.

It is then important to take image formation into account in the thick system. The source then sees the aperture through the entrance pupil, as an image of the field diaphragm in the object space (see section 2.4.4), while the sensor sees it through the exit pupil. These elements then define the etendue [BIG 14].

If the system is aplanatic[6], then equation [4.28] becomes:

$$\mathcal{E} = \pi\tau\mathcal{L}\sin^2(\gamma_{max}) \qquad [4.33]$$

where γ_{max} is the maximum angle of the rays passing through the exit pupil that arrive on the sensor. For symmetric optical systems (that is to say, having optical components symmetrically identical in their assembly), the exit pupil is confused with the diaphragm, and if the angles are small, it yields $\sin(\gamma_{max}) \sim D/f = 1/N$. If this is not the case, formula [4.33] may differ from formula [4.28] in a ratio of 0.5 to 2.

4.3.2.4. *Equations for the source points*

If the object is very small, its geometric image will not cover the whole of the photosite and the previous equations are no longer valid (that is, for example, the case of pictures of a star). It is then essential to abandon the geometrical optics image formation model to take into account the diaphragm diffraction which defines a finite source size and expresses the manner in which its energy is distributed over the photosites (see section 2.6).

4.4. Radiometry and photometry in practice

4.4.1. *Measurement with a photometer*

For the photographer, the measurement of the luminous flux traveling from the scene has always been an important operation. Several concurrent techniques for performing these measurements have been proposed. All these techniques are based on the use of photoelectric cells with calibrated dimensions making it possible to determine a luminous flux generally filtered in the visible wavelength spectrum (objective photometry) and weighted by the luminous efficiency curves (subjective measurements) (also referred to as luxmeter). These measurement systems are either integrated into the camera (they measure energy directly in the image plane, taking into account the attenuation by the objective and the diaphragm), or external to the camera, and it is then necessary to take these elements into account when adjusting the settings of the camera. It should be noted that these differences in principles have been effectively dealt with by the ISO standards which precisely stipulate how these measurements should be made to suit photographic cameras [ISO 02, ISO 06a].

6 A system is aplanatic if it preserves a stigmatic image of any object point neighboring a stigmatic point in a plane perpendicular to the optical axis [DET 97].

In photographic studios, the luminous flux incident on the object is measured in order to balance the sources (spotlights and reflectors). To this end, the cell is placed between the object and the source, turned toward the source (position A in Figure 4.10). This ensures that the various regions of the object are well balanced. The settings are then reconsidered by means of average hypotheses about the reflectance of the object, or by systematic tests.

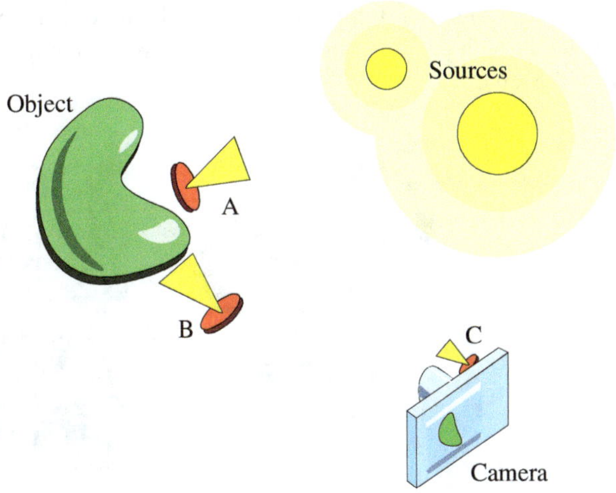

Figure 4.10. *Positions of the light measurement points with a photometer: in A, between the source and the object to ensure a good light balance between the various objects of the source. In B, between the object and the camera to adjust the exposure time. In C, on the camera to determine all of the incident flux*

In the studio under different circumstances, the light intensity reflected (or emitted) by a particular area of the source is measured by placing the cell between the object and the camera, close to the object (position B in Figure 4.10). This allows the exposure time for this area to be directly and accurately determined, at the risk that other parts of the image may be poorly exposed. When the object is not available before taking the shot, it can be replaced by a uniform and Lambertian area usually chosen as a neutral gray with a reflection of 18 %. This value is chosen because it corresponds to the average value of a perfectly diffuse 100 % reflectance range when measured in a large number of scenes. It lies in an area of high sensitivity to contrast perception and thus ensures a good rendering of nuances [ISO 06a].

These two measurements are most often made with rather broad cells (a few cm^2), either with a flat surface (emphasis is then clearly given to the flux originating from the perpendicular directions), or with a spherical cap covering the cell (which is often smaller) which makes it possible to have an integration of the fluxes traveling from all directions (see Figure 4.11 at the center).

In order to reproduce this type of experiment when the object cannot be accessed directly (e.g. outdoors), a photometer can be used, equipped with a very directive lens which can be pointed toward the area to be measured (see Figure 4.11 on the left).

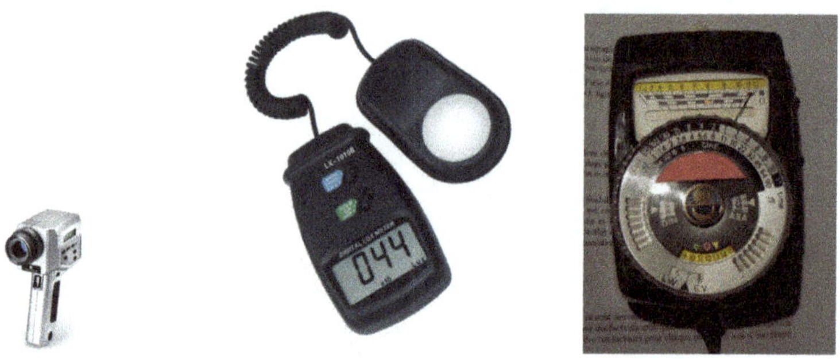

Figure 4.11. *On the left, a photometer allowing directional measurements to be carried out on a very narrow angle (typically 1°). In the center, a light meter capable of measuring incident fluxes in a half-space. On the right, an example of a cell displaying the old DIN and ASA units (the ISO unit is equal to the ASA unit)*

Numerous old cameras had photoelectric cells coupled with the camera objective lens, which allowed the measurement of the full value of the flux emitted by the scene, or an average value around the axis of view. Compared to modern systems, the positioning of the measurements was quite inaccurate. The standards stipulate that a measurement of the overall energy flux must integrate the measurements on a disk $\mathcal{A}$ whose diameter is at least 3/4 of the smallest dimension of the image.

Let A denote the surface of $\mathcal{A}$, the measured energy is then the average of the energies at each point of the disk:

$$\bar{E} = \frac{1}{A} \int_{\mathcal{A}} e(x, y) dx dy \qquad [4.34]$$

It should be noted that the geometric mean could have been measured. This mean value would correspond better to the perception of the visual system, since it would have integrated the energy log (or the optical flux densities), according to the formula:

$$\bar{\bar{E}} = \exp\left(\frac{1}{A}\int_{A}\log[e(x,y)]dxdy\right) \qquad [4.35]$$

This more complex formula has not been retained. In the case of uniform images, $\bar{E} = \bar{\bar{E}}$, but for more complex images, it still yields $\bar{E} \geq \bar{\bar{E}}$. For example, if the image consists of two equal ranges and with energy e_1 and $e_2 = 2e_1$, $\bar{E}$ is equal to $\frac{3}{2}e_1$, while $\bar{\bar{E}}$ is equal to $e_1\sqrt{2}$.

4.4.2. *Integrated measurements*

Modern devices have taken advantage of the integration of sensors to provide greater numbers of measurements of the luminous fluxes emitted by the scene. The measurements are carried out on a large number (between 10 and 300) of generally very small-sized cells, spread over the entire field of the photo. The measurements of these cells are combined in order to provide, on demand:

– either a narrow field measurement on a specific object;

– or a weighted combination of a small number of cells selected in a portion of the image;

– or a combination of all cells to decide an overall exposure.

Some systems also allow the displacement of the axis of view of all the cells within the image. The choice of the measurement strategy can be defined by the user (supervised mode) or automatically determined by the camera after an "analysis" of the scene, which gives the possibility to decide on the best choice (automatic mode). The decision may also deviate from the choice of the average value of the energy (as proposed by equation [4.34]) to adopt more subtle strategies closer to the geometric distribution of equation [4.35], or strategies excluding the extreme values that are in fact saturated, in order to better compute the intermediate values.

It can be seen that the modern camera allows measurements to be obtained very quickly that the analog camera could not. The decisions which are taken to weigh the various cells depend on the expertise of manufacturers and are not available to the general public.

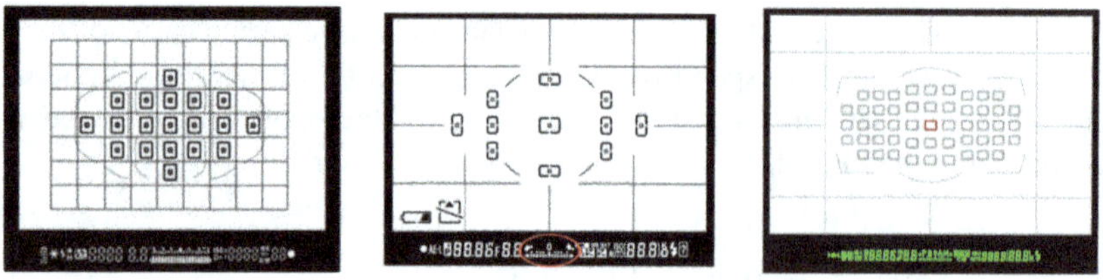

Figure 4.12. *Three examples of arrangements of the energy sensors in the fields of view of digital cameras*

4.5. From the watt to the ISO

With a digital camera, the photographer has several parameters available to adapt the picture to the incident luminous flux:

– the exposure time dt;

– the diaphragm aperture D and, as a result, the f-number: $N = f/D$;

– the sensitivity of the sensor that we refer to by S (this feature does not exist for analog cameras that are loaded with film with given sensitivity for all of the shots).

We have covered the role of the exposure time and the diaphragm in the energy balance. How does the sensitivity come into play?

4.5.1. *ISO sensitivity: definitions*

4.5.1.1. *In analog photography: film sensitivity curve*

The sensitivity of photographic films is governed by the chemistry of their composition and developing processes. History has imposed DIN sensitivities (German standard in logarithmic scale) or American linear standard (ASA) to characterize this sensitivity. It is now standardized by the ISO sensitivity [ISO 03] which is in practice aligned with the ASA standard and unified for films and solid sensors. Generally, the price of high sensitivity comes at the expense of a low film resolution[7] which presents a coarser "grain" (lower limit of the film resolution). A film has a fine grain for a sensitivity of 100 ISO (or 20 DIN) or below. It is then known as "slow". It is known as "fast" for a sensitivity of 800 ISO (or 30 DIN) and beyond. The film has a strongly

7 Resolution will be the subject of section 6.1.2.

nonlinear behavior which, in particular, results in the lack of response at low energies and in saturation when it is overexposed. Doubling the ISO sensitivity is equivalent to exposing the film twice, either by opening the diaphragm by a $\sqrt{2}$ ratio or by exposing with twice the time.

We will choose the example of the black and white negative film[8], but the reasoning would be similar with a positive film, with paper or with a color emulsion: only a few changes of vocabulary about the quantities being manipulated need to be introduced. Film is characterized by its sensitivity curve, H&D curve (Hurter–Driffield) [KOW 72], which expresses its blackening degree according to the amount of light it receives. Such a curve is shown in Figure 4.13: it expresses the logarithm (in base 10) of the transmittance (which is called optical density) with respect to the logarithm of the exposure ξ the film has received.

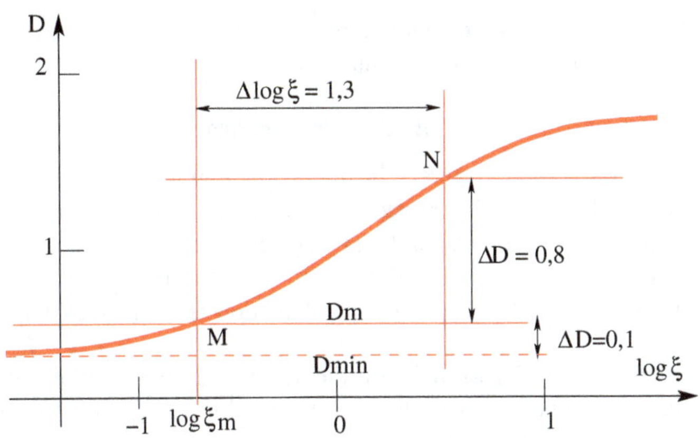

Figure 4.13. *Sensitivity curve of a negative film: blackening of the film according to the log of the exposition energy ξ. The blackening is expressed by the optical density. The film usage linear range is located around the inflection point. To determine the ISO sensitivity of the film, a particular H&D curve is selected among all those that can be obtained by changing the developing conditions. The one that passes through the point N is retained once the point M is defined. The point M is the first point that exceeds the density of the non-exposed film by 0.1. The point N is the point of the curve whose density is that of M plus 0.8 for an energy of $10^{1.3}\xi_m$. The ISO sensitivity is then given by: $ISO = 0.8/\xi_m$; ξ_m is expressed in lux.second*

8 Film or negative paper becomes even darker as it receives increasingly more energy.

The blackening curves depend on many factors (in particular on the conditions for developing film: bath temperatures, product concentrations, etc.). The standard ISO 9848 [ISO 03] defines the ISO index which allows the ordering of the film sensibilities.

Starting from the family of the blackening curves of a given film, the minimum transmittance value of the film is first determined (when it has been developed without receiving any light): that is D_{min}. The curve that passes through the two points M and N, known as standard contrast points, is then selected (it should be noted that there is only one for a given film) (see Figure 4.13), identified by:

– M has a density$D_m = D_{min} + 0.1$ (and an abscissa $\log \xi_m$);

– N has for coordinates: $\log \xi_n = \log \xi_m + 1.3$ and $D_n = D_m + 0.8$.

The ISO sensitivity is then defined by: $S = 0.8/\xi_m$ (with ξ_m expressed in lux.second), rounded to the nearest integer of the ISO table.

4.5.1.2. *In digital photography: sensor sensitivity*

The situation is very different in digital photography because, once the architecture of the sensor is defined (site geometry, choice of materials and coatings), the gray level corresponding to a given pixel count can be affected in two ways: by the analog gain of the conversion of the charge into current, and then by the gain of the digital processing chain downstream (see Chapter 3).

Nevertheless, in order to maintain the same quantities as those that were familiar in analog photography, the ISO has tried to transpose (with more or less success) the existing standards. Thus, a digital ISO standard [ISO 06a] has been redefined which allows the use of familiar acronyms in adjustments that can now be performed on an image-basis and on any camera.

This standard considers three slightly different but nevertheless very similar quantities to characterize the camera response to a flux of light. They are needed because the cameras operate very differently: some do not give access to essential precise definition settings. Their differences are generally ignored by the users and manufacturers willingly misinterpret them for commercial purposes under the same name of ISO sensitivity in commercial documents intended for the general public:

– the standard output sensitivity (SOS), which takes into account the entire image manufacturing chain in a monolithic manner;

– the recommended exposure index (REI), which translates the nominal operating conditions depending on the manufacturer, but does not guarantee image quality;

– the ISO speed that specifies exactly what is obtained on the sensor output when it is exposed to a given luminous flux. The ISO speed is the closest to what is expected of a standardized sensitivity.

For these quantities, the standard is always concerned with conditions as fixed as possible regarding the temperature, the type of illumination, the camera settings (exposure duration, amplifier gain and white balance), as well as the experimental conditions. It is usually proposed to define default values for a daylight illuminant (D55) or for a tungsten illuminant (in which case, the letter T is added to the given value).

Practical considerations about the precise determination of the sensor sensitivity can be found in [BOU 09] in Chapter 9.

4.5.1.3. *ISO speed*

The standard provides two different ways to determine the ISO speed $\mathcal{S}$ when taking pictures, either by promoting the high values of the image or by considering its low values, and in this second case, it offers two different quality levels. These various definitions will apply to very different cameras because between a professional camera and an affordable cell phone, it is not possible to find common operating ranges that allow the definition of a standard protocol. The ISO speed $\mathcal{S}$ is defined as the weighted inverse of an exposure ξ judiciously chosen and expressed in lux.second. The formula type is:

$$\mathcal{S} = \frac{10}{\xi} \qquad [4.36]$$

where the value 10 is chosen for compatibility with existing standards.

The two definitions are the result of the following procedures:

– the saturation method quantifies the sensitivity that characterizes the best image quality possible. It applies equation [4.36] to an irradiance ξ_{sat} which is just below the saturation. The sensitivity is then defined by: $\mathcal{S}_{sat} = 78/\xi_{sat}$. The value 78 is determined such that a picture of an 18% gray[9] gives a desaturated image of 12.7%, that is to say $18.\sqrt{2}\%$). The value $\sqrt{2}$

9 We have explained in section 4.4.1 the reasons for the choice of this 18% value.

is introduced to assume a semi-diaphragm below saturation. It is this method which guarantees the best reproduction of very bright areas;

– the method based on noise concerns on the one hand, very good quality images and in order to achieve this a signal-to-noise ratio of 40 to 1 is required, and on the other hand images of acceptable quality to which it imposes a 10 to 1 ratio. To this end, the curve expressing the signal-to-noise ratio depending on the illumination level is used (Figure 4.14) and the operation points N and M and the corresponding exposure are considered: ξ_{40} and ξ_{10}. Therefore, it can be derived that:

$$\mathcal{S}_{10} = \frac{10}{\xi_{10}} \quad \text{and} \quad \mathcal{S}_{40} = \frac{10}{\xi_{40}} \qquad [4.37]$$

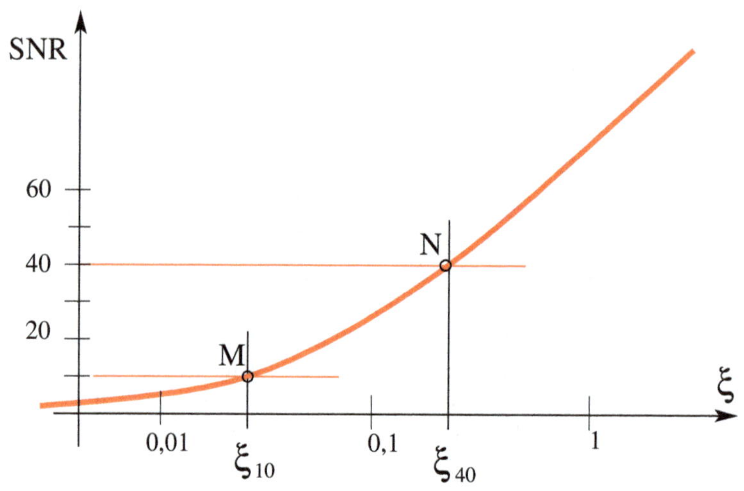

Figure 4.14. *For a digital camera, the sensitivity can be defined from the signal-to-noise ratio curve according to the log of the exposure of the sensor. Two definitions are possible: one for quality systems uses the abscissa of the point N which corresponds to a signal-to-noise ratio of 40, the other, for cheaper cameras, uses the point M with an SNR of 10. Curves such as this one are usually determined experimentally from constant grayscale test patterns*

For a color sensor, ξ is determined by weighting the irradiance noise: $Y = 0.21R + 0.72V + 0.07B$ and in the chrominance channels $Y - R$ and $Y - B$.

4.5.2. *Standard output ISO sensitivity SOS*

The standard output sensitivity[10], considers the whole camera system as a black box. It requires an input image (in practice an 18% gray test pattern) and determines the exposure on the sensor which allows a fixed output level to be obtained. The general formula for a signal whose maximal dynamics is O_{max} is:

$$\mathcal{S}_{SOS} = 10,000/(461\xi_{SOS}) \quad [4.38]$$

where ξ_{SOS} is the exposure which allows the level $0.461.O_{max}$ to be obtained, the term 0.461 originating from the compatibility with the other definitions. Given a signal coded on 8 bits, ξ_{118} allows the obtention of the output level 118 ($118 = 256 \times 0.461$), and the formula becomes:

$$\mathcal{S}_{SOS} = 10/\xi_{118} \quad [4.39]$$

If the exposure measurement ξ is not available (that is placing a sensor on the photodetector), the exposure measurement L of the pattern can be derived by the formula:

$$\xi = \frac{\pi T \nu \cos^4(\theta) L t F^2}{4A^2 i^2} \quad [4.40]$$

or from its simplified form in the case of an infinitely distant object and from standard losses in the lenses:

$$\xi = \frac{65 L_a t}{100 A^2} \quad [4.41]$$

4.5.3. *Recommended exposure index*

The recommended exposure, or REI, is a reference value given by the manufacturer to clarify a good quality operating point of its system. It is allowed full flexibility to assign the sensitivity as it decides to consider the operating range that it deems desirable. It is its "vision" of the sensor sensitivity. This solution has been left to the manufacturer who now has numerous solutions (material or software) to provide very good quality images with an average nominal sensor sensitivity.

10 Standard output sensitivity, or ISO SOS.

To fix the REI, an exposure value ξ_{REI} expressed in lux.second is provided, of which the REI sensitivity is derived by the usual formula:

$$\mathcal{S}_{REI} = \frac{10}{\xi_{REI}} \qquad [4.42]$$

The REI is particularly useful to manufacturers of accessories so that they can adapt their cameras (flashes or photometers) based on the nominal operating conditions.

4.5.4. *Exposure value*

Derived from the energy measurements, the exposure has also received a standardization framework.

The exposure is the product of the irradiance by the exposure duration[11]. Bridging radiometry on the one hand and the sensitivity of the detector on the other hand, the exposure value $\mathcal{Y}$ measures the light received by a scene during the exposure and expresses it in photographer units. To this end, the exposure value is defined from the lens aperture, exposure time and sensitivity of the sensor by:

$$\mathcal{Y} = \log_2 \frac{N^2}{\delta t} + \log_2 \frac{\mathcal{S}}{100} \qquad [4.43]$$

The ISO sensitivity $\mathcal{S} = 100$ is thus used as a reference since it cancels out for $\mathcal{Y} = \log_2 \frac{N^2}{\delta t}$. By convention, the exposure value is then zero for a lens aperture open at $N = 1$ and an exposure time of 1 s.

Each increment of 1 in the exposure value corresponds to a decrease by a factor of 2 of the received energy.

The exposure values have given rise to tables experimentally obtained to indicate the standard parameters of a good photo (snow, sand scenes, under a blue or a covered sky, etc.). However, these tables have fallen into disuse with the narrow coupling of photo-sensitive cells and digital sensors.

11 Exposure and irradiance express a same property, the exposure is reserved to the received energy, while the irradiance rather designates the energy emitted.

5

Color

On the one hand, the concept of color is the result of a subtle operation between the physics of electromagnetic radiations and on the other hand, the psycho-physiology of perception. This double inheritance justifies the complexity of the explanatory mechanisms of color. It is also not surprising to learn that color, since the antiquity, has caught the attention of both the greatest philosophers and physicists. We will quote, in particular, Aristotle, Plato, Pythagoras, Locke, Young, Newton, Goethe, Dalton, Rayleigh, Helmholtz, Lorentz, Maxwell, Schopenhauer, Husserl, Land and many others[1].

It is therefore a difficult task to expose, in a few lines, the main elements of the reproduction of colors by the photographic camera, and we can be certain that we will miss important developments. This is all the more true as the field continues to unravel many unknown as modern techniques for examining biological structures and of perception mechanisms are beginning to appear. A number of excellent books are fully dedicated to color to which we gladly refer the reader who wishes a comprehensive approach of the field [HUN 95, KOW 99, WYS 82].

Faced with the immensity of the topic, what are the objectives of this chapter?

They could be reduced to two formulas:

– How can a colored field be effectively and accurately acquired?

1 On the relationships between philosophy, physics and color we may refer to [KEM 90].

– How can this recorded information be restored to account for the perception that the photographer had thereof?

In order to fulfill these objectives, we should first of all explain how radiation originating from light sources interacts with the objects of the universe and is responsible for the information that will give rise to color stimuli: the green of a leaf, the red of an apple (section 5.1.1). We will have to explain how the human visual system transforms this radiation into a particular perception, referred to as "color", an intransmissible perception, but curiously very widely shared. This will be the subject of section 5.1.2. We will see that this stage is accompanied by a strong information reduction, due to the trivariance of chromatic perception. This trivariance gives the scientist and the technician (as well as the painter) a significant amount of freedom to acquire and represent any color. In order to adopt a universal language shared beyond subjective impressions, the community has equipped itself with representation and measurement tools. Namely the reference observer in the first place and then the chromatic spaces. We will discuss these points in section 5.2.

The broad variety of color spaces reflects the multitude of envisaged applications: to precisely define and measure a color point, to predict the outcome of mixtures, to quantify the distance perceived between colors, to reproduce colored areas on a screen, on paper, on canvas, etc. These steps are fundamental for the photographer, but also for many other players in the business world who wish to take part in the standards. Our task will thus be to keep only the most relevant contributions to photography. Throughout this chapter, we will introduce important concepts such as chrominance, hue, and saturation. We will also introduce spaces familiar to the photographer: RGB, Lab, XYZ, sRGB, CMYK, Adobe RGB, etc. It will allow us, *in fine*, to investigate some difficult problems combining technology and perception subjectivity, such as the effects from changing illuminants, the role of spatial masking and of chromatic persistence, that will naturally lead to the complex issue of white balance faced by all photographers (section 5.3).

We will then access tools to address, in an engineering perspective, chromatic acquisition as seen from the point of view of the photodetector, that is of the choice of a technology capable of associating chromatic information to each pixel. We will distinguish true color acquisition (with three components by pixel), a rare and expensive solution, and spatial multiplexing such as achieved by the Bayer filter array (section 5.4). We will then describe the consequences of multiplexing on the signal quality, and since this solution is very widely adopted, we will deal with the problem of demosaicing (section 5.5) which allows obtaining an image of quality from a subsampled image.

5.1. From electromagnetic radiation to perception

5.1.1. *The color of objects*

The light incident on an eye or on a photographic lens is usually described by a continuum of wavelengths overflowing that largely extends beyond the visible spectrum (which is roughly between 400 and 750 nm). The distribution of wavelengths is specific to the light source on the one hand, and to the interactions suffered by the photons with the various objects in the scene on the other.

When the source is directly visible in the image, its spectral content is affected only by the absorption by the propagation medium. That is the reason why the sunset is seen redder than the Sun at zenith as its rays pass through a thicker atmosphere layer whose absorption depends upon the wavelength according to the Rayleigh diffusion model.

Most of the objects in the scene are however not primary light sources, but secondary sources, that is to say that they return part of the incident light according to very varied mechanisms (more interesting topics about this subject can be read in [CAL 98]).

Most often, the dominant mechanism is a selective wavelength absorption. Thus light penetrates the superficial tissues of a plant leaf; there it meets chlorophyll molecules, a pigment with numerous double bonds whose resonance frequencies lie in the visible spectrum, especially in the large and short wavelengths. Chlorophyll therefore absorbs a lot of light (for the benefit of the plant), but less in the intermediate green wavelengths. The green color is thus propagated towards the external environment, giving the vegetation its color. A similar mechanism explains the red color of blood, the brown of chocolate or the orange of some fruits. But other phenomena can be brought forward as for example:

– the propagation in transparent media with variable indices according to the wavelength resulting in the diffusion of light (prisms), in selective reflections (diamonds) or in energy concentrations within limit angles (rainbows);

– the dispersion of the electromagnetic field in homogeneous environments, especially in the presence of dipoles (Lorentz or Clausius–Mossotti models) explaining the color of oxides or of colored glass;

– the diffusion and the dispersion by particles of size comparable to the wavelength: Rayleigh and Mie theories explaining the color of the sky, but also Melamed or Kubelka–Munk models to account for the way the dusty powders used for printing or for plastic coloring appear like;

– the interference processes involved in very thin layers (oil films) or in stratified environments (wings of a butterfly, varnish, mother-of-pearl);

– the fluorescence processes confined in the quantum dots of nano-crystal giving rise to the colors of certain stained glasses, of doped glasses or of biological markers.

Photography images are created by two types of objects:

– light sources (known as primary), which emit their own photons: these are stars (Sun, stars), lamp bulbs, fires, in very rare cases stimulated emissions of fluorescence[2] or of phosphorescence[3];

– passive objects that reflect part of the light they receive. When these objects reflect large amounts of light, they may be seen as sources of light (designated then as secondary). This is the case of the sky (excluding the Sun), the Moon and many bright surfaces.

Ultimately, the incident light on a camera therefore consists mainly of light sources, of which certain wavelengths are potentially attenuated by the materials constituting the objects in the scene. It is by examining the path of this light, from the sources to the objective, that the color recorded on the photo can be explained.

If the aim was to represent accurately, for example by regular sampling, the wavelength dependence in the visible spectrum $[\lambda_{min}, \lambda_{max}]$ of a luminous flux, it would be necessary to apply the sampling theorem [MAI 08a, VET 14] by determining the bandwidth of the signal from its fastest variations. Experience shows that a small number of samples is often sufficient because spectra emitted by incandescent sources (the most frequent)

2 Fluorescence is the emission of a material subjected to illumination by a photon in a wavelength other than that of the stimulation.

3 Phosphorescence concerns materials that continue to emit light despite their exposure to radiation having stopped, by a relaxation process of the excited levels.

and the absorptions by materials are rarely very narrow or very steep. A sampling effected every 10 nm is often sufficient for many applications; 8–20 measurements should be enough to very precisely reconstruct the composition of the incident light [RIB 03, TRU 00]. It is quite different with discharge lamps and light-emitting diodes, as well as for some synthetic materials, that can have much more chaotic behaviors.

The representation of the content in wavelength by a dozen samples constitutes what is commonly called a multispectral representation of light. When the number of measurements is very widely extended (for example, with 64 or even 256 channels, some of which possibly out of the visible spectrum), a hyperspectral representation is being achieved. These two imaging modes do not fall under the scope of this book and will not be discussed here where we will only be focusing on examining the trichromatic representations.

5.1.2. *Color perception*

5.1.2.1. *The LMS system*

The human visual system works very differently from a sample analysis. It analyzes light with two types of sensors: the cones and the rods[4]. The rods are responsible for a monochrome vision in low light, and the cones for the color vision. When considering photography, it is primarily with the intent of mimicking the daytime vision and we will therefore only focus on the cones. There are three kinds of cones, differently sensitive to wavelengths and designated by LMS, for *long*, *middle* and *short*, these terms qualifying the maximum of their sensitivity, respectively to long wavelengths (in the yellow-green range), medium wavelength (in the green) and short wavelengths (in the blue) (Figure 5.1). It should be noted that the rhodopsyne of the rods presents its maximum between cones S and M. Note also that if human beings have three types of cones[5], many animals just have two such as the cat, sometimes only one as the rat, while the pigeon has five of them.

4 We will discuss here only the more general vision of colors, leaving aside vision anomalies such as the various deficiencies of color perception covered by the generic term of color blindness. Similarly, we will only mention the characteristics shared by a large majority of observers that gave rise to the standard observer and will leave aside the nevertheless frequent variations encountered in non-deficient populations.

5 With the exception of color-blind people who only have two types of cones, or in some cases a single one.

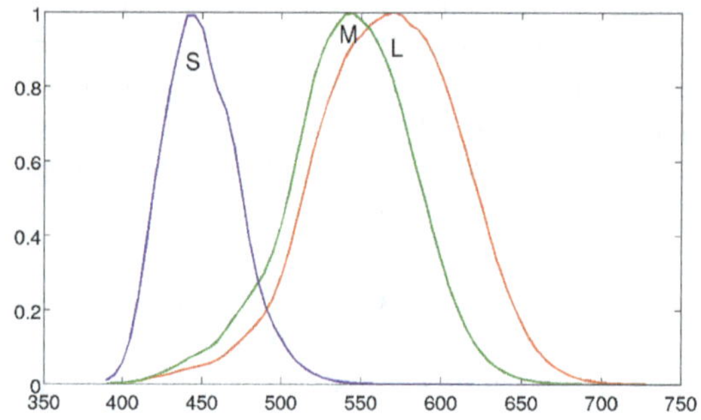

Figure 5.1. *Normalized curves of the response of the three types of cones, with respect to the wavelength, in nanometers, for an angle of view of two degrees. The close proximity between the responses of the cones M and L can be denoted*

Of the three signals coming from these various sensors and arranged in a rather complex manner, visual trivariance makes its appearance, recognized well before the cones and the rods are identified, since already expressed by Young in 1801, and then formalized by Grassman in 1853: "any color stimulus can be reproduced by an additive mixture of three primaries properly chosen".

Visual trivariance is at the base of the main image reproduction techniques in photography, in film and in television (while printing was likely to make use of more complex approaches, sometimes using up to seven primaries). We will come back to this a little further in the text, as it greatly contributes to understanding how standards such as sRGB have been defined and which are universally adopted today. However, we will first of all briefly clarify how LMS signals detected by the retina are processed by the brain [VIE 12, PAL 99], which explains the effectiveness of our color perception and justifies the complexity of color processing in photographic cameras. A more complete description of this visual system as well as analogies with the photo camera are available in [CHA 07a].

The cones are distributed on a fairly regular grid at the back of the retina. There are about twice as many L cones as the M and ten times more than the S. The signals picked up by the photoreceptor cells are divided according to three paths and several processing layers; the neighborhood of the cells in the retina is preserved along the various pathways until the visual areas of the cortex, allowing retinotopic maps to be built in these areas. These mappings are real images of the observed world where the proximity of the objects in the visual field is respected. The three pathways separate certain features very early. The parvocellular pathway builds, from the signals L and M, luminous contrast signals as well as yellow and green chromatic dominant signals to which the magnocellular pathway also contributes. Conversely, the koniocellular pathway develops a picture of the blues from the S signals. These pathways are gathered in the upper layer where chromatic contrast images are created: green-blue opposition on the one hand, green-yellow on the other. A model of the construction of the chrominance signals in the first layers of vision is available in [VAL 93].

5.1.2.2. *Visual pathways and vision areas*

These signals are then processed at the level of the visual areas within the cortex where the concept of color that we assign to objects is developed. We will mention a few important elements involved at this level:

– the first is the notion of color constancy. Although the luminous flux carries in an indissociable fashion, in the contribution of the illuminated object and that of the source that illuminates it, the observer can differentiate one from the other. For example, if the illuminant changes, it does not attribute this development to some change in the color of the object but effectively to the change of lighting, thus emphasizing a real chromatic adaptation [BUR 12]. This is the kind of processing that the “white balance” feature will try to achieve in photographic cameras, with unfortunately often less success.

– a second very important point, especially for the photographer, is the memory of colors, that is both persistent and culturally very important, but also inconsistent because it can be forgotten. It is particularly important for the reproduction of flesh tones (in portraiture), but also for the rare colors of flowers or birds. The mechanisms of this memory are not yet elucidated.

– Last element, that of chromatic masking in which vivid hues “extinguish” nearby chromatic regions, or more generally, where the chromatic perception of a region depends on the nearby regions and even more strongly as they are chromatically more contrasted. This effect can be explained on a short distance by the local integrations of the first cells of the visual pathways, but it can be extended far beyond the receptive fields and cells of the magnocellular and parvocellular pathways and involves a comprehensive adaptation of our interpretation of the whole of the scene.

5.2. Color spaces

An LMS colorimetry has recently been developed from the knowledge available about LMS curves, but it is not used very often [CIE 06, VIE 12]. A large number of other representations have been proposed for almost a century, relying on specific and reasoned choices concerning for example:

– the reference white;

– the three primary colors;

– the orientation of the axes (in particular the choice of assigning an axis to the irradiance);

– the units on the axes (and in particular a linear or logarithmic distribution on the irradiance axis).

The first works undertaken in order to define standards for the transmission of color images, both for photography and for cinema and television, did not take into account the knowledge mentioned above on the human visual system, still unknown at the beginning of the 20th century. They were based on the trivariance expressed by Grassman distributions and the linearity of the perceptive space. They have focused on emphasizing the identification of two generative models:

– an additive model, which builds a color by adding the contributions of three primaries;

– the other subtractive, which by chromatic filtering, subtracts some primaries from a white source.

Photography on film has concurrently used these two models for color films, according to whether positive films (directly producing visible images such as slides), or negative films (requiring the transfer to a medium to render the image) were being employed. Digital photography also uses the two types of synthesis for image acquisition (in quite distinct commercial areas), but the additive synthesis has been chosen as an exchange standard and the user is confronted only with RGB signals (prior to sending their pictures to professionals such as printing or graphic designers that will again be able to reuse the subtractive synthesis).

Subtractive synthesis (using cyan, magenta and yellow masks on the sensor (see section 5.2.6) is preferred on very small sensors (for example, telephones or certain compacts) because the energy balance (measured in photons per site) is better (in the range of 30%) [BOU 09] since each filter allows two components to pass though instead of stopping two of them. Nevertheless, the signal is therein very quickly converted towards an additive space, usually sRGB (which we will discuss later) and the user does not have the opportunity to process these signals. It is therefore on the additive schema that we will focus from now on.

5.2.1. *The CIE 1931 RGB space*

The CIE has defined a reference RGB space (CIERVB 1931)[6] from the following elements:

– the three primaries are monochromatic, with a wavelength of: 700 nm for red, 546 nm for green and 436 nm for blue; pure frequencies have been chosen so as to render a wide color range by positive summations. Their position has been determined in order to cover the areas of high sensitivity of the visual system;

– a reference white is chosen, denoted E or W_0, with constant wavelength power density and equal to 5.3 10^{-2} W/nm; it is the white of equal energy.

An observer has also been defined, which is characterized by its color matching functions (or chromatic), expressing the way which any wavelength is decomposed into three components.

Color matching functions are virtual functions, obtained experimentally, by chromatic range equalization. A pure wavelength is compared to a mixture of the three primaries and the observer aims to equalize the two stimuli, the pure frequency on the one hand, the mixture on the other hand, by judiciously balancing the proportions of the mixture. In general, the balance (a pure wavelength = a suitable mixture of the three primaries) cannot be obtained (for reasons which will be explained in the following lines). One of the primaries must then be added to the unknown wavelength to achieve the balance (a pure wavelength + a primary = a suitable mixture of the two other primaries). In the mixture, the part of this primary added to the white color is then negatively quantified (a pure wavelength = a mixture of two primaries – a portion of the third primary). The resulting curves are presented in Figure 5.2, denoted: $\overline{r}(\lambda)$, $\overline{v}(\lambda)$, $\overline{b}(\lambda)$. As stated, they present almost everywhere at least a negative component, but these components are very small except in the green region where they reflect a strong negative red contribution.

6 Another definition was given by the CIE in 1964 (CIE 1964 standard observer) which does not fundamentally alter the results presented here.

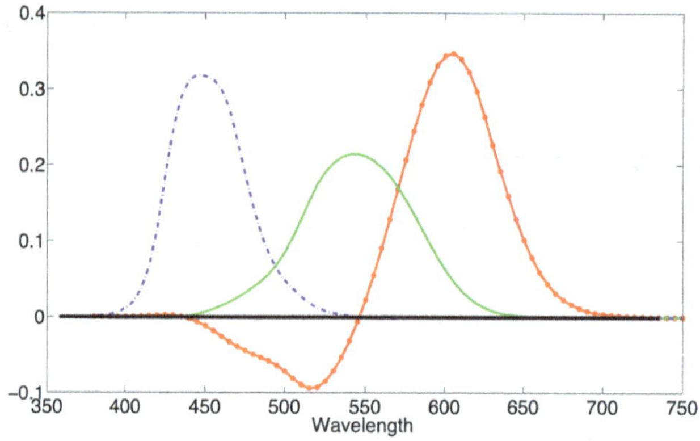

Figure 5.2. *Matching functions of the CIE reference observer – in dashed lines: $\overline{x}(\lambda)$, in solid $\overline{y}(\lambda)$, and with points $\overline{z}(\lambda)$ which presents a significant negative lobe [CIE 06]*

The trichromatic components R, G and B of a stimulus $\Phi(\lambda)$ are given by:

$$\begin{array}{l} R = A \int \Phi(\lambda)\overline{r}(\lambda)d\lambda \\ G = A \int \Phi(\lambda)\overline{g}(\lambda)d\lambda \\ B = A \int \Phi(\lambda)\overline{b}(\lambda)d\lambda \end{array} \quad [5.1]$$

where the integral is taken over the whole visible spectrum, and the normalization value A is equal to 1 if $\Phi(\lambda)$ is expressed in watts, and 683 if it is expressed in lumens (683 lumens per watt corresponds to the maximum sensitivity of the eye (see Figure 4.7)). The coordinates normalized by dividing the sum $R + G + B$ can also be used, they are then written in lowercase:

$$\begin{array}{l} r = \frac{R}{R+G+B} \\ v = \frac{G}{R+G+B} \\ b = \frac{B}{R+G+B} \end{array} \quad [5.2]$$

It is the prototype of all the representations from the real primaries.

The examination of this representation leads to the following conclusions: in the RGB coordinate system, the colors available by positively combining the three primaries, belong to a cube (Figure 5.3 on the left); neutral tones, of increasing brightness, spread over the first diagonal, from black (0,0,0) to white (1,1,1); the Maxwell triangle is the diagonal triangle defined, in this cube, by the equation $R + G + B = 1$; the pure wavelengths form a curve joining the red and the blue colors, passing through the green and the yellow; this curve is open in its direct path between red and blue (there is no pure wavelength corresponding to the purples, mixtures of red and blue).

In any diagonal plane of equation $R + G + B = K$, there is a curve of pure wavelengths of intensity depending of K. This curve for K ranging from 0 to 1 describes a surface external to the cube (except along the three axes, for colors corresponding to the three primaries).

In a diagonal plane (for example for $K = 1$), two variables suffice to represent any color. A chromatic diagram is thus constructed that allows the brightness of the image to be ignored. We traditionally choose R and G as axes (Figure 5.3 on the right). In this diagram, the colors available with primaries R, G and B are always located in the triangle (0:1, 0:1). The site of pure wavelengths is external to this portion of the space. Simple in design, this diagram is still used very little; the one built from the XYZ space being preferred to it (see Figure 5.4).

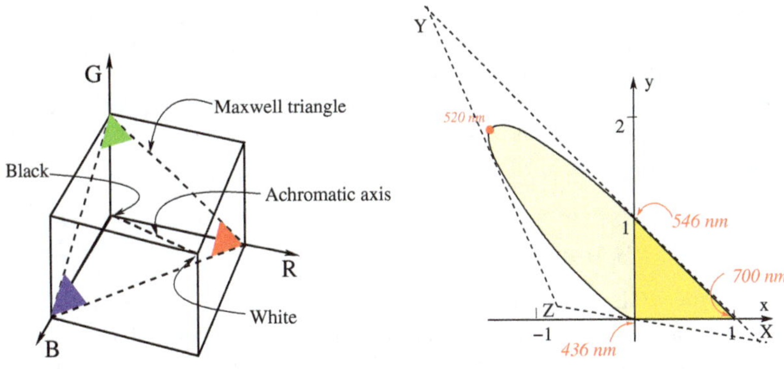

Figure 5.3. *On the left, the cube of RGB colors and the Maxwell triangle. On the right, the chromatic diagram (x,y) built on constant brightness RGB (R+G+B = K): the origin is placed in B and only the normalized R and G coordinates are retained. The site of pure frequencies involves a very important part of negative abscissa. The triangle XYZ is the one that has been selected to define the axes of the CIE 1931 observer in order for any real color be obtained by a positive combination of three terms. For a color version of this figure see www.iste.co.uk/maitre/pixel.zip*

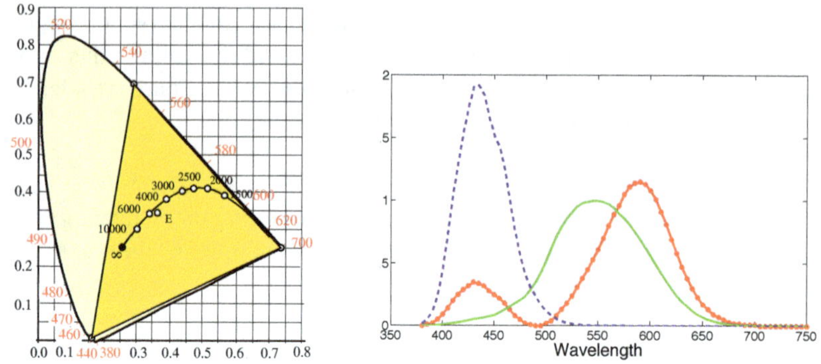

Figure 5.4. *On the left, the chromaticity diagram of the CIE 1931 observer. It is distinguished from that of Figure 5.3 because all the colors have here positive coordinates. The curve external to the triangle is the pure frequency locus, ranging here from 380 nm (deep blue) to 700 nm (red); it is limited in its lower part by the line of purples. The triangle is that of the colors achievable by positive sums of the three RGB primaries chosen by the CIE. The diagram can explain that the negative component of the three primaries* $\overline{x}$, $\overline{y}$, *and* $\overline{z}$ *is strong only for the green-yellow range (see Figure 5.2), but theoretically effectively for any pure wavelength. The Planckian locus is represented by the line crossing the range and graduated in temperatures (Kelvin). The constant energy white point (in wavelength)* W_0 *is represented by the letter E. On the right, the matching functions associated with this XYZ space. It should be noted that they are always positive*

5.2.1.1. *The XYZ representation*

From the RGB space, the CIE has derived (in 1931) a space which is still being used as a reference today because it is based on particularly practical properties since very inspired by human vision. It is however an artificial space that does not correspond to any human observer.

Thus, an axis has been chosen to express the luminance of the image. Then the other two axes have been taken such that any decomposition of a colored stimulus is positive (which is not the case of the RGB space, as we have seen, the pure frequency locus is outside the RGB cube). To this end, we have chosen a particular tangent instead of pure colors as new axes (see Figure 5.3 on the right).

Therefore, it is possible to define a chromaticity diagram from the XYZ space (Figure 5.4). In this diagram, variables x and y are chosen from the two dimensions X and Z of the XYZ space. It is the representation universally adopted to represent a color point, despite some residual flaws that we will discuss later.

Using the colorimetric functions $\overline{x}(\lambda)$, $\overline{y}(\lambda)$, and $\overline{z}(\lambda)$ of Figure 5.2:

$$\begin{array}{l} X = A\int_{visible} \Phi(\lambda)\overline{x}(\lambda)d\lambda \\ Y = A\int_{visible} \Phi(\lambda)\overline{y}(\lambda)d\lambda \\ Z = A\int_{visible} \Phi(\lambda)\overline{z}(\lambda)d\lambda \end{array} \qquad [5.3]$$

where, as in equation [5.2], A is equal to 1 or 683 depending on whether the flux is expressed in watts or lumen, by a change of base we can switch from RGB to XYZ:

$$\begin{bmatrix} X \\ Y \\ Z \end{bmatrix} = \begin{bmatrix} 0.4887 & 0.3107 & 0.2006 \\ 0.17620 & 0.81298 & 0.010811 \\ 0.00 & 0.01020 & 0.98979 \end{bmatrix} \begin{bmatrix} R \\ G \\ B \end{bmatrix} \qquad [5.4]$$

and vice versa from XYZ to RGB by:

$$\begin{bmatrix} R \\ G \\ B \end{bmatrix} = \begin{bmatrix} 2.37067 & -0.9000 & -0.47063 \\ -0.51388 & 1.42530 & 0.08858 \\ 0.055298 & -0.14695 & 1.00939 \end{bmatrix} \begin{bmatrix} X \\ Y \\ Z \end{bmatrix} \qquad [5.5]$$

On the other hand, swapping spaces from the LMS space to the XYZ space can be achieved by modifying the variables:

$$\begin{bmatrix} L \\ M \\ S \end{bmatrix} = \begin{bmatrix} 0.15514 & 0.54312 & -0.03286 \\ -0.15514 & 0.45684 & 0.03286 \\ 0 & 0 & 0.01608 \end{bmatrix} \begin{bmatrix} X \\ Y \\ Z \end{bmatrix} \qquad [5.6]$$

It is currently a lead that is being explored to redefine colorimetry which would free itself from the primary conventions.

5.2.1.2. *Limitations of the RGB space*

The RGB space is not the best space to reproduce the human visual perception because:

– it is difficult to simply attach a color to a point of this space;

– the three RGB components are highly correlated perceptually (decreasing the green component causes the redder hue to come forward);

– it is difficult to separate the intensity and chromaticity notions.

To avoid these inconveniences, it is possible to switch to different spaces whose representation is closer to the human perception of colors, and in which variables can be better decorrelated.

Nonetheless, before leaving the RGB space, we are going to consider limitations that it shares with many other color spaces which we will see further on in the book.

5.2.1.3. *Some properties of trichromatic spaces*

Here we introduce, a few terms which will be useful and that reflect important properties of trichromatic spaces:

– *Metamerism*:

The trivariance of human vision explains the existence of metamerism. Two colors are metamers when they are represented by the same color point but are physically composed of different spectral distributions. Metamerism reflects an irreversible transformation of vision since it is no longer possible to distinguish these two points represented by the same triplet. It is however a valuable asset since it allows the reconstruction of a given apparent form without knowledge of the underlying spectral composition, by a simple adjustment of the three primaries (Figure 5.5 on the left).

– *Gamut*:

With the primaries defined, it is possible to determine all the chromatic space swept by a system of image reproduction, by varying each primary from zero to its maximal energy. Such a volume is called the gamut, that is the range of the representable colors using the primaries. It becomes a sensitive issue when it is necessary to represent colors that are outside the gamut. If the color point to be reproduced is outside the gamut, it will be often toned-down (that is projected on the border of the volume), therefore represented by a nearby less saturated hue, unless is is preferred to change all of the colors by reducing the contrast and the saturation, in order to bring all the color points in the accessible space (Figure 5.5 on the right).

– *Tone mapping:*

The operation that consists of matching a range of colors to the gamut of a device (monitor, printer) is called tonal mapping (see section 10.3.1). It is the subject of very intense studies in photo-printing industries.

– *Planckian locus:*

Black bodies (defined in section 4.1.2) emit radiation according to their temperature, following Planck's law. Each temperature corresponds to a spectrum and to a color point. The Planckian locus crosses the color space from the deepest red to a blue point approximately at $(x = 0.24, y = 0.23)$ (Figure 5.3). It can be graduated in temperatures and reference white points denoted as Dnn, nn can found thereon expressing the temperature in hundreds of Kelvins (see section 4.1.2.1). The equal energy point E, frequently used as the white reference, is also located near this locus and close to the temperature 5,600 K (Figure 5.4 on the left). Since several reference whites are located close to the Planckian locus but not exactly on it, iso-temperature color curves are defined on both sides of this locus. It should be noted that these curves, plotted on the CIE 1931 diagram, are not orthogonal to the Planckian locus.

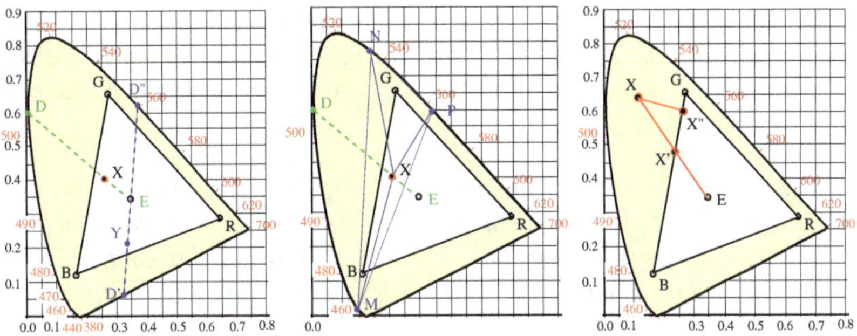

Figure 5.5. *Color space with three RGB primaries and a reference white noted E. Left: point X has for dominant wavelength a blue-green one at 505 nm (point D). The point Y, on the other hand, has no dominant wavelength (the point D' corresponds to a purple which is not a pure radiation). By convention, the wavelength of the complementary (point D") at 562 nm is attributed thereto. In the center: example of metamerism. The point X, single color point in RGB representation, could have been created either by a mixture of blue-green and white (linear interpolation on the line DE), or by a mixture of blue (at 460 nm), of green (at 532 nm) and red (at 564 nm) (barycentric interpolation in the triangle MNP), or even by an infinite number of other combinations of different radiations. They will all be seen as a same color X by the observer. Right: the point X is out of the gamut. It can be represented in the RGB space by point X" (the nearest to X) or by point X' (same hue as X, but desaturated). For a color version of this figure, see www.iste.co.uk/maitre/pixel.zip*

5.2.2. *Other chromatic spaces*

To introduce these different areas, we have to define, literally and mathematically, the characteristics of color: hue, saturation, and chroma [FLE 62, SEV 96], properties well-known to photographers, but also intermediate quantities borrowed from common sense that must find a precise reference in colorimetry. We will follow the definitions of [SEV 96].

5.2.2.1. *Definitions of chromatic quantities*

Brightness can be distinguished from luminance because a bright object (therefore of strong brightness) may be dimly lit and thus have a low luminance. The terms "bright" or "dark" and "brightness" must not be related to the luminance, but to the transmission factor or to the reflection factor of the body.

Brightness is "*a characteristic of a visual sensation, according to which a surface appears to diffuse more or less light relatively to that received. Brightness is evaluated relatively to the perfect diffuser illuminated under the same conditions*".

The luminance of a color range is proportional to its illumination. Luminance is a "*physics quantity that characterizes a surface emitting radiation in a particular direction. It is the ratio of the flux emitted by the solid angle and the apparent area of the surface from the emission direction*".

The hue is defined as the dominant wavelength of a source; it is a "*characteristic of the visual sensation that can be described by qualifiers such as red, yellow, etc.*". All colors have a dominant wavelength, apart from purples (mixtures of red and violet) that are represented by the complementary dominant.

The concepts of chroma and saturation are often confused, their formal expressions will be defined subsequently (equations [5.8] to [5.10]) and their notations will be respectively C and S.

Chroma is "*the coloring level of a surface, evaluated relatively to the light it receives. The chroma of a given surface is a perceptive characteristic independent of the illumination level. For a constant chroma surface, the chroma increases with the brightness of the surface, unlike the saturation*".

Saturation is the purity factor (proportion of pure chromatic color in the total sensation) or the "*coloring level of a surface, evaluated relatively to its brightness. The saturation of a constant chromaticity surface is a perceptive characteristic independent of its brightness*".

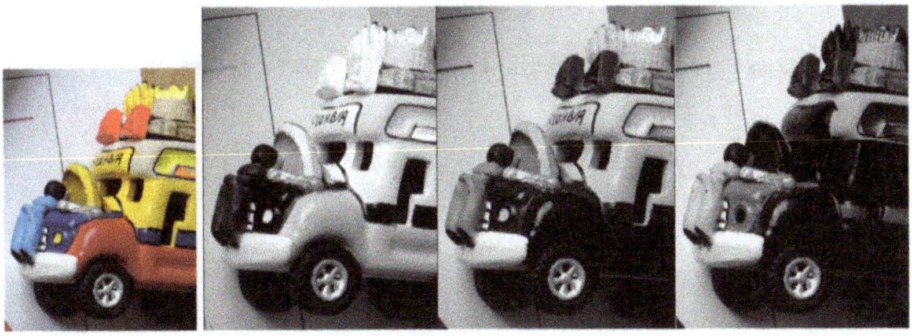

Figure 5.6. *The color image is decomposed into its chromatic channels: red, green and blue which are represented here in grayscale. For a color version of this figure, see www.iste.co.uk/maitre/pixel.zip*

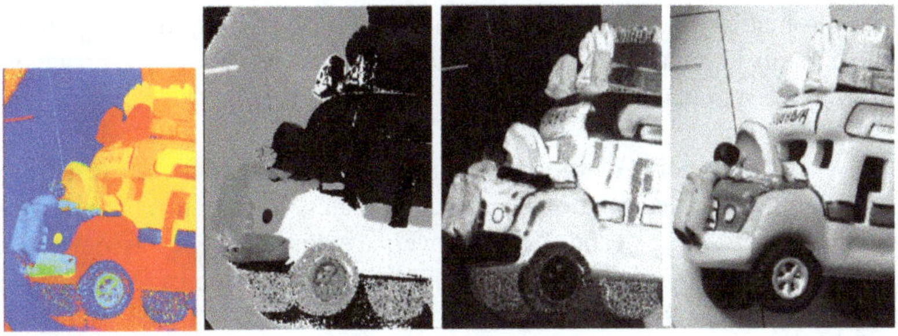

Figure 5.7. *The color image of Figure 5.6 is decomposed here into its chromatic components. On the left, the hue is represented in hues: that is the pure wavelengths from yellow to red. To its right, the hue is represented in grayscale by an angle comprised between O (represented in black, but therefore coding the yellow) through the green (absent from this image and the blue which appears thus in gray) and 2π (represented in white, but coding the red color). On the right, luminance image (this is often the one that is chosen to transform a color image into a black and white image). On its left, the representation of saturation (faded hues appear in black). For a color version of this figure, see www.iste.co.uk/maitre/pixel.zip*

A color range is described by indicating its luminance, the hue and the saturation, or more precisely, the dominant wavelength and the purity factor. In addition, the *Association Française de Normalisation* (AFNOR) has recommended the use of the following words:

– a color is vivid if it is both bright and saturated;

– a color is pale if it is bright and faded (that is little saturated);

– it is deep if it is dark and saturated;

– it is toned-down if it is dark and faded.

5.2.3. *The Lab space*

The Lab space, adopted by the CIE in 1976, bears the name of CIELab. It is a space that intends to be uniform, which means that the differences of color in this space are, as a first approximation, equal to the color differences perceived by an observer.

This space exploits the fact that the luminance of a radiation is independent of its chromaticity. It makes the approximation of the logarithmic response of the eye to this luminance (Weber–Fechner law), by a power of $1/3$, very close, and by a line segment in dark areas.

In Lab, for a point of the space the following components are chosen: L the lightness, a and b, the chroma coordinates. These are obtained from the components, X, Y and Z. Moreover, a white reference is defined of coordinates Xn, Yn, Zn determined depending on the type of the illuminant being chosen [SEV 96]:

– illuminant A: Xn = 109.85, Yn = 100, Zn = 35.58.

– illuminant C : Xn = 98.07, Yn = 100, Zn = 118.23.

– illuminant D65 : Xn = 95.04, Yn = 100, Zn = 108.88.

The Lab coordinates are defined by equations [5.7]:

$$
\begin{aligned}
L &= 116\left(\frac{Y}{Yn}\right)^{\frac{1}{3}} - 16 \\
a &= 500\left[\left(\frac{X}{Xn}\right)^{\frac{1}{3}} - \left(\frac{Y}{Yn}\right)^{\frac{1}{3}}\right] \\
b &= 200\left[\left(\frac{Y}{Yn}\right)^{\frac{1}{3}} - \left(\frac{Z}{Zn}\right)^{\frac{1}{3}}\right]
\end{aligned}
\qquad [5.7]
$$

For argument values less than 0.008 856, the power function $x^{\frac{1}{3}}$ is replaced by the line segment: $7.787x + \frac{16}{116}$.

The orthogonal plane to the axis of brightness is the chromatic plane containing the two axes a and b. It should be observed that the axis a carries the green-red antagonism and the axis b the blue-yellow antagonism in a quite similar fashion to what is found in the early stages of the visual pathways, as described in section 5.1.2.1.

In the Lab space, the variables a and b both vary from –299 to 300, thus taking 600 values. A space denoted by La*b*, utilizes values of a*, b* ranging from –127 to 128, therefore coded on 8 bits. This is the most used in image processing.

It is often preferable to characterize the color of a color point by its hue H and its chroma C, rather than by the values a and b. In order to define H and C, the cylindrical coordinates are used: the chroma is then the distance from the point to the achromatic axis, the hue is defined as the angle made by the direction of the white to the point with the axis a:

$$C = \sqrt{a^2 + b^2} \qquad [5.8]$$

$$H = \arccos \frac{a}{\sqrt{a^2 + b^2}} \qquad \text{if} \quad b > 0 \qquad [5.9]$$

$$= 2\pi - \arccos \frac{a}{\sqrt{a^2 + b^2}} \qquad \text{otherwise} \qquad [5.10]$$

5.2.3.1. *Properties of this space*

The Lab space is almost uniform, which means that a gap between two colors is roughly equal to the gap perceived by humans. The L component is decorrelated from the chromatic components, unfortunately, chroma depends on lightness. The switch from RGB to Lab causes a loss of numerical precision due to the successive changes of spaces, in particular to the approximation caused by the exponent $\frac{1}{3}$.

5.2.4. *Other colorimetric spaces*

Many other color spaces have been proposed to solve users' specific problems.

The spaces named YCrCb or YC1C2 propose linear transformations from an RGB space to offer both linear brightness and emphasis of the chromatic antagonisms. It is this kind of transformation which is recommended during transmissions with JPEG or MPEG encoding.

Space U'V'W' has been built on XYZ. Its component V' is the luminance, but the component W' is equal to $1/3(-X + 2Y + Z)$. Although linearly derived from XYZ, it offers a better homogeneity (the distances are therein closer to the distances perceived by the observer). It was created to provide a simple tool for computer graphic designers and for image processing. It has been improved by the creation of the LUV space, built similarly to Lab with an almost logarithmic luminance, but on U'V'W' and not on XYZ, which thus consists of a good approximation of a perceptually homogeneous space, often preferable to Lab.

The IHS space (*intensity, hue, saturation*) was created by the image processing industry for its ease of use and for the simplicity of the interpretation of the variables. The components of a point of this space are the intensity (I), hue (H) and the saturation (S) and are defined, switching to the YC1C2 space, by the formula:

$$\begin{bmatrix} Y \\ C1 \\ C2 \end{bmatrix} = \begin{bmatrix} 1/3 & 1/3 & 1/3 \\ 1 & -1/2 & -1/2 \\ 0 & \sqrt{3/2} & -\sqrt{3/2} \end{bmatrix} \begin{bmatrix} R \\ G \\ B \end{bmatrix} \qquad [5.11]$$

Then, after choosing $I = Y$, the values of H, and S are defined as follows:

$$\begin{array}{llr} \text{if } R = G = B \implies S = 0 & \text{and} & H = 0 \\ \text{otherwise } S = 1 - \frac{3\min(R,G,B)}{R+G+B} & \text{and if } C2 > 0 \Rightarrow & H = \arccos(\frac{C1}{\sqrt{C1^2+C2^2}}) \\ & \text{otherwise} \Rightarrow & H = 2\pi - \arccos(\frac{C1}{\sqrt{C1^2+C2^2}}) \end{array} \qquad [5.12]$$

But in this space the important property of the reversibility of the transformation is lost due to the term min which is used to define the saturation. It is therefore a workspace rather suited for performing measurements than for representing an image therein.

5.2.5. *TV spaces*

In order to guarantee the universal reproducibility of colors, analog television systems have adopted display and transmission standards, concurrently developed.

The American system NTSC presents an RGB space connected to the XYZ space by direct matrices:

$$RGB\_XYZ_{NTSC} = \begin{bmatrix} 0.607 & 0.174 & 0.020 \\ 0.299 & 0.287 & 0.114 \\ 0 & 0.066 & 0.117 \end{bmatrix} \qquad [5.13]$$

The European systems PAL and SECAM, present the same RGB space connected to the XYZ space by a different matrix:

$$RGB\_XYZ_{PAL/SECAM} = \begin{bmatrix} 0.429 & 0.343 & 0.178 \\ 0.222 & 0.707 & 0.071 \\ 0.019 & 0.132 & 0.939 \end{bmatrix} \qquad [5.14]$$

Certainly, inverse matrices do exist to return to the XYZ space from the video signal.

5.2.6. *The sRGB space*

This space has acquired a significant importance in the field of photography because it has very quickly established itself as the *de facto* standard. Under the pressure of the digital imaging industry, it has contributed to defining a representation mode for the acquisition, the exchange, the display and the processing for the general public (recognized by IEC6 1962-1 standard). It is today the default operation mode of photo cameras; more sophisticated formats, often proprietary, are available to professionals: Adobe-RGB, ACI-RGB, ProPhoto-RGB, etc.

The sRGB space is based on the CIE RGB space, but instead of taking pure frequencies for primaries, it makes use of color points representative of the photophores of the most commonly used video screens. The high-definition television standard (recommendation UIT-R, BT, Rec 709) is employed as reference. It provides the reference white point in the XYZ space: the D65 white point (see section 4.1.2) ($X_w = 0.3127,\ Y_w = 0.3290$). It is a very slightly bluish white. The three primaries are also fixed: red ($X_R = 0.64,\ Y_R = 0.33$), green ($X_G = 0.30,\ Y_G = 0.60$) and blue ($X_B = 0.15,\ Y_B = 0.06$). These primaries are not very saturated (less than the pure RGB frequencies), but in practice, they offer much greater brightness, which results in a narrower gamut (reproducing less correctly the pure frequencies), but more significant at high luminances (Figure 5.8 on the left), therefore better restoring the contrasts of the majority of the shades that are not very saturated when there is a limitation to 8-bit dynamics.

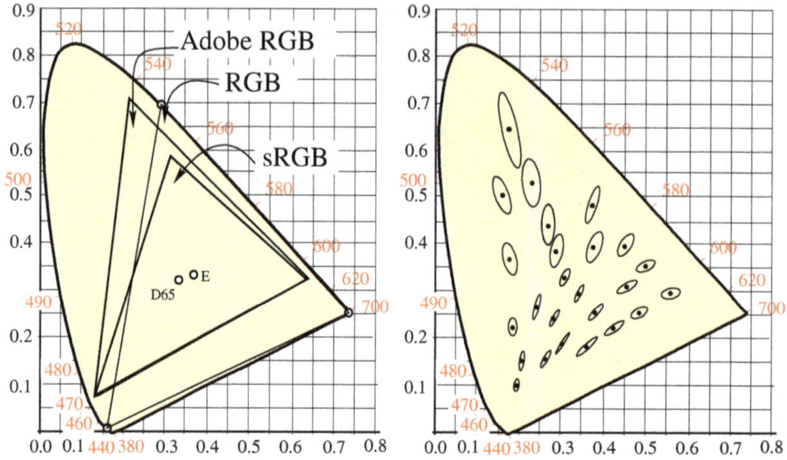

Figure 5.8. *On the left, three triangles of primaries in the CIE1931 space: those of RGB, sRGB and Adobe RGB. The sRGB primaries are less saturated than the pure frequencies of RGB. The choice of the white is the equal-energy white E for RGB, the D65 for the other two. The Adobe RGB uses a deeper green than the sRGB, giving a wider gamut in the blue-greens. On the right, the MacAdam ellipses express the tolerance of the visual system to a chromatic deviation, it is very strong in the green, very low in the blue*

The luminance is inferred by the formula: $L = 0.2126\ R + 0.7152\ G + 0.0722\ B$. This formula correctly highlights the importance of the green channel in the appreciation of the gray hue, importance which will be exploited in digital cameras and will explain Bayer's mosaic structure.

The palette available by sRGB is satisfactory for most applications of regular photography because the vast majority of natural colors occupy the central portion of the sRGB diagram. When it is desirable to reproduce scenes richer in saturated blue-greens, color spaces such as the Adobe-RGB can be used since it has adopted the same white, the same red and blue primaries, but a green at $X_G = 0.21$, $Y_G = 0.71$, giving a triangle spreading more in the upper part of the diagram. This improvement nevertheless comes at the expense of a little green desaturation of the ordinary hues that may sometimes appear less contrasted.

Image display screens, as well as printing systems, most often exploit a specific chromatic space, allowing them to make the best use of the technologies used. It is particularly important that the color points that are sent to them be perfectly identified. It is the role of the standardizations such as sRGB to define the parameters necessary to a good color reproduction. It is also the role of color management software programs that have become indispensable components of image reproduction terminals.

A switch from sRGB to XYZ can be achieved by the formulas:

$$\begin{bmatrix} X \\ Y \\ Z \end{bmatrix} = \begin{bmatrix} 0.4125 \; 0.3576 \; 0.1804 \\ 0.2127 \; 0.7151 \; 0.0722 \\ 0.0193 \; 0.1191 \; 0.9503 \end{bmatrix} \begin{bmatrix} Rs \\ Gs \\ Bs \end{bmatrix} \quad [5.15]$$

and vice versa from XYZ to RGB by:

$$\begin{bmatrix} Rs \\ Gs \\ Bs \end{bmatrix} = \begin{bmatrix} 3.2404 & -1.5371 & -0.4985 \\ -0.9692 & 1.8760 & 0.04155 \\ 0.0556 & -0.2040 & 1.0522 \end{bmatrix} \begin{bmatrix} X \\ Y \\ Z \end{bmatrix} \quad [5.16]$$

5.2.6.1. *Switching from RGB to process color*

So far, we have examined only the additive synthesis which is the basis for digital photography. The image output on a printer makes use of subtractive representations of the printing industry, relying on the CMY primaries (*cyan, magenta, yellow*), or on CMYK process color (K for *key*). In principle, the transition from an additive representation to a subtractive representation is simple since it relies on the notion of color complementarity. In practice, as in every subject related to color, it emphasizes subtleties that we will not mention.

Starting from three values $\{r, g, b\}$ in the range $[0, 1]$, a switch to the coordinates $\{c, m, y\}$ in the subtractive CMY space can be obtained by the relation:

$$\begin{bmatrix} c \\ m \\ y \end{bmatrix} = \begin{bmatrix} 1 \\ 1 \\ 1 \end{bmatrix} - \begin{bmatrix} r \\ g \\ b \end{bmatrix} \quad [5.17]$$

The CMYK process color is obtained by adding a black channel that will result in purer grays and deeper blacks. The determination of this channel can be effected in various ways: the minimum of $\{c, m, y\}$, the distance to the origin $[0, 0, 0]$ from the nearest point to the origin in RGB, where the distance to the origin of the plane containing the minimal brightness point, etc. It results for example in the following formulas:

$$n = \min\{c, m, y\} \tag{5.18}$$

$$\begin{bmatrix} c' \\ m' \\ y' \end{bmatrix} = A \begin{bmatrix} 1\,0\,0\,-1 \\ 0\,1\,0\,-1 \\ 0\,0\,1\,-1 \end{bmatrix} \begin{bmatrix} c \\ m \\ y \\ k \end{bmatrix} \tag{5.19}$$

where A can be equal to 1 or $1/(1-n)$ depending on whether it is sought to expand the contrast or not.

5.2.7. *ICC profile*

We have just seen in the few previous examples that the life cycle of a photograph drives its user to very frequently change their color space representation. This task was made easier by a decision of the International Color Consortium to choose a pivot space (named PCS: *profile connection space*) to allow any device (camera, monitor, scanner, printer) to make its workspace known. This is carried out using the ICC profile of the device that indicates how the images on which it works in this pivot space are to be transformed.

The profile connection space that has been chosen is the La*b* of the CIELAB that we have seen above, with a D50 white (of coordinates X = 0.9642, Y = 1.0000, Z = 0.4289). The transition from a RGB space of a device 1 to a R'G'B' space (or to any other colorimetric space with 4 or n components) of a device 2 is done in two steps: first of all, the transformation of the transition from RGB to La*b* is applied by using the ICC profile of 1, then by inverse transformation of the ICC profile of 2 to La*b*, the image is obtained in the second referential.

The transformations can be done by using transcoding tables with the RGB components, by polynomial transformations or by piecewise approximations. In photography, the starting space is imposed by the sensor. By default, the arrival space is the sRGB space, in particular for images encoded in JPEG. However, the end user can choose a wider space (which contains sRGB), such as to not saturate the very pure colors (see Figure 5.4): AdobeRGB or ProPhoto.

When the device uses the native format (RAW format), there is no default space and the space transformation instructions are either directly applied within the device or in the proprietary software on the host computer, or stored in the EXIF file to be used by a future user.

5.2.8. *Chromatic thresholds*

An important question remains unresolved, which concerns the difference between two colors perceived by an observer. Numerous works have been devoted and in particular MacAdam's experimental ones: [MAC 42, MAC 43]. They have relied on the notion of JNDC (*just noticeable differences*) which, as their name suggests, characterize the smallest differences that the observer perceives. It appears that represented under a fixed luminance, for example in the CIE 1931 space, this area surrounding each color point is really represented by an ellipse whose size and orientation are heavily dependent on the color point (Figure 5.8 on the right). The ellipses are very small in the blue, expressing a very strong sensitivity of the visual system. They are much larger in the green where the eye tolerates great differences. These ellipses, known as MacAdam's, allow the construction of the empirical metrics of the color space.

It is difficult to mathematically express the differences in the way colors appear (see for example the discussion in [LEG 68], Chapter 20). We have indicated that the Lab space (or L*a*b* space) had been developed to provide metrics to the color space. This leads to the expression of a difference of the form $\Delta E(X_1, X_2)$ between two color points X_1 and X_2 by the Euclidean formula:

$$\Delta E(X_1, X_2) = \sqrt{(L_1^* - L_2^*)^2 + (a_1^* - a_2^*)^2 + (b_1^* - b_2^*)^2} \qquad [5.20]$$

a formula known as CIE76. It is experimentally found that $\Delta E(X_1, X_2) \sim 2,3$ corresponds roughly to 1 JND variation.

Nevertheless, this form is rather imprecise. This is particularly true for strong chromatic differences $\Delta E(X_1, X_2)$. It has thus been modified by a more complex form (CIE94) using a new space L*C*h*, then again (CIEDE-2000), using adjustment terms [SHA 05]. On the other hand, these forms still remain approximate and in the same manner as it was during Helmholtz's time, who tackled them in the 19th Century, there is still hope to find some Euclidean metrics to represent the perceived distances in any situation.

5.3. The white balance

To the best of our present knowledge, the most general transition from a space to another color space should be done using the CIECAM02 color appearance model (CIE 8-01 Technical Committee) and more specifically its submodel, point transportation CIECAT02 taking into account the primaries $[R, G, B]$ and the starting white point $[E]$, the primaries $[R', G', B']$ and the arrival white $[E']$ and the illuminant $[L]$. This model, which has the merit of also taking into account the adaptation of the observer to the context for each area of color, is very complex, non-linear and spatially variant. It is in practice not used in photography. For switching spaces between primaries, a linear model is preferred that allows swapping from a $[R, G, B, E]$ base to a $[R', G', B,' E']$ base. When shooting with a given camera, the color filters are obviously fixed on the sensor and the measurement primaries are determined regardless of the content of the scene and the lighting. The recommended arrival space (sRGB) is also known and shifting from one space to another is carried out by matrix operations.

Naturally in this operation, the influence of the illuminant is not separated from the hue of the object. However, it is known that this influence is significant. Without adjusting, the camera will record a yellow or a blue range depending on whether the source temperature is warm or cold, while the range being imaged is neutral white. The on-site observer will adapt themselves to the ambient light and will surely view the white area in both cases, but if confronted to the picture later, and without the context of overall adaptation, they will no longer recognize the color which it perceived and will perhaps be surprised with the overly pronounced hue on the wedding dress.

The objective of white balance is to maintain the color point close to an observation under neutral lighting despite the presence of a colored illuminant. It is one of the few cases where it is sought to move away from accurate physical measurements to mimic the distortion due to the human visual system. This is somehow an attempt to recreate the chromatic adaptation of biological circuits of image processing, by means of photographic cameras.

Independently of the change of space associated with the shift from the primaries of the camera to the primaries of the reference (e.g. sRGB), and after this change of primaries has been carried out, it is intended in this particular case to correct all of the hues by using white balance in order to bring the reference point, here the white color of the dress, towards a desired color point. The shifting formulas brought about by white balance are entirely determined from the information defining the white point, that is from three unknowns (the color point $[E]$ of the starting white point).

Von Kries [VON 79] has proposed a very simplified model of color point transformation. It links in a diagonal manner the three initial components R, G, B to the three arrival components R', G', B' through the intermediate of the coordinates e_R, e_G, e_B of the initial illuminant $[E]$ and the coordinates e'_R, e'_G, e'_B of the arrival illuminant:

$$\begin{bmatrix} R' \\ G' \\ B' \end{bmatrix} = \begin{bmatrix} e'_R/e_R & 0 & 0 \\ 0 & e'_G/e_G & 0 \\ 0 & 0 & e'_B/e_B \end{bmatrix} \begin{bmatrix} R \\ G \\ B \end{bmatrix} \quad [5.21]$$

This model is in practice very satisfying under rather general conditions of usage, which cover a large number of situations encountered by the photographer [WES 82]. It allows the calculations to be quickly made (therefore in particular in the camera) and gives good results. It is universally used to make the white balance and most software only provide the user with the three factors that determine the matrix (of which one, that of the green channel, being anyway usually set to 1).

The operator has several strategies available to achieve this balance between the primaries, enabling a closer approximation of the hues that a reference illumination would provide.

5.3.1. *Presettings*

Most cameras have settings available to allow displaying the type of lighting before exposure. They are generally defined using the reference illuminants A, B, C, Dxx, E and F described in section 4.1.2. They allow the definition of a white point (often by projection on the Planckian locus) which is therefore matched to the D65 of sRGB (or sometimes to other white points that can be chosen) between the measurements and the exposure. These illuminants can be symbolically represented by ordinary life situations: “broad daylight”, “overcast”, “shadow”, which could give rise to slightly different choices according to manufacturers. Thus “broad daylight” corresponds to color temperatures ranging from 5,200 to 5,500 K, "overcast" to temperatures higher than 6,000 K, "shadow" higher than 7,000 K. Incandescent lights correspond to black body temperatures (often much higher than the actual temperatures) from 2,500 to 3,200 K. Fluorescent lighting are found ranging from 3,500 to 4,500 K.

The settings thus proposed are obviously average and the obtained results are most often questionable.

Generally, these choices do not generally affect files saved in RAW format (see section 8.2) that are identical regardless of the selected setting. They are used when converting to JPEG and are stored in the ancillary files such as EXIF, to be used during subsequent processing. When they are used to create image files they are then irreversible since quantization will affect very differently the different dynamics. However, it should be noted that these settings adjusted on the camera often allow avoiding a first conversion of the RGB space specific to the camera to the sRGB space with a default white, followed by a later conversion on computer with software programs for color correction, a correction that is always done with loss of precision in the levels.

5.3.2. *Color calibration*

Unlike these automated approaches, the most accurate way to manage color consists of performing a chromatic calibration that allows that all matrix operations performed by the camera behind the scene be guaranteed. It consists of placing inside the scene a color test pattern whose color points are perfectly defined[7].

The transformations of the chromatic space are then performed which enables the color points of the various areas of the images to be brought towards the target values. This approach ignores the corrections made by the camera and can be combined with any illuminant.

If the test pattern works in reflectance, the obtained calibration will integrate the lighting source unless its spectral distribution be explicitly inserted in the calibration formulas. If the pattern behaves by transparency or by emission, then sources will be taken into account in the corrections.

---

7 Test patterns can be commercial products such as, for example, GretagMacbeth or ColorChecker Passport patterns from X-Rite, or color swatches whose properties have been measured by a spectrophotometer.

As mentioned above, and *a fortiori* in the case of rigorous calibration, there are great advantages in working with native files (RAW), that is to say before any intervention on the signal. The corrections usually assumed by the camera and which concern measurement noises as well as possible bias, mosaicing, vignetting, and eventually chromatic aberration, etc., must then be taken into account. Fortunately enough, there is software available making use of the EXIF file that accompanies the image (see section 8.3.2), to perform these complex operations and transform the image signal into a more portable and more interoperable format, as for example TIFF/EP (see section 8.2.4.1) or DGN (see section 8.2.4.2).

However, these operations can also be carried out on more conventional files (such as JPEGs, so usually after shifting into the sRGB space), at the expense of a notorious loss of quality.

5.3.3. *Gray test pattern usage*

Some cameras have white point adjustment features, using the designation of a reference area before exposure. The white point thus identified is converted to D65 during the transformation of the primaries space of the camera to the final sRGB space.

If the camera does not include this facility, the white areas (or neutral gray) can be *it a posteriori* identified, when further processing on computer. This method is however very sensitive to the choice of particular area and we then become potentially exposed, as indicated above, to accumulate conversion errors if RAW files are not available.

5.3.4. *Automatic white balance techniques*

Automatic determination of white balance can be done during exposure (many cameras offer this mode by default) or after the shooting, in post-processing mode. Naturally, the complexity of the decisions may be greater in this second case.

It is difficult to know how algorithms operate in commercial cameras. It is likely that they use advanced versions of the ones we present now (see also [GIJ 11, MAZ 12]).

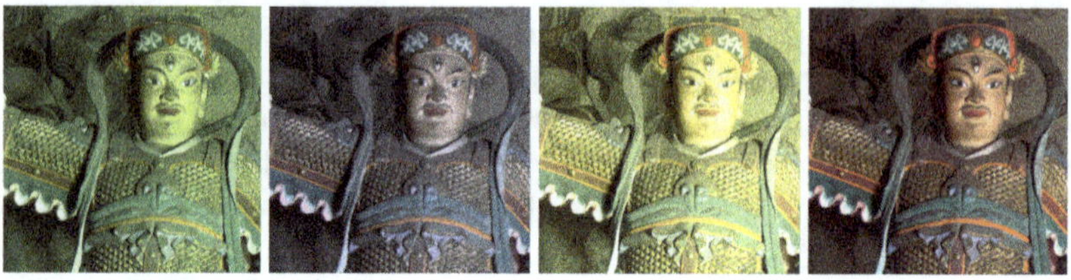

Figure 5.9. *White balance. From left to right: original image, whose parameters (illuminants, camera settings) are unknown, as recorded on chip (during viewing the primaries are assumed to be those of the sRGB space); white balance by the von Kries model, under the gray-world hypothesis, then under the white-page hypothesis, finally on the right, correction made with the (unknown) algorithm of the camera. The very wide variety of renderings can be observed in this difficult situation where the light sources are artificial and the colors strong. For a color version of this figure, see www.iste.co.uk/maitre/pixel.zip*

Most often, it is considered that there is only one light source in the picture. The determination of this source (i.e. of its three components $R_{source}, G_{source}, B_{source}$, which brings back the equation [5.21] to $R' = R/R_{source}, G' = G/G_{source}, B' = B/B_{source}$) can be done according to different strategies based on a few hypotheses (see Figure 5.9):

– under the white-page hypothesis, it is assumed that there is a surface perfectly white in the image; this surface thus gives the color point of the illuminant. Furthermore, since it presents the highest luminance characteristics, it is easy to detect it. This method, as with the following two, relies on statistical studies of numerous natural images which give them a reasonable validity for every day life scenes [GOL 08];

– the gray-world hypothesis [BUC 80] assumes that hues are distributed evenly around the gray axis (on average, the world is grey). The illuminant is then determined as the maximum of the hue distribution (or the average over a small area around this maximum), either in the whole of the image, or after selecting only the slightly saturated pixels, or after having segmented the image to remove the colored areas;

– the gray-edge hypothesis [CHA 12a, GIJ 11] extends a little the gray-world assumption, and thus its generality, at the expense of a greater instability in the computations by statistically studying the variations of colors between neighboring pixels: $\partial R/\partial x, \partial R/\partial y$, etc.;

– the gamut method [FOR 90] tries to determine the illuminant from extreme values that are reproduced in the image and to match it with other gamuts observed further with multiple known illuminants;

– learning methods recover and generalize this statistical approach which consists of accumulating the three-dimensional (3D) chromatic histograms for a wide variety of known illuminants, and then of recognizing, among this very vast collection, the distribution that is the closest to that of the unknown image. When the cloud of points is reduced to its axis of inertia, we are very close to the gray-world situation. When it is reduced to its extremum, this corresponds to finding once more the white-page hypothesis.

Other approaches abandon the hypothesis of an illuminant unique to the entire image and seek to adapt to more illumination sources by locally studying the distribution of the light returned by the scene. This is the case for example of the method known as dichromatic which deals with the problem on a per object basis, provided that each object be constituted of non-Lambertian materials. Under this hypothesis, the light that they reflect towards the sensor is constituted of two components: one is diffuse (Lambertian), whose color is greatly influenced by the object (see section 5.1.1), the other specular, which exactly reproduces the spectral content of the source. Statistical analysis of the color diagram sometimes allows these two components to be isolated (we are looking for an elbow shape in the axis of inertia of the cloud) and the light source that illuminates it to be identified, object-wise.

The most advanced methods are based on human color vision models, necessarily complex as we have seen in the introduction to this chapter. Following the remarks exposed in [PRO 08], the approaches must account for the differential properties of the visual system that determine a perception as opposed to the immediate neighborhood of the fixed area; they must also account for the spatial variations of vision which are enabling the perception of color to evolve in the field of vision according to the context; finally, they must be non-linear in accordance with the physiological responses.

Thus, the proposed methods are quite expensive in terms of computations: the ACE model, [RIZ 03] or the RACE model [PRO 08], itself based on a particular version of the Land and McCann Retinex vision model [LAN 71], allows reporting and accounting for both the white-page and the gray-world hypotheses, but locally applied.

5.3.5. *The Retinex model*

We will return to this property of perception that a color depends on the other colors that surround it. Long considered as originating in the higher areas of the brain (hence the name Retinex, a portmanteau word of retina and cortex), this property can in fact be traced as early as the first levels of vision, near the retina. It uses a particular type of visual cell called double-opponent cells, which respond differently depending on whether the luminous field presents a color contrast between the center of the cell and its periphery. These cells are mainly sensitive to the red-green chromatic contrast but the same can also happen with blue-yellow contrast. Long discussed, the presence of double-opponent cells in primates is now established. It leads to a very complex model of vision because it prompts to take into account in a very localized manner, the visual field in its entire spatial and chromatic aspect.

With the family of the Retinex models, detailed in [LAN 93], Land has attempted to come forward with algorithmic answers to the simulation of this property and thus to account for some reactions of the visual system, such as simultaneous contrast chromatic illusions, ignored by earlier approaches. The role of the Retinex model was mainly notable in computer vision rather than in physiology where biologically better inspired models are preferred.

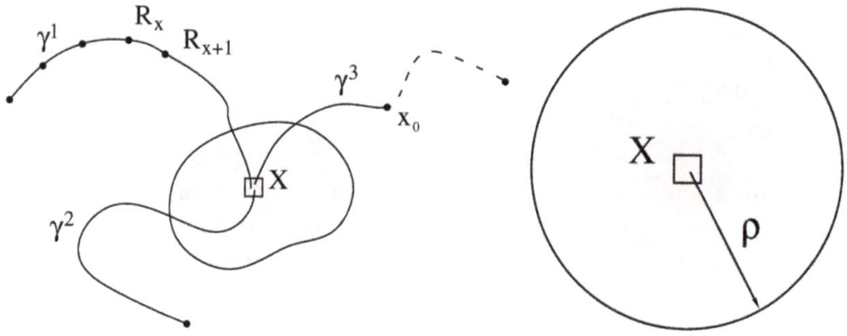

Figure 5.10. *Left: in the Retinex model, paths are chosen ($\gamma^1, \gamma^2, \gamma^3$), which lead to point X whose appearance is to be determined. The lightness ratio of successive points is determined on each path (here in the red channel R). As long as the ratios $R(x+1)/R(x)$ are greater than 1, they are summed up, determining the lightness by averaging all the paths according to equation [5.22]. If the ratio becomes less than 1, the point being reached (here x_0) becomes the new origin of the path (here γ^3). Right: the calculation area of the neighborhood of a pixel X in the case of a multiscale Retinex*

In the Retinex model from 1971 [LAN 71], Land defines the "lightness" L of each pixel (with three components L_R, L_G, L_B) depending of its neighborhood. To this end (Figure 5.10), he randomly draws paths that lead to the pixel being considered and on these pathways examines the evolution of the ratio of the signals R, G or B in two successive points, as well as the accumulated product of these ratios. If these ratios are close to 1, it sets them to 1 (thresholding process). If the accumulated product, after being deviated, goes back to 1, it considers the current pixel as the new origin of the path (zeroing process, which in fact amounts to consider only the paths whose origin is a local lightness maximum). The lightness L in a channel (here we choose the red measurement channel R) is then calculated from all paths κ_k resulting in the pixel, by the formula:

$$L_R = 1/N \sum_k l^k \qquad \text{with} \qquad l^k = \sum_{x \in \kappa_k} \delta \left| \frac{R_{x+1}}{R_x} \right| \qquad [5.22]$$

where N is the number of paths and δ is equal to 1 if the ratio is greater than a threshold and 0 if it is close to 1. The final image is obtained by combining the three lightnesses L_R, L_G, L_B in each pixel thus calculated.

This model suffers from numerous inaccuracies in the definition of the parameters that thus open the door to many variants (several by their authors themselves). A rigorous mathematical interpretation was made thereof in [PRO 05], where its abilities to solve difficult white balance problems are also demonstrated by exemplifying. However, it proves rather complex to implement.

The simplest Retinex model is rather far different than this model and loses many of the expected properties. It makes the assumption that an image contains objects that perfectly reflect red light, others green, and others blue. The objective is to find the maximal values of the Red, Green and Blue channels (denoted R_{max}, G_{max} and B_{max}). This triplet therefore defines the spectral content of the illuminant. Each pixel in the image is changed by standardizing: $R' = R/R_{max}$, $G' = G/G_{max}$, $B' = B/B_{max}$. A white balance that comes close to that which we have mentioned previously is thus obtained.

This algorithm can be applied to the whole image only under the hypothesis that a single source is illuminating it. Otherwise, the image must be partitioned into sub-regions. It will then be assumed that each subregion is illuminated only by a single source and the algorithm will subsequently be applied to the sole subregion. Problems can obviously arise during the partition into subregions as well as when connecting the regions.

A family of algorithms has come closer to the original idea of the Retinex by proposing an approach by filtering with emphasis on the notion of context surrounding a pixel. This is for example what has been proposed in [JOB 97b].

To this end, the lightness in red channel (in the Retinex sense) in a pixel x is defined by:

$$L_R(x, y) = \log(R(x, y)) - \log(R(x, y)) * \phi(x, y, \rho) \qquad [5.23]$$

where $R(x)$ is the intensity of the red channel of the image in (x, y), $\phi(x, y, \rho)$ is a low-pass filter which therefore takes the neighborhood ρ of the pixel into account. The log function is here involved to express the Weber-Fechner's law of perception. A similar computation is carried out on the other two channels. Various filtering functions have been proposed, decreasing according to the distance r to the center of the window: Gaussian functions or in $1/r^2$, extending the area influencing the value of the pixel on rays ρ from 30 to 300 pixels according to the importance of the details of the image. Finally, a multiscale Retinex model was proposed [JOB 97a], which calculates $L_R(x, y)$ in the equation [5.23] by a linear combination of filtering obtained with variable values of ρ:

$$L_R(x, y) = \sum_k \alpha_k \left[\log(R(x, y)) - \log(R(x, y)) * \phi(x, y, \rho_k)\right] \qquad [5.24]$$

The different variants of Retinex generally lead to a significant improvement in the readability of the images, especially when they contain highly contrasted areas. They often lead to chromatic re-balancing which makes them more agreeable. They behave however somewhat critically in certain configurations when highly saturated colors are in the neighborhood. They then require heuristic adjustments limiting the magnitude of the applied corrections that make their automation very sensitive [PET 14].

5.4. Acquiring color

The objective of a color sensor is to transform the luminous flux traveling from the scene, which generally depends continuously on the wavelength in the visible range (between 400 and 750 nm), into three chromatic components, since the human visual system is trivariant (see section 5.1.2.1). In conventional photography, "color films" are composed of three multicolor and chromogenic layers in three wavelength ranges corresponding more or less to retinal sensitivity. Film therefore reconstitutes a colored luminous flux in each "pixel", similar to the incident flux (if the film is "positive" as this is the case with slides) or complementary to the incident flux (if the capture is "negative").

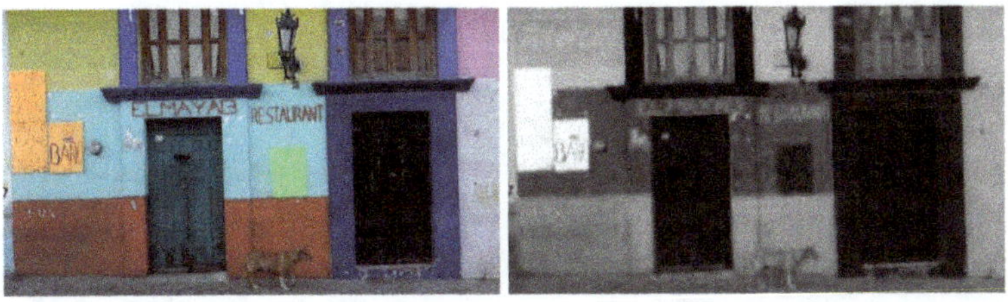

Figure 5.11. *Decomposition of an image in its three RGB components. The original image is on the top left. The red channel on its right is represented in grayscale for an easier comparison with the green and the blue channels in Figure 5.12. For a color version of this figure, see www.iste.co.uk/maitre/pixel.zip*

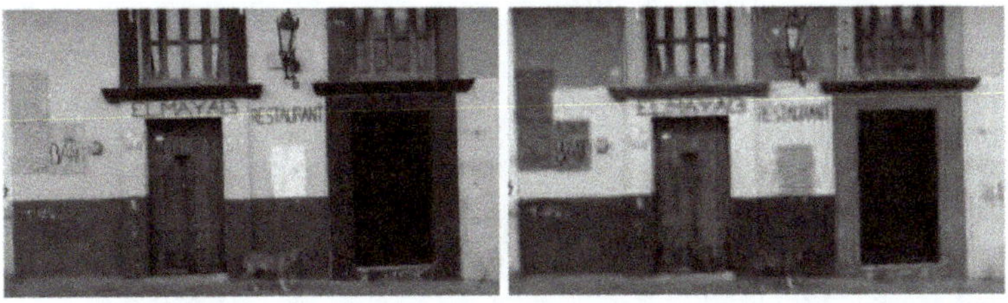

Figure 5.12. *Decomposition of an image into its three RGB components, the green and blue channels (continued from Figure 5.11). For a color version of this figure, see www.iste.co.uk/maitre/pixel.zip*

After passing through the chromatic filters $\phi_R(\lambda)$, $\phi_G(\lambda)$ and $\phi_B(\lambda)$ (see Figure 5.23), the picture, before being converted by the sensor, is constituted of three channels denoted i_R, i_G and i_B:

$$i_R(x,y) = \int_{\lambda=0.4\mu}^{\lambda=0.8\mu} i(x,y)\phi_R(\lambda)d\lambda \qquad [5.25]$$

and similarly for i_G and i_B.

The image signals $i(x,y)$ or their components $i_X(x,y)$, for $X \in \{R,G,B\}$, due to their great diversity are not easily modeled, but among their known properties, it can be observed (see [MAI 08a]) that their power spectrum decreases very quickly and follows a distribution rather similar to a Lorentzian model:

$$\frac{\alpha I_O^2}{\alpha^2 + u^2 + v^2} \qquad [5.26]$$

where α characterizes the exponential decrease of the autocorrelation of the image (the larger α is, the finer the details of the picture will be). The variations of α are typically between 0.1 and 5 for ordinary images. This signal is in addition filtered by the optical system. If we consider that the diaphragm ensures most of the filtering (no aberration, no default exposure settings), and in the case of a circular diaphragm, this signal has a power density spectrum of:

$$|I(u,v)|^2 = \frac{\alpha I_O^2}{\alpha^2 + \rho^2} \cdot \frac{J_1^2(a\rho)}{a^2 \rho^2} \qquad [5.27]$$

with $J_1(z)$ the Bessel function of the first kind, $\rho = \sqrt{u^2 + v^2}$ and a the diameter of the diaphragm.

It is therefore a highly decreasing function and one whose bandwidth is limited.

In the case of a solid sensor, historically (in television for example), in contrast to what has been done in photography, it was decided to separate the optical flux into three channels (by dichroic beam splitters), to receive each channel on a sensor (vidicon, plumbicon, etc.) and to transmit the three signals separately, either on separate cables, or by multiplexing them on the same carrier, but without recombining them electronically. The combination into a colored luminous flux is then ensured by screens that juxtapose stimuli in a very dense manner in the three primaries. When observed at a distance, the image composed of the three primary gives, considering the limited resolution of the human visual system, a colored luminous flux very close to the original.

Digital photography has evolved from these two ancestors (film and television) but has also developed in its own way. Similarly to color television, it has been driven to come up with solutions to measure the three components of the luminous flux. Similarly to film, it has aimed at reconstructing an immediately visible image.

An extremely attractive idea is to design solid sensors which, such as photographic color emulsions, give the three components of the luminous flux in each photosensitive site. Such an idea has given rise to sensors now available commercially, but which still remain a very small minority on the photographic market. We first present it in section 5.4.1.1 as well as other yet very futuristic ideas. We will then describe in section 5.4.2.2 the most widespread solution that does not follow this lead but is based on the juxtaposition of sensors sensitive to different wavelengths.

5.4.1. *"True color" images*

We thus denominate a digital image calculated from a matrix that we will assume as infinite, as the sites having the dimensions $\eta \times \eta$ and their repetition step $\mu \times \mu$ (with naturally, $\mu \geq \eta$) (see Figure 5.13), starting from an image $i(x, y)$ as it appears just before the color filters and the sensors. It is an image that continuously depends on the space variables ($\{x, y\} \in \mathbb{R}^2$), as well as on the wavelength ($\lambda \in \mathbb{R}^+$), which has undergone a filtering of the spatial frequency by the optics (whose diaphragm is limited). It is therefore a signal with spectrum bounded in all directions; we will denote this bound $\mathcal{B}$.

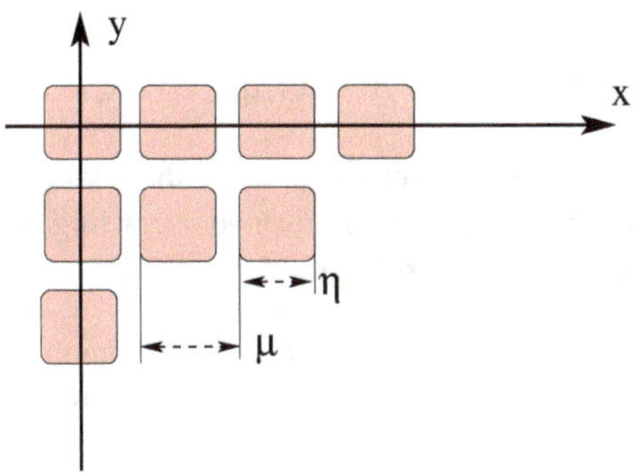

Figure 5.13. *The photosensitive areas of the sensor have a size of $\eta \times \eta$, they are separated by a guard zone in order to constitute a periodic matrix of step μ (with $\mu \geq \eta$)*

We will proceed to the detection. Let o_R, o_G and o_B denominate the detected components. In the case of a "true color" image, each site renders three signals that will undergo identical processing:

$$o_X(x, y) = \left[i_X(x, y) * \text{rect}(\frac{x}{\eta}, \frac{y}{\eta}) \right] ш (\frac{x}{\mu}, \frac{y}{\mu}) \quad \text{for } X \in \{R, G, B\} \qquad [5.28]$$

where $*$ expresses the convolution ($c(t) = a(x) * b(x) = \int_{-\infty}^{\infty} a(x)b(x-t)dx$) and the function Ш is the Dirac comb with a step a : Ш $(x/a) = 1/a \sum_{p=-\infty}^{\infty} \delta(x - pa)$.

It should be noted that it is at this point that the rotational symmetry around the optical axis which characterizes optical systems and that film cameras preserve is lost[8]. In effect, two directions are introduced corresponding to the regular disposal of the photosites on the matrix of sensors. These two directions are generally the horizontal and vertical directions, which are very widely favored in nature. But that choice is not unique as will be seen in section 5.4.2.3. It should also be observed that a second reason will cause this rotational symmetry to be lost: the shape of each photosite. This form is theoretically independent of the repetition step of these photosites. In practice, however, and for reasons of ease of manufacture, the photosites are usually rectangular (or square), and aligned in the direction of the axes of the matrix[9].

Under these assumptions (regular matrix with orthogonal axes, square photosites aligned with the matrix), we will calculate the frequency content of a "true color" image:

$$O_X(u,v) = [I_X(u,v)\ \mathrm{sinc}(\eta u, \eta v)] * \text{Ш}\ (\mu u, \mu v) \text{ for } X \in \{R, G, B\} \quad [5.29]$$

which expresses that the spectrum $I_X(u,v)$ of the analog image is low-pass filtered by $\eta \times \eta$ sized detectors, and repeated according to a $1/\mu \times 1/\mu$ due to the discrete structure of the pixel grid.

Can a digital "true color" system verify the Shannon sampling theorem? Yes, if the signal that reaches the sensor has a limited frequency $\mathcal{B}$, such that:

$$\mathcal{B} < 1/\mu$$

It is noteworthy that the integration window (of size $\eta \times \eta$) of each photosite comes appropriately to attenuate the signal spectrum (cardinal sine term of equation [5.29]) within the repetition range of the spectrum:

8 Strictly speaking, the diaphragm (common to digital and film cameras) is not exactly circular, but takes the form of a polygon (with six, seven, or nine sides), regular, often curvilinear. It is very close to a circle.

9 It will also be necessary to consider the effect of the microlenses placed on the matrix, of a possible anti-aliasing filter, as well as of filters separating infrareds and ultraviolets and take them into account in order to obtain an exhaustive study of the sensor.

$[-1/2\nu,\ 1/2\nu\ ;\ -1/2\nu,\ 1/2\nu\ ]$. If ideally we can assume conditions are met where μ equals η (no guard zone between two photosites), then the sinc function cancels out just at the zero of the parasite first-order, which does not cancel aliasing but reduces it significantly.

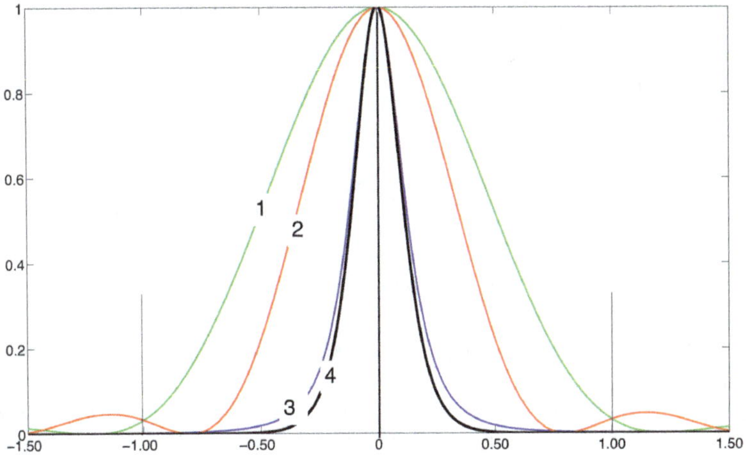

Figure 5.14. *The spectrum of the image signal (curve 1), maximal at frequency zero, the filter imposed by the diaphragm (curve 2) and that imposed by the integration on the sensor (curve 3). If the period of the sensor defines useful frequencies in the band $u \in [-1, 1]$ we have arbitrarily chosen a diaphragm that cancels out at $\pm$ 1.22 and a sensitive cell such as $\mu = 1.25\eta$. The signal resulting from the two filterings prior to periodization by the sensor matrix is represented by the curve 4. For a color version of this figure, see www.iste.co.uk/maitre/pixel.zip*

5.4.1.1. *The Foveon from Sigma*

This CCD sensor consists of sites, each corresponding to a pixel and issuing an RGB signal. The principle of chromatic selection is based on the differences of the absorption power of silicon based on the wavelength. The sensor is therefore conceived as a stack of three layers, each delivering a signal specific to a chromatic range. Thus the blue light is largely absorbed and the first microns being crossed will receive an electric charge which will mainly depend on the blue component of the signal. The green signal will contribute to the next deeper and thicker layer. Finally the bottom of the site will detect the red component over a still more important depth. In the case of the X3 Sigma sensor, a site typically has a surface of $7 \times 7\,\mu m^2$ and a depth of $5\,\mu m$. In Figure 5.15, the absorption curves of the wavelengths are represented according to the depth of penetration.

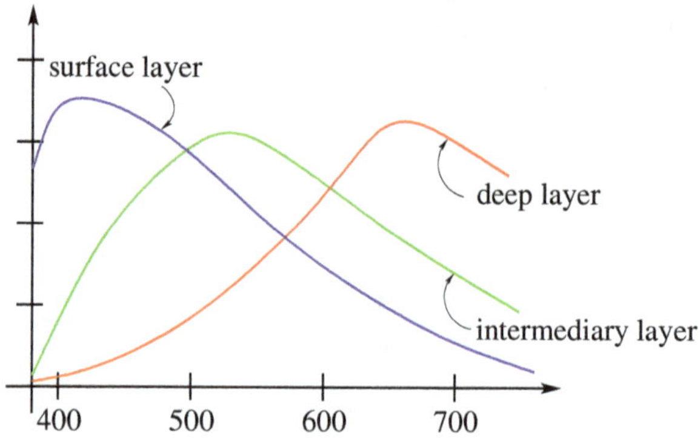

Figure 5.15. *Relative sensibility of the three layers of a Foveon Quattro-type sensor depending on the wavelength (in nm) for the three depths of the embedded layers. Such a sensor provides true color images (three values for each pixel); it takes advantage of the different penetration of the various wavelengths in silicon. Thus, the photodiode placed in the surface layer is very sensitive to blue radiation, while that of the deepest layer is mainly affected by red radiations*

Let's consider that the adopted principle allows the RGB signal to be available in all pixels. It is thus sometimes said that this sensor has 15.4 million "true" pixels, in contrast to other photo cameras that just have, in each photosite, one value R, G or B, and which must therefore combine them to reconstruct an image. The images obtained are deemed to be of excellent quality and the fineness of the color reproduction, greater than that which a CCD of the same size allows using a Bayer filter. We will come back to this point further in the book.

In 2010, Foveon marketed a 15.4 million-pixel sensor. In 2014, another sensor of the same company, Quattro, made use of a slightly different scheme. The first photodiode layer, near the surface, is very dense (with a step of 4.35 μm), while the two deeper layers have a double step of 8.7 μm. The signal is reconstructed by taking advantage of the high resolution of the first layer, the same way as carried out in satellite imagery, by enriching the spectral channel of high resolution informations of the panchromatic channel. The Quattro offers twenty million pixels in the topmost layer and five in each of the two other layers. The reconstruction of 3×20 million pixels is done by image processing, on the one hand to separate the chromatic components from quite saturated raw signals, on the other hand to resample the three channels with the same step.

5.4.1.2. *The other solutions for "true color" images*

Other avenues are being explored to enable true color detection in layers of silicon. Thus, in [SEO 11] the theory of a sensor consisting of silicon nanofibers stacked vertically in a monocrystalline silicon substrate is presented. The color selection is then made, not by the depth of penetration as in the Foveon sensor, but by the diameter of the fibers exploiting the selective excitation of the propagation modes. [PAR 14] presents the first images obtained with a sensor designed on this principle, using nano-fibers whose diameter varies from 80 to 140 nm, placed on a regular frame with a step of 1,000 × 1,000 nm.

We have also mentioned, at the beginning of Chapter 3, the advances made to take advantage of the specific structures of quantum dots often exploiting, microstructured graphene, but at an even smaller scale, in order to simultaneously achieve the detection of the image and chromatic selection.

Quantum dots are electronic structures presenting very narrow potential wells and confined in spaces of a few nanometers (hence their name). Considering their dimensions, they offer optical properties relevant for both emission and for light absorption as well as remarkable mechanical qualities. Nanostructured graphenes are very fine layers of graphene (a few atoms thick) initially obtained by crystal growth (known as Stransky–Krastanov growth). They are well suited to integration inside more complex electronic structures. Colloidal systems on the contrary make obtaining liquid solutions possible from the dissolution and the recrystallization of various materials such as lead or cadmium sulphides which are transformed into films. The size of the domains thus created makes it possible to select the absorbed wavelengths that continuously cover the entire visible spectrum and beyond and therefore enable a color selection essential to photography. The association of the two types of production is at the heart of photographic applications.

But these sensors are still rare. The first quantum dot cameras just began to make their appearance on the mobile phone market in 2015[10]. They consist of a nearly continuous photosensitive layer regularly sampled on its underside by connections to the VLSI in charge of the digital image formation.

---

10 These first photodetectors are marketed under the name of QuantumFilm by the Californian company InVisage.

Finally, other solutions are envisaged that are completely moving away from the previous approaches. As an example, we can mention the idea of returning to an R, G and B channel separation as in television, but at the microscopic level of each photosite. A Nikon patent is protecting this idea (presented schematically in Figure 5.16). In this solution, a microlens concentrates the luminous flux on an aperture. The luminous beam is then split by dichroic mirrors and sent on three specialized detectors separately processing the R, G and B channels. It can be observed that "true color" comes at the expense of a rather complex integrated optical circuitry that will be placed under the microlens only if it covers a far wider field than the elementary photodetector.

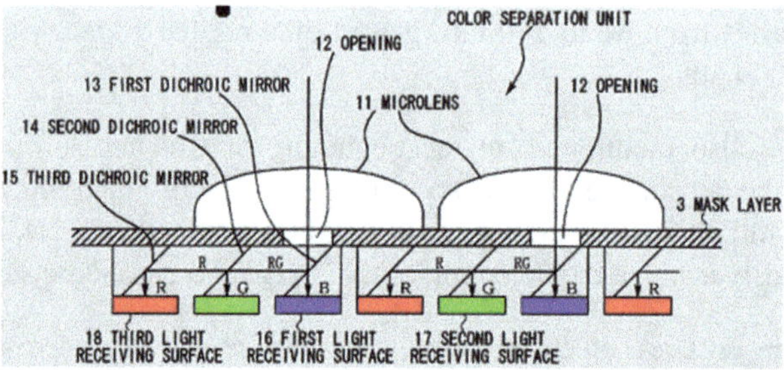

Figure 5.16. *The operating principle of a "true color" sensor by separation of the colored flux in each site. This idea has not yet been implemented. It was presented in a 2007 Nikon patent (patent USPTO 7,138,663)*

5.4.2. *Chromatic arrays*

Storing three colors in each pixel is not the solution that has been overwhelmingly chosen by photographic camera manufacturers. They have, instead, adopted a very symmetrical solution from the one that was prevailing for video screens and which has favored the juxtaposition of sites, each specializing in the detection of a color. This solution works with both the CCD and CMOS technologies that each have a larger wavelength response than the range of the visible wavelengths. It suffices then to cover a site with a chromatic filter to detect a component in this site; the neighboring site is masked by another filter. The three-color image will be rebuilt by signal processing.

Various filtering configurations have been proposed that we will detail later. We will first mention the simplest form, by bands, and then that that was retained for most consumer cameras: Bayer filtering, which we examine next.

5.4.2.1. *Striped arrays*

Striped arrays (or striped maps) are simply obtained by juxtaposing alternating red, green and blue stripes of sensors (see Figure 5.17 on the right). This array has the lowest periodicity (3×1) and is the most simple to implement since it results in identical reconstructions for all the rows and all the bunches of three pixels. We will also highlight that it processes the three R, G and B colors in a similar way.

Starting from the reference equation [5.28], this configuration leads to a signal which is equal to:

$$o_X(x,y) = \left[i_X(x,y) * \mathrm{rect}(\frac{x}{\eta},\frac{y}{\eta}) \right] Ш\,(\frac{x}{3\mu},\frac{y}{\mu}) \text{ for } X \in \{R,G,B\} \qquad [5.30]$$

whose frequency content equals:

$$O_X(u,v) = [I_X(u,v)\ \mathrm{sinc}(\eta u, \eta v)] * Ш\,(3\mu u, \mu v) \text{ for } X \in \{R,G,B\} \; [5.31]$$

where it can be seen, by approximation with equation [5.29], that if the vertical spatial frequencies (according to Oy) are processed as those of a "true color" system, the horizontal spatial frequencies (according to Ox) are assigned a bandwidth $1/3\mu$, three times smaller, therefore exposing the image to unpleasant phenomena of spectrum folding (or *aliasing*)[11]. It is actually the regular result of downsampling of the image plane since it was decided to assign only one site in three to the measurement of each RGB signal.

As we have said the three R, G, B channels are processed in the same fashion, undergoing the exactly same filtering[12] and potentially presenting the same defects.

---

11 It depends naturally on the manufacturer to switch "horizontal" and "vertical", by rotating the sensor. It can also be oriented according to the diagonal, but at the cost of a complexification of the electronic circuits, if the intention is to keep an image with a rectangular format according to the ordinary axes.

12 Up to a dephasing term $\exp(-2j\pi\mu u)$ that we ignore, which expresses that the measurements are not made at the same point. These problems will be widely covered when we will discuss demosaicing.

5.4.2.2. *Bayer filtering*

The array filter was proposed in 1976 by B.E. Bayer, an engineer at Kodak. It consists of favoring the green channel, to which 50% of the site is allocated, compared to the red and blue that just have 25% each of the sites. Their disposition is arranged in a basic 2×2 pixel pattern across the matrix (see Figure 5.17 on the left), therefore slightly less effective on average than the previous. This choice is based on the observation that the green channel is the closest to a luminance channel to which the human eye is particularly sensitive to fine details. As a result, the green channel will be sampled with a rate of 50%, whereas the red and blue channels will be assigned a rate of 25% (to be compared with the rate of 33% globally assigned to the three components in striped filtering).

The chromatic components R and B are strongly undersampled and accordingly will give *a priori* channels with a frequency twice lower, in accordance with the properties of the visual system that distinguishes much finer luminance details. It can be expected that Bayer filtering, although a little less efficient with the radiometric flux, provides nevertheless a better quality perception to the human observer matching its properties.

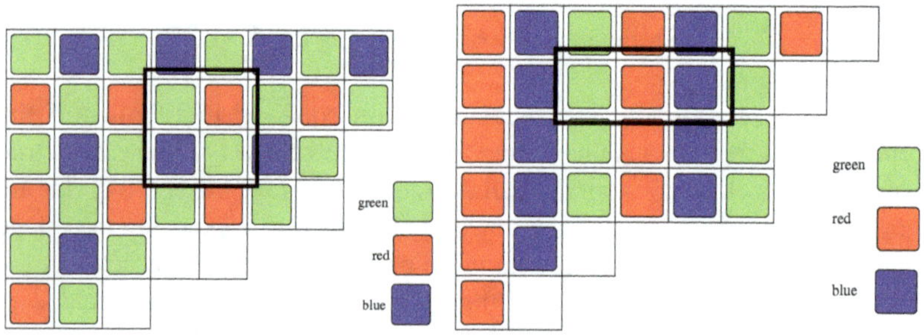

Figure 5.17. *On the left, Bayer filter array; half the sites are sensitive to the green component, a quarter to the blue, a quarter to the red. The repetition element step of the pattern is 2×2. On the right, a chromatic stripe selection array. A third of the pixels is assigned to each color. One direction is heavily under-sampled while the other is correctly sampled. The pattern is 3×1. The configurations of the basic patterns of the periodicity are traced in black. For a color version of this figure, see www.iste.co.uk/maitre/pixel.zip*

The signal of the blue channel, or of the red channel is written:

$$o_X(x, y) = \left[i_X(x, y) * \text{rect}(\frac{x}{\eta}, \frac{y}{\eta}) \right] Ш \, (\frac{x}{2\mu}, \frac{y}{2\mu}) \text{ for } X \in \{R, B\} \qquad [5.32]$$

and its frequency content is:

$$O_X(u,v) = [I_X(u,v)\ \mathrm{sinc}(\eta u, \eta v)] * \text{Ш}\ (2\mu u, 2\mu v) \text{ for } X \in \{R, B\} \quad [5.33]$$

while the green channel is written as:

$$\begin{aligned} o_G(x,y) = &\left[i_G(x,y) * \mathrm{rect}(\tfrac{x}{\eta}, \tfrac{y}{\eta})\right] \text{Ш}\ (\tfrac{x}{2\mu}, \tfrac{y}{2\mu}) \\ &+ \left[i_G(x,y) * \mathrm{rect}(\tfrac{x}{\eta}, \tfrac{y}{\eta})\right] \text{Ш}\ (\tfrac{x+\mu}{2\mu}, \tfrac{y+\mu}{2\mu}) \end{aligned} \quad [5.34]$$

$$\begin{aligned} o_G(x,y) = &\left[i_G(x,y) * \mathrm{rect}(\frac{x}{\eta}, \frac{y}{\eta})\right] \\ &* [1 + \delta(x+\mu, y+\mu)]\ .\text{Ш}\ (\frac{x}{2\mu}, \frac{y}{2\mu}) \end{aligned} \quad [5.35]$$

and its frequency content is (up to a phase term $\exp(-j\pi\mu(u+v))$):

$$O_G(u,v) = [I_G(u,v)\ \mathrm{sinc}(\eta u, \eta v)]\,.\cos(\pi\mu(u+v)) * \text{Ш}\ (2\mu u, 2\mu v) \quad [5.36]$$

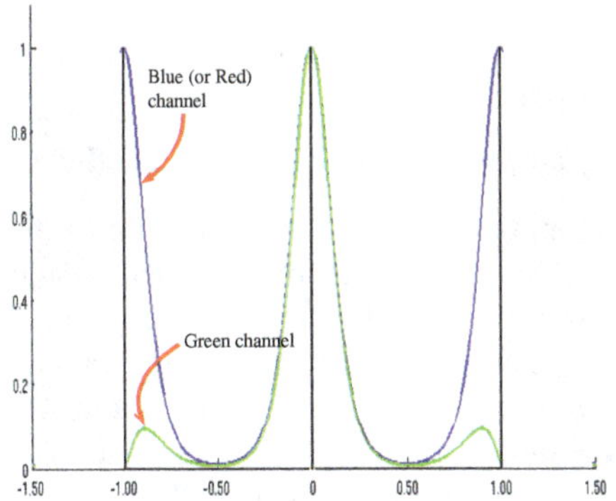

Figure 5.18. *Given a dimension, the power spectral density of an image created by a sensor using a Bayer mosaic filter: green channel on one hand, blue channel (or red) on the other hand. The periodization of the photosites corresponds to a bandwidth between -0.5 and +0.5. The blue channel, sampled every two pixels, undergoes a very strong aliasing that the green channel barely exhibits. For a color version of this figure, see www.iste.co.uk/maitre/pixel.zip*

As expected, it is observed that blue and red frequencies will undergo a strong aliasing in both directions, u and v. The green signal benefits from twice as much samples and from favorable filtering (the zero of the cosine appropriately cancels out the first aliasing for $u = 1/\mu$). It therefore results in an (almost) satisfactory signal in the green channel, while it is significantly under-sampled in the R and B channels (see Figures 5.18 and 5.19).

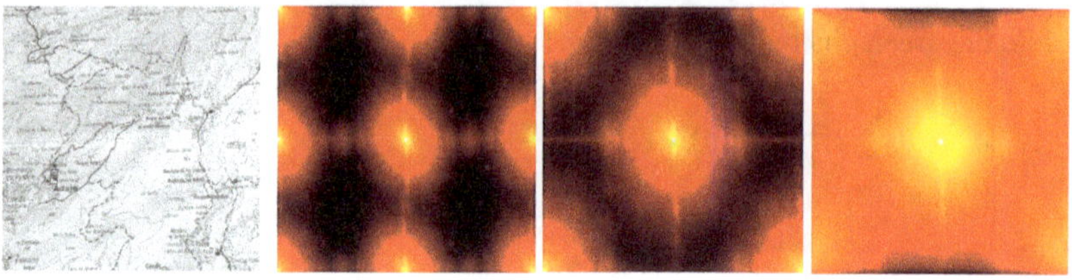

Figure 5.19. *On the left, the green channel, reconstructed by interpolation, of a RAW image. The three images on the right represent the frequency spectrum module of the blue channel before interpolation, then of the green channel before interpolation, and finally of the green channel after interpolation. To facilitate the readability of these spectra, they have been filtered by a low-pass Gaussian filter, and represented in logarithmic scale. It can be emphasized that aliasing is twice as strong from the blue channel compared to the green channel and that the role of the interpolation in rejecting noise orders of the green channel outside the bandwidth thus reduce aliasing. For a color version of this figure, see www.iste.co.uk/maitre/pixel.zip*

5.4.2.3. *Other sensor arrays*

Several avenues have been explored to provide other arrays of sensors:

– Kodak has proposed solutions in which three masked R, G and B photodetectors and a non-masked (therefore panchromatic) sensor is utilized. These solutions can tolerate fairly small periods (typically 2×2) but the solutions developed have rather explored research combining the Bayer approach (a green channel twice as rich) and panchromatic measurements and result in large periodicities (4×4), allowing the reconfiguration of the sensor to suit the needs of the end user (see Figure 5.20);

– Fujifilm has proposed quite a complex CMOS matrix (called EXR), which groups neighboring sites detectors of the same color (see Figure 5.21 on the left) oriented at 45 degrees from the horizontal [FUJ 11]. The photodetector can also be dynamically reconfigured to adapt to three specific situations:

- the situations where a very high resolution is looked for, where all pixels are treated independently,

- situations where there is very low luminosity, where the pixels of the same color are grouped in pairs in order to reduce the noise,

- high dynamic situations (HDR images), where two almost similar images are created, but offset by one pixel allowing access to strong dynamics;

– Fujifilm has also introduced a second "pseudo-random"-type of array called Trans-X with the aim of reducing the effects of spectrum aliasing without resorting to polarizing filters (see Figure 5.21 on the right). We will note, from the analysis of these examples, that the reconfiguration properties of the sensor are at the heart of the projects of new matrices, as well as photometry and colorimetry considerations;

– in 2003, Sony proposed a sensor with four primaries, with a similar configuration to Bayer's, but with one of the green pixels replaced by an *emerald* pixel (that is green-brown). This sensor, which seemed originally developed for surveillance applications, has also been installed on a consumer camera (the DSC F628). The images, even though of good quality, have not been made subject to particular media coverage despite the originality of their capture principle, which suggests that they do not differ noticeably from images captured with a conventional Bayer filter array.

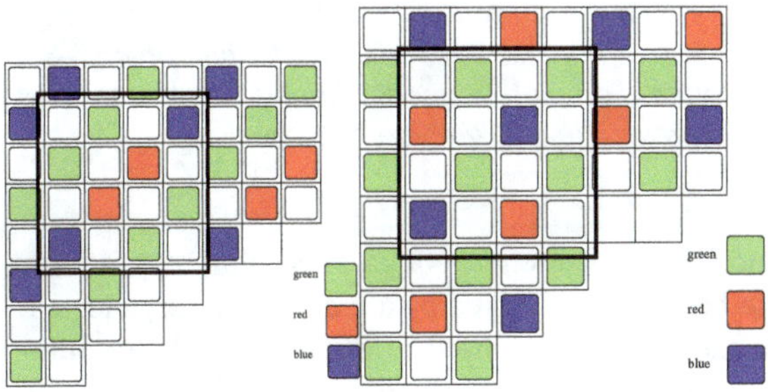

Figure 5.20. *Two configurations offered by Kodak to replace the Bayer filter array: sites in white have a panchromatic sensitivity supposed to give a better smoothness to the image. For a color version of this figure, see www.iste.co.uk/maitre/pixel.zip*

5.4.2.4. *CMYK arrays*

So far, we have left aside chromatic selection matrices based on subtractive color synthesis, that is using the complementary colors of the RGB colors of additive synthesis: the cyan, magenta and yellow hues.

However, a number of devices exist based on this principle. Their most striking feature is that they present a better energy balance than the sensors masked by RGB filters at the cost of a slightly less accurate colorimetry. They are therefore reserved for very small sensors for which the energy balance is critical: very compact devices, e.g. cell phones.

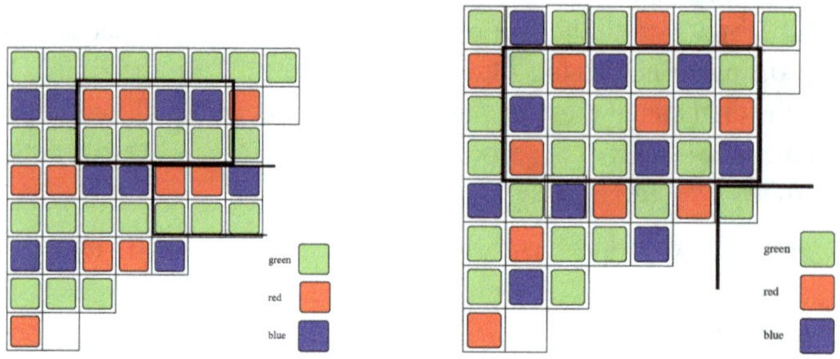

Figure 5.21. *Configurations proposed by Fujifilm using original filter arrays: left, the ESR configuration uses a basic 4×2 pattern (in a quincunx) originating from a reconfigurable matrix (in the case, the array is rotated by 45° with regard to this representation, the neighboring pixels of the same color are coupled 2×2 to provide signals with stronger dynamics or a better signal-noise ratio). On the right, the X-Trans array (6×3 basic pattern on a hexagonal mesh) which made its appearance in 2013 and whose first objective is to overcome the moiré phenomena on high-frequency periodic textures. For a color version of this figure, see www.iste.co.uk/maitre/pixel.zip*

The adopted model is often that of process color, on a block of 2×2 pixels, the fourth site being often masked by a green filter very close to the maximal sensitivity of the eye. Widely used in printing, CMY encoding then adds a fourth black channel to allow for better contrast in dark shades. It is the CMYK space (K for *key*).

CMYK signals are rarely accessible to the end user, they are the subject of a conversion towards a standard, usually sRGB, before recording them (see section 5.2.6).

5.4.3. *Chromatic selection of the arrays*

The photosensitive elements used in photography (CCD or CMOS) provide extended sensitivity over the whole visible range, defined by the forbidden band gap of the component (see section 3.2.1) and extending into the infrared. The response of the sensor is determined by the material, silicon, and by the processing that it undergoes. In all cases, it covers quite extensively the visible range (Figure 5.22) and even often requires that filters be used, cutting out the infrareds (warm filter) and sometimes the ultraviolets (cold filter), to isolate the visible spectral band (see Figure 3.9).

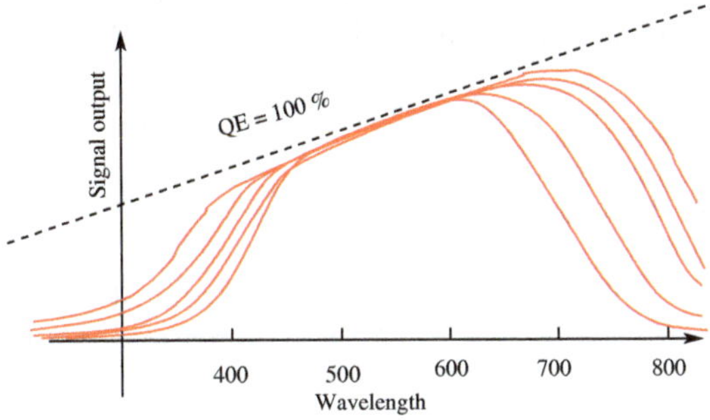

Figure 5.22. *Silicon sensitivity curves (in arbitrary units) according to the wavelength. These curves cover the visible range quite extensively, but they depend on the geometry of the junction and on doping. The quantum efficiency of 100% introduces a linear wavelength limit. For small wavelengths, the curve moves up if the number of surface carriers is small (low density of surface states) and if the junction is thin. In the case of long wavelengths, the curve climbs up if the lifespan in the substrate is extensive, if the junction is thick and if the epitaxial layer is far from the surface (see Figure 3.2)*

Color filters can be placed on the path of the image, between the sensor and the warm and cold filters, to select the desired channels (RGB or CMY).

The chromatic sensitivity of the sensor in the three R, G, and B channels is then the product of the spectral sensitivity of the constituent with the spectral transmittances of the chromatic filters and the various optical elements assessed (IR filter (ϕ_{IR}), UV filter (ϕ_{UV}), microlens (ϕ_{mL}), anti-aliasing filter (ϕ_{AA})). Equation [5.25] becomes:

$$i_R(x,y) = \int_{\lambda=0,4\mu}^{\lambda=0,8\mu} i(x,y)\phi_R(\lambda)\phi_{IR}(\lambda)\phi_{UV}(\lambda)\phi_{mL}(\lambda)\phi_{AA}(\lambda)d\lambda \quad [5.37]$$

and similarly for i_G and i_B.

The chromatic filters ϕ_R, ϕ_G, ϕ_B (whose manufacture is briefly described in section 3.3.3) present a great diversity of chromatic profiles, between manufacturers on the one hand, but also for the same manufacturer, within its range. Information about the response of the filters is generally unavailable to the user who must carry out a calibration to access them. He must then compromise with all the filters interposed in front of the sensor: $\phi_{UV}, \phi_{IR}, \phi_{mL}, \phi_{AA}$ and unless there is specific information on these elements, he will have to incorporate the responses of these filters in the determination of the chromatic responses ϕ_R, ϕ_G, ϕ_B. This is what we will subsequently do, combining equations [5.25] and [(5.37] letting:

$$\Phi_R(\lambda) = \phi_R(\lambda)\phi_{IR}(\lambda)\phi_{UV}(\lambda)\phi_{mL}(\lambda)\phi_{AA}(\lambda) \qquad [5.38]$$

and similarly for Φ_G and Φ_B.

The curves Φ_R, Φ_G and Φ_B should be determined for values regularly spaced by λ, typically 10 nm, reducing the three integrals (5.37) to linear equations.

Ideally, the calibration is effected using a spectrophotometer that sweeps the wavelength spectrum and makes an extended image of it (for example, using an integrating sphere). For each wavelength λ of known energy, the user measures the three values Φ_R, Φ_G, Φ_B, that thus integrate not only the parameters of all the filters, but also those of the objective lens. It is a long and delicate operation which can only be done in laboratory with specialized equipment (spectrophotometer, radiometer to determine the energy of the source, integrating sphere to ensure the uniformity of the observed range).

More simple approaches are often preferred that can be performed during a single exposure using calibrated testing patterns of color ranges[13]. The data from the test patterns manufacturers are then used as ground truth and the unknown spectral sensitivity values are determined by inverting the system of the measurements obtained (equation [5.25]). If the main objective is to obtain the curves every 10 nm between 400 and 800 nm, it is necessary to have forty different colored ranges available at least to reverse the system. In practice, it is essential that many more be available to regularize the system [HAR 98].

Test patterns are most often reflection ranges (the illuminant should then be previously determined with the help of gray test pattern), but it can be seen nowadays that fluorescent test patterns or arrays of light-emitting diodes are emerging which makes it possible to prescind from the light of observation. These inversion techniques are however quite approximate because they are very sensitive to measurement noise and allow that sufficiently smoothed

13 The color test patterns referred to as *color checkers* are the same than those used for white balance in section 5.3.

curves only be obtained at the price of a very strong regulation of the system, using several *a priori* with sensors [HAR 98, JIA 13].

If green filters bring forward fairly regular distributions, blue and especially red filters offer a wide variety of profiles as shown in [JIA 13] after the study of 28 consumer cameras (Figure 5.23). Several authors have focused on statistically modeling these profiles in order to analytically derive the response of a sensor from a small number of measurements only. Polynomial decomposition approaches in Fourier and Gaussian series have thus been proposed. The approach which decomposes spectral responses by principal component analysis (PCA) of the previous 28 cameras has given the best results, leading to a very faithful representation with simply two vectors per channel, the first being the average curve of Figures 5.23, the second offering more fluctuating spectral characteristics.

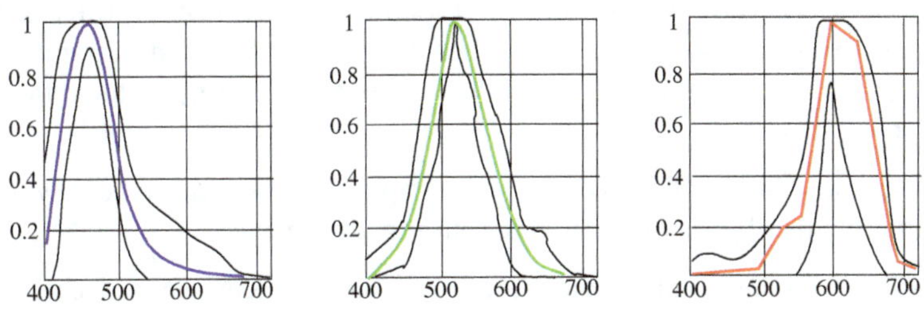

Figure 5.23. *Variability of the color filters: the fine curves represent the extreme values of the sensitivities of the sensors for the R, G and B channels of 28 consumer sensors (curves obtained from the results of [JIA 13]), that is to say that any sensor presents curves entirely comprised between these extreme values. The thick lines represent the average curve obtained from the 28 responses (these are thus the first basic vectors of a principal component analysis)*

However, it will be in vain that similarities with the response of the human visual system LMS (curves of Figure 5.1) will be looked for, in addition to the fact that they will not be found in theoretical studies that aim to define the optimal sensor array minimizing the reconstruction errors of a sample statistically representative in color and in luminance [PAR 06].

5.5. Reconstructing color: demosaicing

We have seen the very particular structure of the image signal on output of a Bayer-type sensor. The various colors are sampled according to offset grids, at frequencies still insufficient to enable an accurate reconstruction according to the rules of signal processing. Admittedly, the use of anti-aliasing filters allows decreasing the deviation from the Nyquist–Shannon conditions, but unless the performance of the sensor is significantly degraded, the normal conditions of analysis of a signal still remain very distant.

Our experience shows us however that this very unfavorable situation does not lead to bad results. The vast majority of images taken under these conditions offer an entirely correct quality, and do not present, in particular, the degradations that should be expected from a notoriously under-sampled signal: the appearance of artificial frequencies in the image and aliasing phenomena on sharp transitions.

This quality is explained for three essential reasons:

– the scenes observed do not often show high frequencies (this is the case for example of portraits), or when they show some, the accuracy of the representation in these regions is not an essential quality element (as for example in hair);

– the shooting conditions or the photographer's choice lead to heavily filter the image either by the diaphragm or settings adjustments;

– the techniques of image reconstruction by demosacing achieve a remarkable job by applying empirical rules.

We will now examine these demosaicing techniques. Note that most operate inside the camera and ensure the processing is carried out at the shooting rate. Other, more complex techiques can only be achieved on a computer. They require native signal (RAW) to be available. Naturally, they give far better results than the previous ones, but the merits of each method are often kept for expert discussions and do not significantly affect the vast majority of the registered pictures.

5.5.1. *Linear interpolation demosaicing*

A first family of approaches considers the three independent channels and restores the blue component (and similarly for the green and the red ones) only from the sole blue samples. We are naturally confronted with a particular case of the conventional problem of image resampling [UNS 95a, VET 14].

5.5.1.1. *Linear interpolations, channel by channel*

Nearest neighbor: first of all, we can observe that zero-order interpolation (to the nearest neighbor) is not a good solution on a Bayer grid, because there is not a unique neighbor since there are four neighbors (at a distance 1) around a missing pixel in the green channel and two (at a distance 1) or four (at a distance $\sqrt{2}$) in the red and blue channels according to the row being considered (Figure 5.24). Anyway, it is a mediocre interpolation, which certainly respects the high frequencies of the signal, but which strengthens the aliasing effects. It is preferable to avoid it if computation capabilities allow.

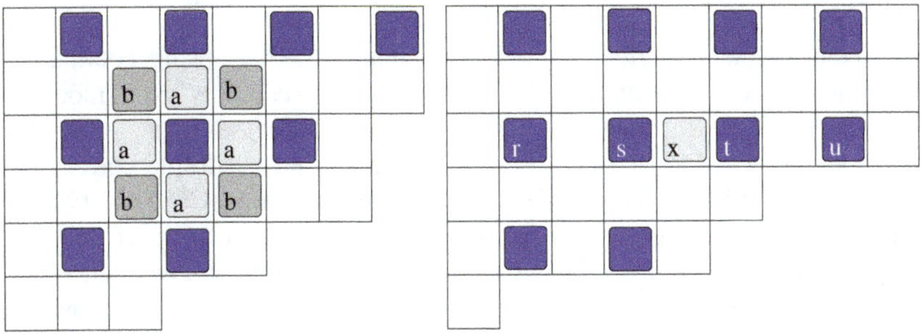

Figure 5.24. *On the left, reconstruction of the blue channel (the same procedure would be used for the red channel). Only blue pixels are displayed: the pixels to rebuild are within a distance of 1 of a blue site (pixels a in light gray) or at a distance of $\sqrt{2}$ (pixels b in dark grey). On the right, the calculation of pixel x by cubic interpolation from four neighbors r, s, t and u*

Linear interpolation: this interpolation is doubly linear: on the one hand, because in contrast to the methods presented in sections 5.5.2 and 5.5.3, it is effected by linear combination of pixels, but also because it is obtained by a linear convolution kernel (in contrast to quadratic and cubic interpolation which make use of higher order kernels). It is equally bilinear because the kernel is linear[14] in x and in y. It replaces the missing green pixel by the average of the four adjacent pixels. It therefore performs a cardinal sine filtering in the frequency domain which attenuates the high frequencies. Red and blue pixels are processed identically:

– the pixels a (Figure 5.24) are the average of the two adjacent pixels;

– the pixels b the average of the four diagonal pixels. It should be noted that these pixels b can be calculated by the same formula as the pixels a once the pixels a have been calculated.

14 Note that bilinear interpolation is linear in x, as well as in y, but not in (x, y) since the two-dimensional function is a hyperbolic paraboloid comprising terms in xy.

Quadratic interpolation: being based on neighborhoods comprising an odd number of points surrounding the pixel to reconstruct row or column-wise, therefore non-symmetrical, these second-degree polynomial interpolations are not employed in image processing.

Bicubic interpolation: it utilizes larger neighborhoods and as a result requires more significant computations.

There are many different order-3 polynomial kernels using continuous interpolation functions and continuous first derivatives. They give smoother images than linear interpolation (disappearance of staircases or jaggies). The cubic spline [UNS 91] (which uses as a function the quadruple correlation rectangular function) and the Catmull–Rom interpolation [COO 67] (which do not explicitly use the knowledge of the derivatives) are often used because they are very suitable for image reconstruction. The first gives smoother images, the second more precise contours. By using the notations of Figure 5.24 on the right, the value in x of the blue pixel can be calculated, knowing the four values B_r, B_s, B_t and B_u, it yields, for the spline interpolation:

$$B_{spline} = [0.0208 \quad 0.4792 \quad 0.4792 \quad 0.0208] \begin{bmatrix} B_r \\ B_s \\ B_t \\ B_u \end{bmatrix} \qquad [5.39]$$

and for the Catmull–Rom interpolation:

$$B_{C-R} = \frac{1}{16}[-1 \quad 9 \quad 9 \quad 1] \begin{bmatrix} B_r \\ B_s \\ B_t \\ B_u \end{bmatrix} \qquad [5.40]$$

An identical calculation allows the pixels B_a, located between two pixels on a vertical, to be interpolated. The pixels placed in diagonal are calculated on a 4×4 pixel square window thus involving 16 coefficients or through the implementation of one-dimensional (1D) interpolations from the grids previously calculated.

The improvement of the quality of the cubic interpolation is generally sensitive in the presence of sudden transitions in the image. Over a unique transition the aliasing effects are attenuated. Over multiple and narrow transitions, they reduce the aliasing phenomena.

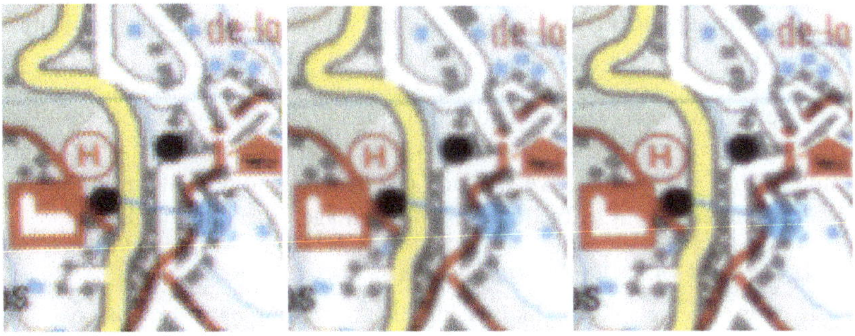

Figure 5.25. *Demosaicing. On the left, nearest-neighbor interpolation: the effects of aliasing are very sensitive and the colors are erroneous on the edges of the objects, but the high frequencies are correctly rendered. At the center, bilinear interpolation reconstruction: checkerboard effects are less sensitive on strong transitions (but still present), the picture is a little fuzzier. On the right, from a local study of the gradients guided by the green channel: the quality of the edges improves significantly (the checkerboard effects disappear)*

5.5.2. *Per channel, nonlinear interpolations*

The presence of contours in images drives the choice of interpolation strategies adapting themselves to the context. A first approach consists of replacing the average by the median to calculate the green channel. Since a missing green site is surrounded by four known sites, the median is then the average of the two central values of the four points. Another idea is to detect the contours in heavily under-sampled images and to reconstruct, by linear interpolation, just the portions of the image exempt from contours. In the presence of a contour, the interpolation will only be made in the half-space where the pixel to be reconstructed is situated with regard to the contour. This strategy allows sharper edges to be obtained and contributes to a better image accutance (see section 6.1.5).

5.5.3. *Interchannel, non-linear interpolations*

Taking account of the three channels to determine the most favorable interpolation mode has allowed for very important progress in the quality of the reconstruction of images during demosaicing. Two points are particularly interesting:

– the correlation between colors is a property shared by a large number of images of ordinary scenes;

– the measurement mode (by Bayer array) introduces a principle complementarity in the samples of the various channels.

Several proposals were made, seeking a compromise between the reconstruction quality on the one hand, and the complexity of the computation on the other. We will cite a few expressions, that can be found either in cameras or in software programs of image processing toolkits.

5.5.3.1. *The Hamilton–Adams method*

It is one of the first methods using all three channels simultaneously. It has been patented by Kodak under the name of its creators [HAM 97]. It only uses row or column signals, leading to 1-dimension interpolations only, but according to one or the other direction (Figure 5.26).

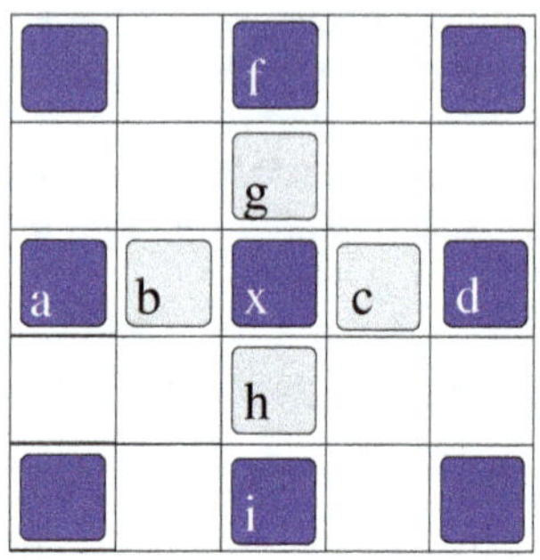

Figure 5.26. *The Hamilton–Adams interpolation utilizes row and column pixels to determine the value of the pixel x in the unknown channels (here the green and the red)*

In order to calculate the unknown green value V_x in a pixel x, where only the blue B_x is known, the horizontal Δ_x and vertical Δ_y gradients are calculated using the first derivative of the green channel and the second derivatives of the blue channel:

$$\begin{aligned} \Delta_x &= |G_b - G_d| + |2B_x - B_a - B_d| \\ \Delta_y &= |G_g - G_h| + |2B_x - B_f - B_i| \end{aligned} \qquad [5.41]$$

The following decisions are then taken:

$$\begin{aligned} &\text{if } \Delta_x > \Delta_y \;\; G_x = (G_b + G_c)/2 + (2 * B_x - B_f - B_i)/4 \\ &\text{if } \Delta_x < \Delta_y \;\; G_x = (G_g + G_h)/2 + (2 * B_x - B_a - B_d)/4 \\ &\text{if } \Delta_x = \Delta_y \;\; G_x = (G_b + G_c + G_g + G_h)/4 \\ &\qquad\qquad\qquad + (4 * B_x - B_f - B_i - B_a - B_d)/8 \end{aligned} \qquad [5.42]$$

The red and blue channels are interpolated according to a simpler scheme once the green is interpolated. Thus, the blue pixel at point b will have the value:

$$B_b = G_b + (B_a - V_a + B_x - G_x)/2 \qquad [5.43]$$

The Hamilton–Adams method gives good quality results for a modest computational cost.

5.5.3.2. *Variable number of gradients method*

This method is often denoted VNG for *variable number of gradients*. Starting from a pixel in a channel, the other two channels are to be determined. To this end, the difference between the known channel and the unknown channels around the pixel will be measured and this difference will be added to the known value. Since it can only be done on fairly uniform neighborhoods, eight areas around the pixel must be first determined and those that are uniform are looked for among those areas (Figure 5.27). For each region, the gradient is therefore calculated in a direction specific to the region, combining the gradients of the various channels. A decision is then made about the homogeneous regions, by means of an empirical rule (it is at this point that the variable number of gradients is involved). In these regions, the differences between the channels are determined. These values are then added to the known channel to form the two unknown channels. It should be noted that the calculations are slightly different when measuring the differences depending on whether the pixel at the center of the window is green or red/blue.

5.5.3.3. *Pattern research*

The first concern is about the interpolation of a green pixel, thus bound by four green pixels. Their average is calculated and those that are lighter than the average are denoted by H, and the others by L. Four classes of configurations are next identified (up to a rotation), according to the representation of Figure 5.28. The first and the last are identified to contours, the second to a line, the third to a corner.

Various strategies are then possible to determine the unknown pixel x. For example, in the case of an edge, the median value of the four neighbors is chosen for x (that is the average of the second and the third of the four values ordered in increasing manner). To determine a corner or a line, a larger domain is used (of 5×5 pixels) and the knowledge available about the homogeneity of the neighborhoods is exploited to choose a robust value (e.g. median) amongst selected pixels.

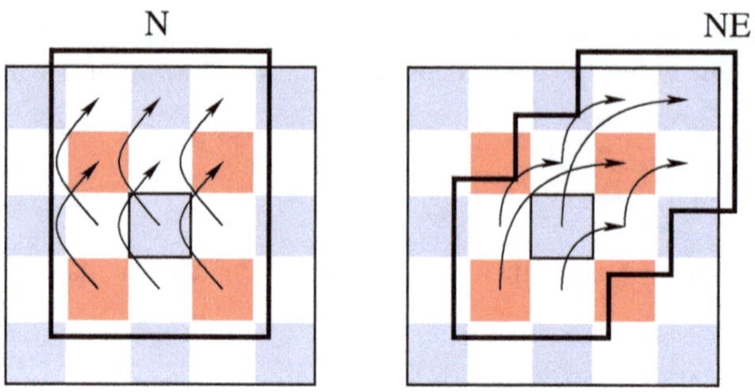

Figure 5.27. *Two regions (North region and Northeast region) surrounding a pixel and used for the calculation of the homogeneity of the pixel neighborhood in the method of the variable number of gradients (the green pixels are left in white). The intensity of the gradient is calculated following the orientation of a weighted average of the differences between the pixels linked with an arrow*

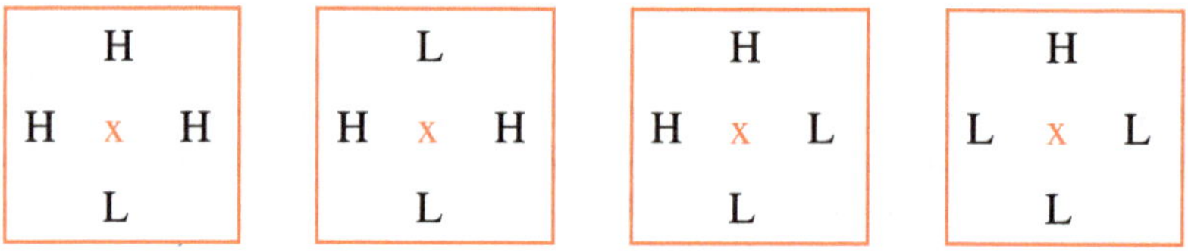

Figure 5.28. *The four possible patterns to interpolate a green pixel: from the left to the right* contour, line, corner *and still* contour. *A pixel denoted as H has a brightness higher than the average of the four pixels of the site while a pixel L has a lower brightness*

In the case of red and blue pixels, a similar but simplified strategy is often adopted based on linear interpolations and medians on well chosen groups.

5.5.3.4. *Interpolation guided by the green channel*

This type of methods take advantage of the very strong correlations between channels. Since the green channel is very similar to a luminance channel, it will be the focus of the first efforts of reconstruction, and then the blue and red channels will be derived from the green signal the same way chrominance signals are derived from the luminance channel in a television.

The interpolation of the green is achieved for example by means of the pattern method.

Next, to determine a blue pixel (noted x), the four diagonal gradients of the green channel are calculated in the neighborhood of x. From these values, a decision is taken whether the region is homogeneous or not and the value x is derived by a combination of the four values of the known blue pixels, weighted by the inverse of the green gradients (or a weighted or frustrated sum, that is deprived of certain terms deemed unsuitable).

This approach, of reasonable complexity, often gives very satisfactory results, far better than those that linear approaches give.

5.5.3.5. *Offline methods*

A large number of methods make it possible to refine the reconstructions presented here, but at the expense of much more complex computations that can be achieved today in the camera. From these techniques, some approaches come forward that place the image signal in a perceptual space (such as Lab), in order to better qualify the distances between nearby pixels. Iterative approaches should also be cited that iterate twice over sites initially unknown to reassign the values defined by taking the calculated context into account. Next, we can mention methods that search for similar configurations elsewhere in the image to take advantage of internal similarities between images (piecewise or *patches* methods). Finally, some techniques are increasingly frequently used, relying on learning methods (here again often by *patches*), from databases of similar images. These methods are described in [BUA 09, CHU 06, GET 12, ZHA 05], and comparisons of results on the website [IPO 14].

In conclusion, it should be noted that nowadays demosaicing techniques are able to obtain great results, which, confronted with ordinary photography situations, leave too little room for future improvements. They quite correctly take into account the effects of aliasing in very textured areas when there is no anti-aliasing filter available. These demosaicing methods are available through several software programs including free software such as RawTherapee [HOR 14b]; they provide the user with the ability to choose which method will be used. The only remaining critical situations, sometimes poorly handled by current techniques, concern very high frequency regions where the three channels show slight correlations and patterns are extremely fine.

In the case of sensors that do not adopt an array filter such as Bayer's, it is possible to transcribe many of the algorithms developed for the Bayer array, but the complex configurations (such as the X-Trans matrix from Fujifilm, see Figure 5.21) require an original investment, often significant, in order to achieve equal performance [RAF 14].

6

Image Quality

The quality of the images produced by a camera is a key concern for photographers, but the subject is a particularly complex one. Quality assessment needs to take account of a variety of measurable technical aspects, associated with the performance of the technologies involved (see Chapters 2 and 3). It also involves perceptual considerations, which are more or less clearly understood; these are often standardized, using the notion of a standard observer (Chapters 4 and 5). However, the notion of quality is based predominantly on highly subjective a esthetic criteria, which are generally cultural, linked to both the scene and the conditions in which it is viewed, and highly dependent on observers themselves. These subjective and cultural criteria have led photography to be considered as an art, with academies, museums, exhibitions, etc.; quality should, therefore, be judged using the criteria of this art. However, since we are deprived of means to translate it when conceiving a camera, this point of view will be evoked only in the last section of this chapter and rather briefly. The rest of the chapter will examine the technical aspects of quality, along with their psychovisual effects via the properties of human perception, as we understand it.

In section 6.1, we will begin by considering the criteria used by engineers in order to measure image quality: the signal-to-noise ratio, resolution, transfer function, sharpness and acutance. These quantities are those used in works on photographic materials; our intention here is to provide readers with the vocabulary necessary to interpret test results using these analytical methods.

We will then present work which has been carried out with the aim of establishing a global measurement of image quality, with or without a reference (section 6.2). These efforts are generally not discussed in works on photography, and tend to be restricted to the field of image transmission (particularly compression) or image processing (particularly in the context of optimizing shape recognition, target detection and tracking).

Finally, we will present a notion taken from information theory (section 6.3) which aims to provide a single framework for regrouping various forms of impairment. This framework is not widely used and remains incomplete; nevertheless, instances of its use have produced some interesting conclusions.

6.1. Qualitative attributes

A wide number of objective criteria may be used in analyzing the quality of a photograph, but not all of these criteria are suitable for use in generalized analysis.

Depth of field, for example, is a quality which we often wish to maximize (for landscapes, sports or group photographs); however, in some instances, it should be minimized (particularly in the case of portraits, but also when photographing flowers, for example, or works of art). Depth of field is almost totally determined by camera lens settings, and specifically by the aperture/focal length ratio. A means of estimating this ratio was discussed in section 2.2. As this quantity is entirely dependent on the scene and the specific intentions of the photographer and is controlled by a single camera setting, it will not be considered as a qualitative attribute in the context of this chapter.

The same is true for aberrations, discussed in section 2.8, which result exclusively from a specific assembly of lenses in specific usage conditions; these errors can be corrected, up to a point, in postprocessing.

Chromatic rendering may also be considered in this way. As we saw in Chapter 5, equation [5.20] may be used, for any point object, to measure the perceptual distance between the true color and the color shown in the image. In theory, this allows us to measure the overall chromatic quality of a photograph. However, this approach requires much more information than is generally available, and is based on additivity and spatial invariance hypotheses which are not verified; in practice, it is only ever used for the purposes of chromatic calibration.

We will, therefore, consider certain simple criteria which may be adjusted in order to model relatively realistic situations. These criteria are used in establishing the predetermined settings offered by a number of cameras, and by toolboxes used in postprocessing tools.

6.1.1. *The signal–noise ratio*

In Chapter 7, we will provide a detailed analysis of the various types of noise which affect image signals. In this context, we will only consider the impact of this noise on image quality, and the way in which this noise may be characterized *a posteriori*. One way of measuring noise as a function of the input signal is to photograph and analyze grayscale test patterns, such as that shown in Figure 6.1, with constant bars. The variation in the level of gray across each bar clearly reflects the noise affecting the image.

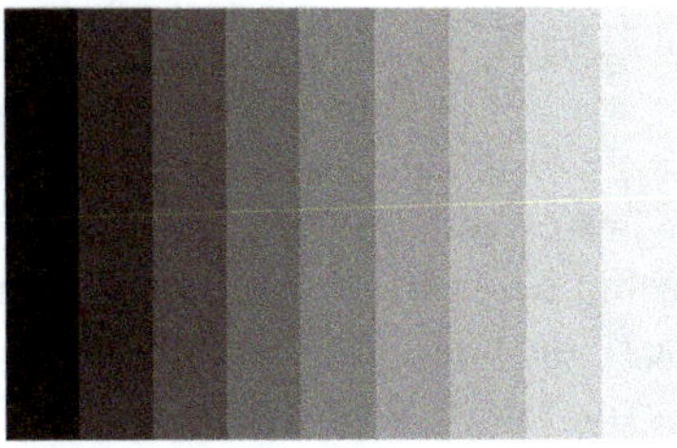

Figure 6.1. *Grayscale test card used to measure noise values for 10 different gray levels*

6.1.1.1. *Definition: signal-to-noise ratio*

An elementary representation of image quality may be obtained using the signal-to-noise ratio (often denoted by SNR). The signal-to-noise ratio is expressed as the relationship between signal power and noise power, converted into decibels (dB). Thus:

$$SNR = 10 \log_{10} \frac{Power(image)}{Power(noise)} \quad dB \qquad [6.1]$$

The power of any given image is the square of its amplitude, which we generally consider to be proportional to its levels of gray. This power is, therefore, strictly limited by N times the square of its maximum intensity: $N2^{2n}$, where N is the number of pixels and n is the number of bits used for the binary representation of each pixel (or $\nu = 2^n$, the number of levels of gray).

The peak signal noise ratio (PSNR) is a more generalized form of this formula, which considers the same noise in relation to the maximum signal carried by the image. Thus:

$$PSNR = 10 \log_{10} \frac{2^{2n}}{Power(noise)} \quad dB \qquad [6.2]$$

hence, for a non-degenerated image, $SNR < PSNR$.

The noise value is generally considered to be unbiased, with a mean value of zero. This hypothesis will be verified in Chapter 7. Its power is, therefore, N times the variance σ^2. Expressing the variance in levels of gray, we obtain:

$$PSNR < 10 \log_{10} \frac{2^{2n}}{\sigma^2} = 20 \log_{10} \frac{2^n}{\sigma} \qquad [6.3]$$

The relationship $D = 2^n/\sigma$ is the dynamic of the imaging system, and is an important parameter when considering image quality. This dynamic is also expressed as the relationship between the maximum number of carriers in a photosite and the variance of the noise (the sum of thermal, reading and sensor non-homogeneity noise, expressed in electrons).

6.1.1.2. *PSNR and quantization*

In an ideal case, with a signal of very high quality (obtained using a very good sensor and strong lighting), the photonic and electronic noise are negligible, and the image is affected by quantization noise alone (section 7.2.5). In this case, for a uniform distribution of gray levels across the whole dynamic, the signal noise ratio is limited by:

$$PSNR < 10 \log_{10}(12 \times 2^{2n}) = (10.8 + 6n) \; dB \qquad [6.4]$$

i.e. for an image coded in 1 byte ($n = 8$ bits), the maximum PSNR is: $PSNR_{max} \sim 59$ dB, and for an image coded in 2 bytes ($n = 16$ bits), $PSNR_{max} \sim 108$ dB.

6.1.1.3. *PSNR and photonic noise*

The photonic noise, which is independent of additional noise introduced by electronic equipment, is random and is therefore accompanied by Poisson's noise. This will be discussed further in section 7.1. If the signal is strong, this noise is negligible; however, in two specific situations it is critical: first, if the scene is poorly lit, but a short exposure is required, and second, if the size of photosites is significantly reduced.

The psychovisual studies carried out and reported in [XIA 05] established a useful, empirical law known as the "thousand photon rule". It states that if fewer than 1,000 photons are received by a site during scene capture, the photonic noise, across uniform ranges with a medium level of gray, begins to be perceptible, with a signal noise ratio of around 100/3 (30dB). In complex scenes, significant masking effects are involved, which have the effect of lowering this threshold and making the results harder to use.

6.1.1.4. *Number of effective levels of gray*

In Chapter 7, we will also see that the level of noise σ is often dependent on the incident energy level $\mathcal{E}$, and should, therefore, be written as $\sigma(\mathcal{E})$. The number of distinguishable levels of gray ν_e is, therefore, expressed by the formula:

$$\nu_e = \int_{\mathcal{E}_{min}}^{\mathcal{E}_{max}} \frac{d\mathcal{E}}{\max(1, \sigma(\mathcal{E}))} \leq \nu = 2^n \qquad [6.5]$$

The curve $\sigma(\mathcal{E})$ is generally obtained using successive captures of grayscale test cards, such as that shown in Figure 6.1.

6.1.1.5. *SNR and sensitivity*

In photography, the notion of the signal noise ratio is intimately linked with the sensitivity $\mathcal{S}$ of the sensor, as we saw in section 4.5. While the notion of noise is contained within that of minimum haze in the definition of ISO sensitivity for film (Figure 4.13), one of the recommended definitions for solid sensors refers explicitly to the curve associating the signal noise ratio with the incident energy (Figure 4.14).

Using definitions [4.37], and based on the hypothesis that there is a linear relationship between the signal noise ratio and the logarithm of the incident energy (this relationship is quasi-verified for reasonable energy levels), we obtain the following signal noise ratio for an incident energy $\mathcal{E}$ expressed in lux-seconds:

$$SNR(\mathcal{E}) = \frac{40 \log_{10}(\mathcal{E})}{\log_{10}(\mathcal{S}_{40})} \qquad [6.6]$$

(a similar equation exists for lower quality equipment, using the measured sensitivity $\mathcal{S}_{10}$ for an SNR of 10). The variation in the signal noise ratio for a given scene while varying the ISO sensitivity between 32 and 100,000 is shown in Figure 6.2.

However, as we have seen, this definition of $\mathcal{S}$ is one of the five standardized definitions of ISO sensitivity. This definition is not the most widely used by manufacturers, and is therefore not universally used in determining noise levels.

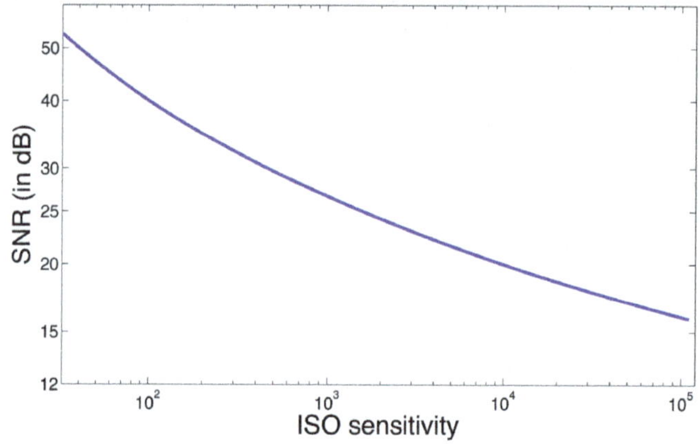

Figure 6.2. *Evolution of the signal noise ratio (in dB) with the same lighting (here, 100 lux-seconds) for ISO sensitivity values ranging from 32 to 100,000 ISO*

6.1.1.6. *PSNR and image compression*

Unfortunately, it is often difficult to determine the precise expression of noise variance, as we will see in Chapter 7. For this reason, the use of expression [6.3] to qualify natural images is limited to cases where detailed knowledge of capture conditions is available. Nevertheless, this expression is important when attempting to express impairment due to a specific treatment, such as compression, where the noise is measured directly as the difference between the original and coded images. Thus, lossy coders produce signal noise ratios varying from 40 dB (coders with very subtle distortion) to 25 dB (high distortion coders)[1] (see Chapter 8).

1 It is impossible to determine what should be considered weakly or highly perceptible without specifying precise examination conditions. These precautions are always taken in real test situations, but will not be discussed here for reasons of simplicity. Interested readers may wish to consult [SAA 03] for further details.

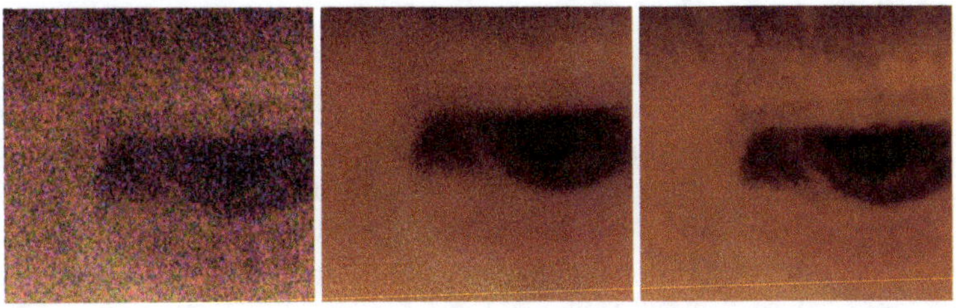

Figure 6.3. *Three different noise levels affecting an image (detail) taken with high sensitivity (6,000 ISO). Left: no treatment applied to the sensor output. The noise is highly colored, due to the Bayer matrix which measures the signal. Center: noise reduction applied during demosaicing, followed by Bayesian filtering using local windows. Right: the same treatment, but using a non-local Bayesian filter. For a color version of this figure, see www.iste.co.uk/maitre/pixel.zip*

6.1.2. *Resolution*

Resolution is that which allows us to perceive the fine details contained in an image. In section 2.6.7, resolution was defined as the minimum distance between two distinguishable points in an image; this definition also makes use of the Rayleigh criterion[2].

This definition may be used to establish several approaches for measuring image resolution based on sensor data alone.

2 The term "resolution" is ambiguous, referring to either a distance or its inverse, depending on the source. In this context, we will apply the definition used in optics, which determines resolution as the smallest quantity (whether linear, as in our case, or angular, as in astronomy or microscopy [PER 94]). This definition is not perfect, as high resolution produced by a high-quality instrument will be associated with a low value. A different definition has been used in other domains (such as image processing, [FIS 05]), where the resolution is described as the number of distinguishable elements per measurement unit (the maximum spatial frequency). This is expressed in ppi (pixels per inch) or dpi (dots per inch). Taking this approach, the resolution is proportional to the inverse of the resolution defined using our method. It reflects a common sense approach (high resolution equates to high quality), but leaves aside notions of etymology, Rayleigh's experiments and standard practice in the scientific community.

6.1.2.1. *Elements affecting resolution*

6.1.2.1.1. Size of photosites

As a first step, it is relatively easy to determine a lower limit for resolution based on knowledge of the sensor. With the relevant information concerning this sensor – its size ($l_x \times l_y$) and the number of sensitive photosites (number of pixels) ($N_x \times N_y$) – it is possible to deduce theoretical maximum resolutions, in the image plane, in both horizontal and vertical directions: $\delta x = l_x/N_x$, $\delta y = l_y/N_y$. These resolutions are expressed as lengths ($[L]$) and are generally given in millimeters, micrometers or fractions of millimeters.

This resolution corresponds to a maximum spatial frequency defined as the inverse of the resolution ($[L^{-1}]$) and expressed in lines per millimeter[3].

Calculated using sensor data alone, the resolution and the maximum spatial frequency are independent of the optical elements and settings used (diaphragm aperture and focus). At this point, we may wish to consider the specific structure of the Bayer matrix, which ignores sensitive photosites of the same color using a factor of $\sqrt{2}$ for green and a factor of 2 for red and blue. The resolution is, therefore, modified in the same proportions. This loss of resolution is partially offset by demosaicing software, but these techniques use specific image properties which cannot, strictly speaking, be generalized for all images (section 5.5).

Taking account of the focal distance f of the selected photographic lens and the distance d from the observed object, it is possible to determine the transverse magnification of the device (see definition [1.3]): $G = f/(f+d)$. This enables us to deduce a lower limit for the distance between two separable objects in a scene (resolution in the object plane):

$$\delta X = l_1(f+d)/N_x f \qquad [6.7]$$

For all far-off objects in "ordinary" photography, this is reduced to the following simple form:

$$\delta X \sim l_1 d/N_x f \qquad [6.8]$$

3 In photography, the expression "line pairs per millimeter" is also used, expressing the alternation of light and dark lines.

6.1.2.1.2. Diffraction

With access to relevant information concerning camera lens settings, it is possible to take account of the effects of lens diffraction in measuring resolution. In the case of a perfectly-focused circular lens, with no errors other than diffraction, formula [2.43] states that any object point, no matter how fine, will necessarily be subject to spread in the form of an Airy disk with a diameter of $\delta x_a = 2.44\lambda f/D$, where f is the focal length and D is the diameter of the diaphragm. For an aperture number f/D of 4 and an average wavelength of 500 nm, the diffraction disk is, therefore, spread over 2 μm, close to the dimension δx of small photosites in compact cameras.

For very high-quality equipment (with large photosites), diffraction affects the resolution of photos taken with low apertures (see Figure 6.4) and in areas of perfect focus. Cameras with small photosites are generally relatively cheap, and images are affected by a variety of other errors, notably geometric aberrations and chromatic aberrations in the lens, which are more problematic than diffraction.

Diaphragm diffraction, therefore, constitutes a final limitation to resolution in cases where other causes of resolution loss have been removed, particularly focus issues, but also movement (photographer and objects), chromatic faults, etc. It is not possible to establish general rules concerning these two final points, but the first issue can be taken into account.

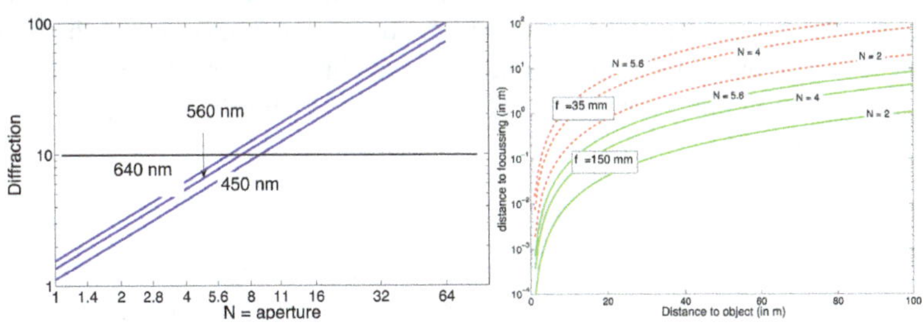

Figure 6.4. *Left: diameter of the ideal diffraction figure of a photographic lens as a function of the numerical aperture N, for the three primary colors RGB (this diameter is independent of the focal distance of the lens in question). Most cameras have photosites of between 1 and 10 μm in size. Right: distances between the object and the focus plane such that the diaphragm diffraction figure is equal to the blur of the focus. Above the curve, the focus error is dominant. For lenses with a focal distance of 35 and 150 mm and numerical apertures of 2, 4 and 5.6*

6.1.2.1.3. Focus error

The expression of the size ε of blurring due to a focus error Δ on an object was given in section 2.2. Formulas [2.9] and [2.10] take sufficient account of low-level blurring in the case of an object in any given position or in the case of an object far away from the lens. These formulas[4] allow us to deduce values for the deviation of the focus Δ, which causes greater blurring than the diffraction figure produced by the optical equipment (Figure 6.4):

$$\Delta > 1.22\lambda \frac{p(p-f)}{D^2} \qquad [6.9]$$

or, in the case of a very distant object:

$$\Delta > 1.22\lambda \frac{p^2}{D^2} \qquad [6.10]$$

If we simplify this last equation [6.10], expressing $\bar{p}$ in meters, $\bar{D}$ in millimeters and taking $2.44\lambda = 1$ (so implicitly $\lambda = 0.41$ micrometers[5]), we obtain the following simple formula:

$$\Delta > \frac{\bar{p}^2}{2\bar{D}^2} \qquad [6.11]$$

This formula now involves only the distance $\bar{p}$ between the object and the camera and the diameter $\bar{D}$ of the aperture. For an object deviation in the focus plane of less than Δ, diffraction should be taken into account in the resolution, if this value is higher than the intersite distance. In all other cases, the focus error is dominant.

6.1.2.2. *A posteriori resolution analysis*

When the relevant information for calculating resolution using the steps described above is not available, a "system approach" may be used, which

---

4 Formulas [2.9] and [2.10] give the distance Δ between the two extreme positions, in front of and behind the focus point, with a blur of ε; here, the mean value $\Delta/2$ will be used to express the displacement of a single point in relation to focus. However, neither of these two extrema is situated at this distance from the focal point, as focusing is based on the average geometric (and not arithmetic) distance between the two points.

5 $\lambda = 0.41$ micrometers is not particularly realistic as a wavelength, as it would be very dark blue, within the ultraviolet range. However, it may be used in giving a conservative definition of the lower bound.

does not isolate individual factors involved in image construction, but rather takes them as a whole in the form of an operational configuration. This is achieved using a juxtaposition of alternating black and white lines (Figure 6.5). The contrast measurement for the image obtained in this way may be used to determine the smallest step δ_p which allows us to distinguish lines. A threshold is sometimes fixed at 5% contrast in order to define this resolution or at 27% by extension of the Rayleigh criterion to any given lines (see section 2.6.7).

This measurement is not immediate, however, as shown in Figure 6.5: aliasing phenomena may affect the contrast measurement, unless care is taken to filter out high frequencies outside of the bandwidth tolerated for sampling by photosites. Hence:

$$\delta_p > \max(\delta_1, \delta_a, \varepsilon) \quad [6.12]$$

As it is difficult to measure resolution in this way, and as particular precautions need to be taken, a different method is often used. This method, determination of the modulation transfer function, requires us to use the same precautions.

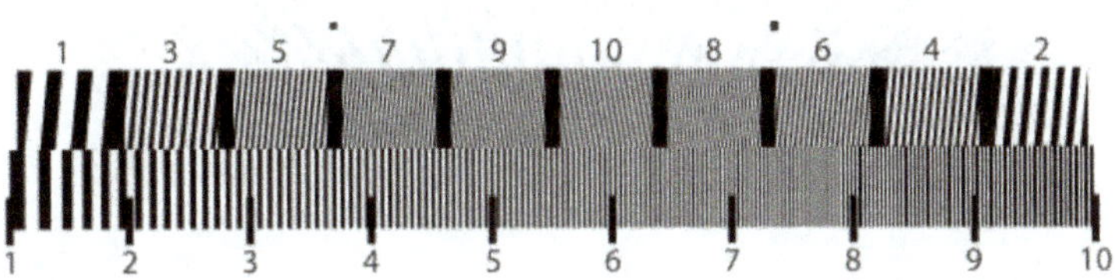

Figure 6.5. *Detail from the ISO-12233 measurement test card, used to measure resolution using alternate black and white bands of variable width, both vertical (lower section) and oblique (upper section). Binary test cards of this type are easy to produce, but contain a continuum of very high frequencies, leading to aliasing phenomena in all photographic reproduction systems. These phenomena are difficult to interpret, as we see here. For this reason, sinusoidal test cards are often preferred (see Figure 6.6)*

6.1.3. *The modulation transfer function*

The modulation transfer function (MTF) was introduced in equation [2.41]. It is denoted by $H(u, v)$, where u and v denote the spatial frequencies associated with x and y.

6.1.3.1. *MTF and test cards*

Conceptually speaking, the MTF is measured using the image taken from a sinusoidal test card of increasing frequency (Figure 6.6), of the form:

$$o(x, y) = e_0(1 + \cos(2\pi\alpha yx)) \qquad [6.13]$$

For an ordinate line y, the input signal is a pure frequency with a value of $u = \alpha y$, and its energy at this frequency is, by construction, equal to $|e_0|^2$. The image obtained is $i(x, y)$, for which the energy at frequency u is carried by the square of the Fourier transformation $I(u, 0)$ of i.

$$H(u, 0) = \frac{|I(u, 0)|^2}{|e_0|^2} \qquad [6.14]$$

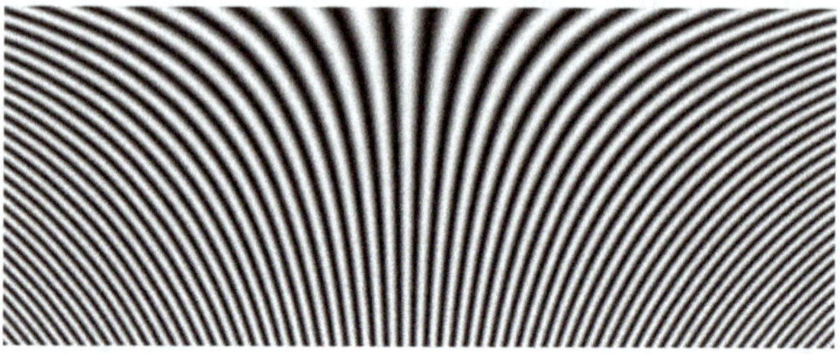

Figure 6.6. *Pure frequency test card used to study resolution in the horizontal direction. The test card is sinusoidal, and the step used decreases in a linear manner from top to bottom*

This MTF can also be observed qualitatively, using image $i(x, y)$ directly; for low frequencies, the signal is transmitted in its entirety, producing high contrast. As the frequency increases, the contrast decreases, reflecting the fall in the MTF. The progressive loss of contrast clearly illustrates the evolution of the MTF.

6.1.3.2. *Indicators derived from the MTF*

The MTF has been used in the development of more concise indicators characterizing the resolution of a camera. The most widely used indicator is currently MTF50 (or MTF50P), which expresses resolution as the spatial frequency for which modulation is attenuated by 50 % (or to 50 % of its maximum value, in the case of MTF50P (as P = peak)) [FAR 06]. This value is used in the ISO 12233 standard [ISO 00].

The final frequency transmitted by the system may be observed at the point where contrast disappears. This corresponds to the resolution limit (or resolution power) seen above, if, for example, the contrast reduction bound is fixed at 5%.

6.1.3.3. *Directionality*

Note that the test card in equation [6.13] behaves similarly in terms of horizontal and vertical lines, and may be used to characterize frequencies in both directions. These measurements are well suited for analyzing the response of a matrix-based photoreceptor. However, we may also wish to consider frequencies in other directions (for example, diagonal frequencies, used in analyzing the role of the Bayer mask). For these purposes, we simply turn the test card in the required direction.

If more precise adaptation to the symmetry of optical systems is required, specific test cards may be used, such as those shown in Figure 6.7, which allow for the characterization of radial frequencies (expressed using a single variable $\rho = \sqrt{(x^2 + y^2)})$ or tangential frequencies (expressed along the length of a circle of radius ρ).

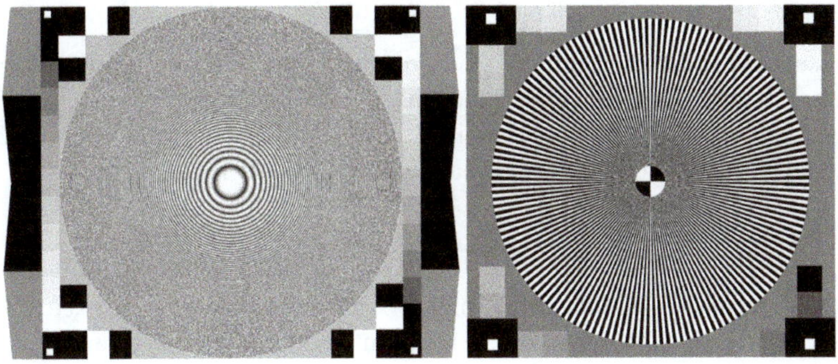

Figure 6.7. *Two additional resolution test cards. Left: test card used to define radial frequencies; right: test card used to define tangential frequencies. These test cards may be binary or sinusoidal. The moire patterns seen on these test cards (like those in Figure 6.5) are due to aliasing phenomena at high frequencies*

Methods using variable step test cards, while widely used in practice, are excessively global; this can be seen if we move the test card in the image. The contrast reduction frequency is generally higher in the center of the image than at the edges, expressing the fact that the MTF is not constant for all points in the image.

6.1.3.4. *MTF and impulse response*

To obtain a more local measurement, allowing us to indicate the resolution limit for all points in the image field, it is better to use the definition of the MTF based on the Fourier transformation of the impulse response, or point spread function (PSF), $h(x, y)$.

Hence:

$$H(u, v) = \mathcal{TF}(h(x, y)) \tag{6.15}$$

A first approach would be to analyze an image of a scene made up of very fine points. Let $\delta(x, y)$ denote the Dirac pulse:

$$o(x, y) = \sum_k a_k \delta(x - x_k, y - y_k) \tag{6.16}$$

and:

$$i(x, y) = \sum_k a_k h(x - x_k, y - y_k) = h(x, y) * \sum_k a_k \delta(x - x_k, y - y_k) \tag{6.17}$$

Isolating each source point, then readjusting and summing the images formed in this way, we obtain the image $\bar{i} = \sum_k a_k h_k(x, y)$, where h_k represents the image of source point k brought into the center of the field. From this, we may deduce an estimation of the transfer function:

$$H(u, v) = \mathcal{TF}(\bar{i}(x, y)) \tag{6.18}$$

This method is unfortunately highly sensitive to noise, as we see when analyzing the signal noise ratio. The results it produces are, therefore, generally mediocre. However, it is used in cases where test cards cannot easily be placed into the scene, or when other, better structures are not available. For example, this is the case in astronomy (which involves very large number of punctual sources) and microscopy.

One means of improving the measurement of the impulse response is to use a line as the source image o_l. This line is taken to be oriented in direction y. The integral of the image i_l obtained in this direction gives a unidimensional signal, which is a direct measurement of the impulse response in the direction perpendicular to the straight line.

$$\int i_l(x, y) dy = h(x, o) \tag{6.19}$$

6.1.3.5. *MTF measurement in practice*

However, the gains in terms of signal noise ratio are not particularly significant, and the step edge approach is generally preferred.

In this approach, the observed object o_{se} is made up of two highly contrasting areas, separated by a linear border. The image obtained, o_{se} integrated in the direction of the contour, is thus the integral of the impulse response in the direction perpendicular to the contour:

$$\bar{i}_{se}(x) = \int i_{se}(x, y)dy = \int h(t - x, o)dt \qquad [6.20]$$

and:

$$h(x, 0) = \frac{d}{dx}\bar{i}_{se}(x) \qquad [6.21]$$

This is the most widespread method. The test card $o_{se}(x, y)$ generally includes step edges in two orthogonal directions, often at several points in the field (producing a checkerboard test card). This allows us to determine the two components $h(x, 0)$ and $h(0, y)$ of the impulse response at each of these points, then, by rotating the test card, in any desired direction θ.

Note that the step edge method presents the advantage of being relatively easy to use on natural images, which often contain contours of this type; this property is valuable when no sensor is available for calibration purposes (a case encountered in mobile robotics or satellite imaging).

6.1.3.6. *Advanced test cards*

Several techniques have improved the line-based impulse response measurement approach, refining the test card in order to give a more precise estimation of the MTF. The Joshi test card, for example, uses circular profiles in order to take account of the potential anisotropy of the sensor [JOS 08].

Other approaches involve inverting the image formation equation for a known image, but based on random motifs; theoretically, this produces a flat spectrum. The spectrum of the obtained image may then be simply normalized in order to obtain the desired MTF. Delbracio [DEL 13], for example, showed that the precision of all MTF measurements is limited by a bound associated with the sensor (essentially due to the effects of noise). The author proposes a test card and a protocol designed to come as close as possible to this bound (see Figure 6.8).

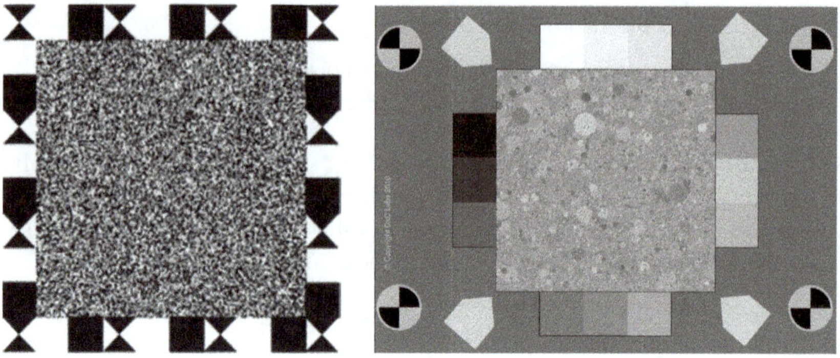

Figure 6.8. *Two examples of test cards used to measure the resolution of imaging systems. Left: the random test card developed by M. Delbracio-Bentancor. Right: the DxO dead leaf test card, taken from [ALV 99], used to characterize texture preservation. The spectrum frequency reduction is expressed in* $1/u^2$

A completely different approach proposes the use of test cards with a random pattern of dead leaves, presenting a polynomial power density spectrum across the full spectrum. These patterns have the advantage of being invariant, not only in translation, rotation and with dynamic changes, but also for changes of scale, due to their fractal properties [GOU 07, CAO 10a, MCE 10]. These test cards currently give the best performance for global estimation of the impulse response. The measurements they provide are only weakly subject to the effects of sharpness amplification techniques, which we will discuss in detail later (see section 6.1.5).

To a certain extent, the resolution problem may be considered to be fully explained using knowledge of the MTF at various points of the image field for all focal distances, apertures and focusing distances, as this information contains all of the material required in order to predict an image for any observed scene. However, this information is bulky and complex, and does not allow us to predict, in simple terms, the way in which this image will be "seen" by an observer. We, therefore, need to consider whether it is better to concentrate on reliable reproduction of low frequencies, to the detriment of higher frequencies, or vice versa.

6.1.4. *Sharpness*

The term "sharpness" is used by photographers to express the quality of frequency content, referring to good representation of fine details, alongside good levels of image resolution. However, this notion does not translate into a measurement which can be applied to an image, and remains a specialist concept, used, for example, to classify multiple photographs of the same scene taken using different parameters. Generally, an image taken with a medium aperture (for example, 5.6 or 8) often presents higher levels of sharpness than those taken with a wider aperture (involving, *a priori*, less diffraction error) or a narrower aperture (with greater depth of field). Sharpness is, therefore, a subjective quality, which, in practice, takes the form of better preservation (less damping) of high frequencies in areas of interest.

Using a given MTF image, it is relatively easy to modify the spatial frequencies carrying these useful data, forcing sharpness beyond that which could be obtained using an ideal system. This type of operation is widespread in photography, and began to be used even before the introduction of digital image processing techniques.

6.1.5. *Acutance*

Acutance, while not universally used, provides a solid foundation for the sharpness criterion as long as care is taken to establish precise observation conditions.

6.1.5.1. *Sensitivity to spatial contrast*

In the specific situation where we are able to define both the observation distance, the luminosity of the image and the luminosity of the lighting, the average sensitivity of the human visual system to the frequency content of an image is well known. This information is summarized [LEG 68, SAA 03] by the spatial contrast sensitivity function (SCSF).

This function is determined experimentally, proposing a variety of visual equalization tasks to observers, using charts with varying contrast and frequency. The curves obtained in this way present significant variations depending on the specific task and stimulus used [MAN 74]. When evaluating image quality, the model proposed by Mannos and Sakrison (Figure 6.9, left) is generally used; this is described by the following empirical formula:

$$\begin{array}{ll} \phi(u_\theta) = 2.6 \times (0.0192 + 0.144 u_\theta) \exp(-0.114 u_\theta)^{1,1} & \text{if } u_\theta \geq u_{min} = 8 \\ \phi(u_\theta) = 1 & \text{otherwise} \end{array} \qquad [6.22]$$

Frequency u_θ is expressed in cycles per degree, and the index θ reflects the fact that the human visual system is not isotropic, and is more sensitive to horizontal and vertical orientations (Figure 6.9, right), according to the formula [DAL 90]:

$$u_\theta = \frac{u}{0.15\cos(4\theta) + 0.85} \qquad [6.23]$$

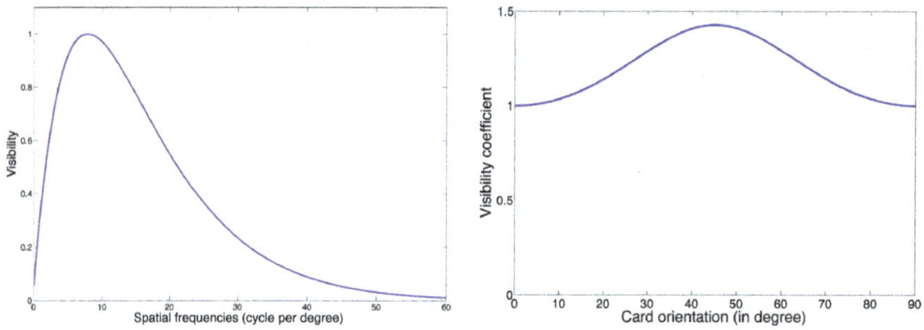

Figure 6.9. *Left: curve showing the sensitivity of the human visual system to spatial frequencies in a horizontal direction, based on [MAN 74]. The frequencies are expressed in cycles per degree. The curve has been normalized so that the maximum has a value of 1 at eight cycles per degree. Depending on observation conditions, the same image may be perceived very different; on a television with a 1 m screen observed from a distance of 1.5 m, for example, we are able to distinguish 150 pixels per degree, whereas on a computer screen observed from a distance of 60 cm we only perceive 40 pixels per degree. Right: effect of test card orientation on the perception of spatial frequencies: a frequency oriented at 45° has the same apparent contrast as a horizontal or vertical frequency which is 1.45 times higher*

6.1.5.2. *Acutance*

This quantity takes account of both the SCSF and the MTF, used to give a single figure qualifying the capacity of an image to transmit fine details, useful for image formation by a user in specific observation conditions:

$$A = \int_{-u_{max/2}}^{u_{max/2}} \phi(u, v)\ H(u, v)\ dudv \qquad [6.24]$$

Thus, for a fixed resolution, the acutance measure gives a greater weighting to frequencies in the high-sensitivity zone of the human eye, based on specific observation conditions. The acutance of an image will differ depending on whether it is observed on a cinema screen, a television or a computer; this corresponds to our practical experience of image quality.

It has been noted that the acutance, determined using a single MTF, only characterizes the center of an image; certain researchers have recommended the use of a linear combination of acutances at a variety of points in the image. However, this refinement is not sufficient to compensate for all of the shortcomings of acutance as a measure of quality; these limitations will be described below.

6.1.5.3. *Limitations of acutance:*

The notion of acutance appears to be based on a solid experimental foundation (measured MTF) and to take account of the human visual system (SCSF); however, it has not been generalized. This is due to the fact that acutance is a particularly fragile measure, easy to modify using artificial means and does not always reflect the perceived quality of an image.

To increase acutance, we begin by exploiting the high sensitivity of image sharpness at certain frequencies. It is easy to produce very high acutance levels in any image by magnifying these frequencies, even if the scene does not naturally contain details in the observer's high-frequency sensitivity zone. This amplification approach essentially affects noise levels, decreasing the signal-to-noise ratio. While the acutance level increases, the resulting subjective quality decreases.

Moreover, certain details which are amplified by the improvement process may be of secondary importance for quality perception, and their accentuation may even be undesirable as they conceal more attractive properties. Professional photographers are particularly attentive to this issue, and make use of all attributes of a photo (lighting, focus, etc.) in order to retain only the esthetically relevant elements of a scene. Increasing the acutance greatly reduces the accuracy of the representation of a scene.

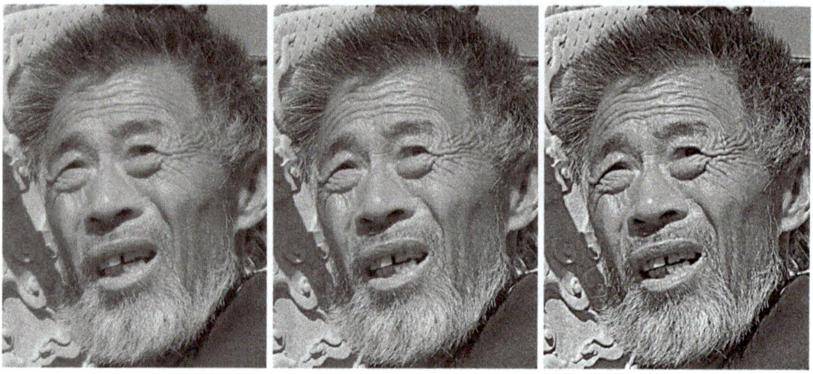

Figure 6.10. *Acutance amplification by unsharp masking (USM). The image on the left is the original; the two images on the right have the same resolution, but have been subjected to increasing amplification of the mean frequencies*

Acutance is nevertheless a highly useful quality component, and experience shows us that careful application of acutance amplification can result in quality gains. This is illustrated in Figures 6.10. Moreover, the property has long been used in film photography, and sophisticated, complex methods[6] have been developed to achieve this goal, notably unsharp masking.

6.1.5.4. *Unsharp masking*

Unsharp masking generally uses a direct combination of pixels around a given point in order to produce moderate amplification of high frequencies in an image. Unsharp masking reinforces image contours, increases the visibility of details and increases legibility, but to the detriment of fidelity. Moreover, it also increases noise, resulting in the appearance of unwanted details in the image; these are generally interpreted as sharpness, but can prove problematic.

Unsharp masking is a two-step process. We begin by evaluating a low-pass filtered version $i_{lp}(x, y)$ of the image $i(x, y)$, before removing a portion of this

6 In photography, unsharp masking is carried out by exposing a photograph through a dimmed, slightly blurred negative of the original, hence the name of the procedure.

filtered image from the original image, and increasing the contrast as required in order to use the full dynamic range. Frequencies not affected by the low-pass filter are, therefore, subject to relative amplification.

The first stage can be carried out using a Gaussian filter, for example, generally in the image plane; in the Fourier plane, this filter is expressed by the formula:

$$I_{lp}(u,v) = I(u,v) \times \frac{1}{\sigma\sqrt{2\pi}} \exp\left[-(u^2+v^2)/2\sigma^2\right] \qquad [6.25]$$

where parameter σ sets the range of the frequencies to be dampened. This parameter is generally determined by the user. This stage consists of calculating the image after unsharp masking i_{usm}:

$$i_{usm}(x,y) = \gamma\left[i(x,y) - \alpha i_{lp}(x,y)\right] \qquad [6.26]$$

where parameter α is used to control image filtering levels, and coefficient γ allows us to control contrast through best use of the available dynamic.

In many applications, formula [6.26] is applied exclusively to contour pixels in order to avoid excessive increases in noise. This is done using a third parameter, δ_{usm}, which establishes the minimum difference between a pixel and its neighbors. If no difference greater than δ_{usm} is found, then the pixel will not be treated.

A more direct approach involves convolution in the image plane via the use of a carefully selected mask, or via iterative techniques based on diffusion equations.

Many cameras systematically apply unsharp masking during image acquisition, often with different parameters depending on the selected capture mode. Photographs of landscapes or sporting events are, therefore, treated using more stringent parameters than portraits, for example. Furthermore, the images produced by low-cost cameras are often subject to aggressive treatments, which improve the perception of the image on the camera's own small screen and hide resolution or focusing errors; however, the quality of printed versions of these images is limited. More complex cameras allow users to adjust unsharp masking settings; in this case, unsharp masking is only applied to images in JPEG format (and not to RAW images) (these are the objects of Chapter 8).

Note that unsharp masking should implicitly take account of observation conditions. It is, therefore, most useful in the context of postprocessing, and it is better to use the unsharp masking tools included in image treatment software than to allow indiscriminate application during image capture.

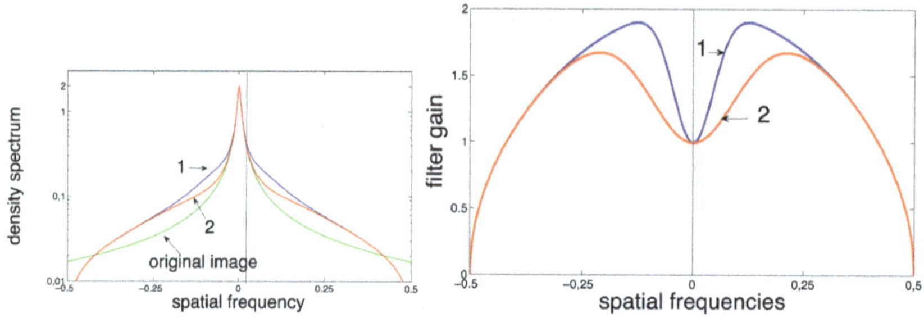

Figure 6.11. *Unsharp masking: power density spectrum of an image (logarithmic scale), before and after filtering, using two different types of unsharp masking filters. Right: transfer functions of the two filters, in the form* $[1 - \alpha \exp(-au^2)]$*, multiplied by a function* $\sqrt{cos(\pi u)}$*, which cancels out at the limits of the bandwidth in order to reduce aliasing effects. For a color version of this figure, see www.iste.co.uk/maitre/pixel.zip*

6.2. Global image quality assessment

Over the last 50 years, the scientific community has used these quality criteria, along with others discussed below, in attempts to qualify image quality (and not the quality of photographic systems, as we would wish). This work has led to the creation of a subdomain of image processing, often known as image quality assessment (IQA) and to extensive publications. The role of IQA is to provide reliable tools for predicting the visual quality of images, applicable to a wide variety of applications, from compression to coding, onboard robotic systems, site surveillance, target detection and tracking in industrial or defense applications, image watermarking for copyright protection purposes, etc. A thorough review of work carried out in this subject is available in [CHA 13].

Work on IQA has notably resulted in the creation of an International Telecommunications Union recommendation (current version ITU-R BT.500, 2013 [ITU 13]) which sets out the conditions for establishing a subjective image quality score for television pictures[7]; this recommendation may also be applied to stationary images.

The recommendation uses two complementary five-level scales, one representing quality, the other representing impairments. The first is essentially used for no-reference evaluations, while the second is used for evaluations based on a reference (see Table 6.1).

	Quality scale		Impairment scale
5	Excellent	5	Imperceptible
4	Good	4	Perceptible, but not annoying
3	Fair	3	Slightly annoying
2	Poor	2	Annoying
1	Bad	1	Very annoying

Table 6.1. *Table showing the quality and impairment scales put forward in ITU6R BT 500-13, [ITU 13]*

The aim of this project is thus slightly different from that which we have considered so far, as no attempt is made to define quality attributes. We also note that these scales give little consideration to images of very high quality, such as those obtained in photography, which generally only use levels 4 and 5.

Finally, note that, while the final evaluation is global ("excellent", for example), the ITU approach, like most of those examined below, is based on the collection of a quantity of local information (such as perceptible impairments) which are then aggregated into a single criterion. This pooling operation is sometimes subject to explicit formulation, but on other occasions, as in the ITU recommendation, remains unformalized.

We will now present a number of methods which operate based on the use of a real reference image (rather than a test card, as in previous cases), before considering some examples of no-reference methods[8]. As we will see, most work in this domain is based on analysis of the human visual system, and methods generally attempt to reproduce biological mechanisms. It is now possible to design evaluation systems which accurately mimick this system of

7 Other recommendations may be found in the final report issued by the Video Quality Experts Group on the Validation of Objective Models of Video Quality Assessment Phase II, issued in 2003 and available at http://www.vqeg.org.

8 Intermediate "reduced reference" evaluation systems may also be encountered, but will not be discussed here.

perception. While these methods have yet to provide an operational solution to the problem of quality estimation in photography, they have produced promising results, and provide an interesting formal framework, which we will discuss below.

6.2.1. *Reference-based evaluations*

A family of measures exists based on the Euclidean distance between two images: i, the image under test and j, the reference image, or, in an equivalent manner, between their cross-correlation functions, whether or not these are normalized or centered. The universal quality index (UQI) criterion [WAN 02] is expressed as:

$$UQI(i,j) = \frac{4\sigma_{ij}\bar{i}\bar{j}}{(\sigma_i^2 + \sigma_j^2)(\bar{i}^2 + \bar{j}^2)} \qquad [6.27]$$

using mean values $\bar{i}$ and $\bar{j}$, variances σ_i^2 and σ_j^2 and the centered cross-correlation:

$$\sigma_{ij} = \frac{1}{N-1}\sum_1^N (i-\bar{i})(j-\bar{j}) \qquad [6.28]$$

This criterion takes values between -1 and 1. It has a value of 1 if the images are proportional. The UQI uses a single term to express three impairment types: loss of correlation, loss of contrast and distortion of luminosity. It has been widely revised and extended in recent years; the structural similarity (SSIM) criterion, for example, adds two constants to formula [6.27]:

$$SSIM(i,j) = \frac{(2\sigma_{ij} + c_2)(2\bar{i}\bar{j} + c_1)}{(\sigma_i^2 + \sigma_j^2 + c_2)(\bar{i}^2 + \bar{j}^2 + c_1)} \qquad [6.29]$$

These constants, c_1 and c_2, are expressed empirically as a function of image dynamics, and result in a regularization of the criterion in uniform zones. They typically take values between 2 and 50, and $c_2 \sim 10\ c_1$. Other authors have extended the UQI to color images. However, the cross-correlation criterion remains rather crude, barely more satisfactory than the signal-to-noise ratio, and its use is extremely limited outside the field of coding applications. Other approaches, which replace this criterion by cross-information measures [SHE 06] or distances in a singular value decomposition (SVD) [SHN 06], are little better.

Many researchers have noted that indiscriminate summing of a local measure is not particularly compatible with human judgments, which are particularly focused on certain structures or zones of specific interest to the observer, often due to their semantic content. For this reason, an additional step may be added before measuring the difference between the reference and test images, consisting of selecting or amplifying zones of interest. The criterion used (SNR, UQI or SSIM) is then weighted using this visibility, or salience, map before summing. The construction of these maps has been widely discussed, with methods based on the use of perceptual methods such as the SCSF (equation [6.22]), on the use of information theory to identify zones of interest or on empirical selection of zones featuring high contrasts and contours. For example, Wang [WAN 11] uses a statistical image model based on a mixture of multiscale Gaussians. Shared information from the reference and test images is calculated locally, and then distributed using a multiscale representation to describe the significance of each difference. Unsurprisingly, in this context, it is generally best to amplify the importance of high-contrast zones within the image, and of zones where there is a large difference between the test and reference images.

A number of authors have modeled these human processes, and attempted to use statistical methods to "learn" them. One example of this type of approach is the machine learning-based image quality measure (MLIQM) algorithm [CHA 12b]. Using the many known criteria involved in perceptual evaluation, the MLIQM algorithm does not, however, involve a fully developed model of human vision. It makes use of a wide variety of measures, including both spatial properties (13 primitives) and spectral properties (12 primitives) of test and reference images. The spatial primitives, which essentially reflect the structures in the image, are calculated on multiple scales using distortion maps, each reduced to a single score. The frequency primitives are calculated over three levels and four orientations; as with the structural coefficients, they are expressed in the form of a single final score. The vector of the structural and spectral primitives is then classified using a support vector machine (SVM), which has previously been "trained" using a range of examples. The results obtained in this way are converted into a score, from 5 to 1 on the ITU scale (see above); these scores present excellent correlation with the scores given by human observers for reference databases.

The visual SNR (VSNR) criterion [CHA 07b] is more complex, being based on richer psychovisual bases, and involves a two-step approach. First, it detects a variety of potential impairments, using mechanisms similar to the first levels of human vision. Distortions between the test and reference images are calculated, band-by-band, after decomposing the image into frequency bands using one-octave steps. At the same time, visibility thresholds are established for each band. This allows us to carry out the second stage, identifying the perceptible components of differences between the image under test and the reference. These components are then subjected to empirical priority rules for frequency bands (low frequencies have a higher weighting only if the structures including the differences are large). When compared to human judgments across a wide base of images and impairments, this criterion presents strong prediction capacities; however, the implementation is relatively heavy.

6.2.2. *No-reference evaluation*

No-reference methods for image quality evaluation have also been the subject of extensive research [FER 09, HEM 10]. *A priori*, this type of approach is better suited to the evaluation of photographic equipment; however, as we will see, examples of this type are not yet fully operational. Most of the proposed methods consider two specific aspects of quality: contour definition, on the one hand, and the absence of noise, on the other hand. Other approaches focus specifically on compression-related impairments: JPEG and JPEG2000. These are less relevant to our specific context. After discussing contour sharpness and the presence of noise, we will consider the way in which the statistics of natural scenes may be used, before considering a more advanced method based on phase coherency.

6.2.2.1. *Noise measurement*

Noise b is traditionally evaluated by studying the neighborhood of the origin of an image's autocorrelation function. Its power is often measured by using the excess at the origin of this function compared to the analytical extension of the surrounding values. Writing the noised image in the form $i_b(x, y) = i(x, y) + b(x, y)$, and considering that the unnoised image has a correlation $C_i(x, y)$, the noise is impulsive and not correlated to the signal, the autocorrelation of the noised image becomes:

$$C_{i_b}(x, y) = C_i(x, y) + \sigma_b^2 \delta(0, 0) \qquad [6.30]$$

Well-verified models of C_i show the exponential reduction either in terms of x and y, or, better, in $\rho = \sqrt{x^2 + \alpha y^2}$, α, taking account of the possibility of anisotropy, and allow us to deduce σ_b by studying C_{i_b}.

6.2.2.2. *Contour sharpness*

The sharpness of contours is an issue familiar to all photographers, and is the reason for the inclusion of a specific element, the autofocus system (these elements will be discussed in section 9.5.2), in photographic equipment. However, the approach taken may differ. Focus search features which act on image quality (this is not universal: other elements operate using telemetric distance measurement via acoustic waves, or using stygometry, i.e. by matching two images taken from slightly different angles), only use small portions of the image (either in the center of the pupil, or scattered across the field, but in fixed positions determined by the manufacturer) in order to obtain measures. These features allow dynamic focus variation in order to explore a relatively extended space and determine the best position. The criteria used to define the best position are very simple: signal variance, or the cumulative norm of the gradients. Simplicity is essential in order for the feature to remain reactive.

Using the IQA approaches described here, however, an exhaustive search may be carried out over the whole field, or we may look for the most relevant zones to use for this measure. However, we only have an image to use in order to decide whether or not a point is good. As we are not limited to real-time operation, longer calculation processes may be used in these cases. We, therefore, wish to characterize contour sharpness at global level, across the whole image. To do this, images are often divided into smaller cells (of a few hundred pixels), and the calculation is carried out separately for each cell. This process generally involves three steps:

– The first step determines whether or not a contour is present in cell k. If no contour is present, then the cell is left aside. Otherwise, the position of this contour is determined using local gradient measures.

– We then define the width $\psi(k)$ of the contour, measuring its extent, the length of the line with the greatest slope.

– Finally, a global quality score is deduced from the various measurements obtained, allocating weightings which may be based on visual perception rules (for example, using salience maps, or leaving aside measurements lying below a specified visibility threshold).

Local measures may then be combined in the following manner [FER 09]: we begin by defining a just noticeable blur (JNB) threshold $\psi_{JNB}(k)$, which is a function of the variance and contrast of cell $\mathcal{C}_k$. The probability of finding a single contour of width ψ within a window k is given by the psychometric function[9]:

$$P(\psi) = 1 - \exp\left[-\left|\frac{\psi}{\psi_{JNB}(k)}\right|^{\beta}\right] \qquad [6.31]$$

Variable β is determined experimentally based on image dynamics. In the presence of multiple independent contours in a cell, the probability of detection will be:

$$Proba(\mathcal{C}_k) = 1 - \prod_{j\in\mathcal{C}_k}(1 - P(\psi_j)) = 1 - \exp\left[\sum_{j\in\mathcal{C}_k}\left|\frac{\psi_j}{\psi_{JNB}(k)}\right|^{\beta}\right]^{1/\beta} \qquad [6.32]$$

and the consolidated measure for the whole image i is expressed:

$$Proba(i) = 1 - \prod_{\mathcal{C}_k\in i}[1 - Proba(\mathcal{C}_k)] = 1 - \exp(-D^{\beta}) \qquad [6.33]$$

Variable $D = -1/\beta\log[1 - Proba(i)]$ and the number l of treated cells are then used to define a sharpness index, $S = l/D$.

6.2.2.3. *Statistics of natural scenes*

The use of statistics measured using a variety of natural scenes is one way of compensating for the lack of references in evaluating image quality. However, few statistics are sufficiently representative to allow their use in this role [SIM 01, TOR 03].

The natural image quality evaluator [MIT 13] (NIQE) method uses statistics based on image amplitude [RUD 94]: an intensity $\hat{i}$ given by the intensity i of each pixel, corrected using a weighted bias $\hat{\mu}$ and weighted variance $\hat{\sigma}$:

$$\begin{aligned} \hat{i}(x,y) &= \tfrac{i(x,y)-\hat{\mu}(x,y)}{\hat{\sigma}+1} \\ \hat{\mu}(x,y) &= \iint_{\mathcal{V}(x,y)} w(x',y')i(x',y')dx'dy' \\ \hat{\sigma}(x,y) &= \sqrt{\iint_{\mathcal{V}(x,y)} w(x',y')\left[i(x',y') - \hat{\mu}(x',y')\right]^2 dx'dy'} \end{aligned} \qquad [6.34]$$

9 This function is monotonic between 0 and 1 as a function of its argument, and is widely used in modeling the results of psychovisual experiments. The function gives us a probability of 63% of finding a contour of size $\psi_{JNB}(k)$.

where $\mathcal{V}(x, y)$ represents a small circular domain around the point (x, y), and $w(x, y)$ is a Gaussian centered on this point. These same elements, known as mean subtracted, contrast normalized (MSCN) coefficients, have been used in a number of other methods, such as the blind/referenceless image spatial quality evaluator (BRISQUE) [MIT 12].

The most robust properties, however, are those based on spectral properties.

The distortion identification-based image verity and integrity evaluation (DIIVINE) [MOO 11] method, for example, uses models of wavelet coefficients of multiscale decompositions of an image. These coefficients are particularly interesting in the context of image coding. The marginal and joint statistics of these coefficients have been shown to be represented successfully using Gaussian mixtures.

The parameters of these models are learned from original images, and then from images subjected to a variety of impairments: blurring, added noise, compression of various types and transmission over lossy channels. Applied to an unknown image, the DIIVINE method begins by determining whether or not an image is defective, and if so, what type of impairment is present; it then assigns a quality score which is specific to this impairment type. This procedure has produced satisfactory results over relatively varied databases, but the full procedure remains cumbersome, as it requires multiple successive normalizations.

6.2.2.4. *Global phase coherence*

Instead of measuring contour spread, it is possible to consider the inverse quantity, i.e. the image gradient and its distribution within the image.

This problem has been addressed in a more analytical manner by examining the way in which transitions in the normal direction of the contour occur, and, more precisely, by studying the global phase coherence [LEC 15]. The underlying idea is based on the fact that a well-focused image is highly sensitive to noise disturbance and to phase modifications of the various frequencies making up the image. These abrupt variations can be measured using particular functions derived specifically for this purpose:

– global phase coherence $\mathcal{G}$, expressed as a function of the probability that a global energy measure (the total variation, TV) of an image i_ϕ, subject to random phase noise will be lower than that of the original image i:

$$\mathcal{G} = -\log_{10} Proba\,[TV(i_\phi) \leq TV(i)] \qquad [6.35]$$

where the total variation of a function $f(x)$ is expressed as:

$$TV[f(x)] = \int_{-\infty}^{\infty} \left| \frac{\partial f(x)}{\partial x} \right| dx \quad [6.36]$$

and that of an image $i(x, y)$ depending on two indices is expressed as:

$$TV(i(x,y)) = \iint |\nabla i(x,y)|[dx\ ,\ dy] \quad [6.37]$$

However, this criterion can only be calculated via a cumbersome simulation process, with random drawings of a number of phase realizations;

– the sharpness index, in which the random phase disturbing i is replaced by convolution by a white noise w:

$$SI(i) = -\log_{10} \Phi \left[\frac{\mu - TV(i)}{\sigma} \right] \quad [6.38]$$

where $\Phi(x) = 1/2\pi \int_x^\infty \exp(\frac{-u^2}{2})du$ and μ and σ^2 are the expectation and variance of $TV(i * w)$;

– the simplified sharpness index S, deduced from the term given above by replacing σ with σ'; the expression of the latter term is more compact, and is obtained by calculating the second derivatives i_{xx}, i_{yy} and i_{xy} and the first derivatives i_x and i_y of i:

$$\sigma' = \sqrt{\frac{1}{\pi} \left[\frac{i_{xx}^2}{i_x^2} + \frac{i_{xy}^2}{i_x i_y} + \frac{i_{yy}^2}{i_y^2} \right]} \quad [6.39]$$

While the complexity of these criteria is greatly decreased, they behave, experimentally, in a very similar manner as a function of the blur present in the image. Although they are well suited to monitoring the treatments applied to images (such as deconvolution), these criteria suffer from an absence of standardization, making them unsuitable for our application.

6.2.3. *Perception model evaluation*

A number of the mechanisms involved in image perception have already been described in this book. These relate to photometric sensitivity (section 4.2.1) or spectrometric sensitivity (section 5.1.2), the persistence of the impression of color and the effects of spatial masking (section 5.3). In addition to these mechanisms, we need to consider multiscale processing, which allows observers to examine scenes in a hierarchical manner, either by considering the general distribution of wide bands, or by considering fine local details, via a continuum of descriptions of variable resolution. Wavelet pyramid, Gaussian pyramid or radial filter decompositions provide a good representation of this property.

Methods based on perception models are particularly useful for expressing degradations which take place during image acquisition. They presume the existence of an "ideal" image, to which the acquired image is compared. The reference image and the image being analyzed are processed in parallel using a "perception model", which is supposed to convert these two images into stimulus fields similar to those delivered by the visual system (Figure 6.12). As we have seen, multiple fields exist, as the cortical columns in the visual system break down images by chromatic content, orientation and spatial frequency. These stimulus fields are then compared, point-by-point, using a metric (which is itself perceptive) to determine what users will actually perceive as being different, and what will remain undetected. Taking account of all these perceived differences, with careful application of weighting, we then assign a score; this may be considered to be the perceived difference between the two images, and seen as a loss of image "quality".

6.2.3.1. *Expressing perceptual difference*

Chandler [CHA 13] provides a relatively general form of the simulated response of a visual neurone to a set of stimuli, suitable for use in measuring the quality of an image in relation to a reference. This method will be described below, using the same notation as elsewhere in this book. First, image $i(\mathbf{x})$ at the current point $\mathbf{x} = (x, y)$ is decomposed into a channel $\tilde{i}(\mathbf{x}, u, \theta, \chi)$, using three types of information:

– the spatial orientation, denoted using variable θ;

– the spatial frequency, denoted by u;

– the chromatic channel[10], obtained by combining signals L, M and S (section 5.1.2), denoted by χ.

Channel $\tilde{i}$ is weighted using the SCSF, $\phi_{u,\theta}$ (equation [6.22]). It is raised to a power p, expressing the nonlinear nature of the neurone, and normalized using a term which reflects the inhibiting effect of neighboring neurones. This term includes a saturation term b and summing across the neurones S connected to the neurone under study. An exponent q (generally lower than p)

10 The article [CHA 13] only considers black and white signals, and the presented model ignores the chromatic components.

is used to express the nonlinear nature of the inhibiting pathway, and a gain g is applied to the whole term. The response $\ddot{i}$ from the channel is thus expressed:

$$\ddot{i}(\mathbf{x}_o, u_o, \theta_o, \chi) = \frac{g\ \left[\phi_{u,\theta}\ \tilde{i}(\mathbf{x}_o, u_o, \theta_o, \chi)\right]^p}{b^q + \sum_{(\mathbf{x},u,\theta)\in\mathcal{S}} \left[\phi_{u,\theta}\ \tilde{i}(\mathbf{x}, u, \theta, \chi)\right]^q} \qquad [6.40]$$

The parameters governing this equation are generally determined by experimenting with known signals, and are adapted for digital images using classic simplifications (for example, frequencies are converted into octaves using wavelet decompositions, orientations are reduced to horizontal, vertical and diagonal directions, the domain $\mathcal{S}$ is approximated using eight neighbors of the central pixel and the variables covered by χ are either the red, green and blue channels, or the red/green and yellow/blue antagonisms).

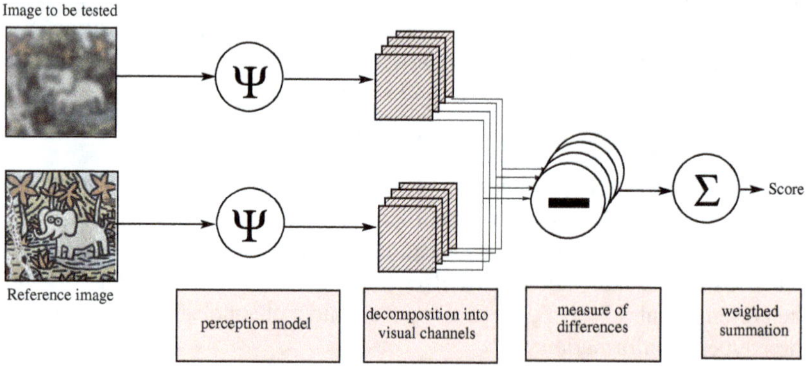

Figure 6.12. *Image quality assessment based on a reference image and a perception model*

6.2.3.2. *Expression of quality*

The model given by equation [6.40], applied to the image under treatment and the reference image, provides us with signals, one for the image under treatment, $\ddot{i}(\mathbf{x}_o, u_o, \theta_o, \chi)$, the other for the reference image $\ddot{j}(\mathbf{x}_o, u_o, \theta_o, \chi)$. In its simplest formulation, the distortion at point $\mathbf{x}_o$ is defined using the differences between these two values. Based on the hypothesis that the perceptual channels are independent, which is relatively well verified, we obtain:

$$d(\mathbf{x}_o, u_o, \theta_o, \chi) = \left[\sum_{u,\theta,\chi} |\ddot{j}(\mathbf{x}_o, u_o, \theta_o, \chi) - \ddot{i}(\mathbf{x}_o, u_o, \theta_o, \chi)|^{\beta} \right]^{1/\beta} \qquad [6.41]$$

Other formulations make use of psychovisual considerations in order to avoid characterizing differences for all points, concentrating instead on smaller areas. Using a simple model, this difference, compared to a threshold, allows us to create a map of perceptible differences. Integrated using a logistical model, this gives us a single quality score, only valid for the specific image in question. Averaged over a large number of images, this score enables us to characterize specific equipment in specific image capture conditions.

Some examples of general models for image quality assessment are given in [WAT 97, BRA 99, LEC 03, LAR 10].

However, work on human perception has essentially resulted in the creation of quality indices for specific applications, which only take account of a limited number of distortions. A wide variety of indices of this type exist [CHA 13]: noise measurement, structural differences, contour thickening, etc.

6.3. Information capacity

Another way of approaching the notion of image quality is to use vocabulary and techniques from the field of information theory, as suggested by Shannon. In this approach, the camera is considered as a channel, subject to the laws of physics and technology, and governed by parameters (aperture, exposure, focus and sensitivity) fixed by the user. This channel is used to transmit a degraded version $i(x, y)$ of a scene $o(x, y)$. The aim in using information theory is to measure message degradation resulting from the transmission process, to define upper performance limits which we may hope to attain in an ideal situation, and to suggest parameter values to optimize performance.

This approach was first explored in the 1960s by physicists, notably for image processing applications in the field of astronomy, but rapidly encountered problems due to the limitations of analog signals. The discrete representation of images produced by digital sensors offers new possibilities, complementary to those examined above. An information-based approach was put forward in [TIS 08] to respond to a key question posed by users when buying photographic equipment: does an increase in sensor resolution improve image quality?

6.3.1. *The number of degrees of freedom*

Digital images are coded using n bits, and each pixel (i, j) in the image is subject to a noise, with an average value which will be presumed to be zero (this will be verified in Chapter 7), and a standard deviation $\sigma(i, j)$ which may vary within the image, for example due to the different gains applied to different sites (see section 3.2.3), to aberrations (section 2.8) or to vignetting (section 2.8.4.1). We thus have $\nu(i, j)$ independent levels within (i, j):

$$\nu(i, j) = 2^n/(max(1, \sigma(i, j))) \qquad [6.42]$$

In this formula, the term $max(1, \sigma(i, j))$ expresses the fact that, if the noise σ is less than a level of gray, then all of the 2^n levels are independent. If the dependence of the noise as a function of the signal is known exactly, we can use the precise number of effective levels ν_e (see equation [6.5]).

In total, the image, therefore, presents ν_1 degrees of freedom:

$$\nu_1 = \sum_{i,j} \nu(i, j) \qquad [6.43]$$

if all sites are independent [FRI 68]; ν_1 is also known as Shannon's number.

This expression of the number of degrees of freedom, in this case measured at the sensor output point, may be compared to the degrees of freedom of the incident wave providing sensor input.

This number is given by the surface of the sensor S divided by the angular resolution of the lens $d\Omega$ (Ω = solid angle of the lens). In the case of a circular aperture of diameter D and focal distance f, and with a wavelength λ, the angular resolution is given by $d\Omega_c = (D/\lambda f)^2$ in the coherent case and $d\Omega_i = (D/2\lambda f)^2$ in the case of incoherent photographic imaging; the latter case will be considered here [FRA 55, FRA 69]. Taking $N = f/D$ as the aperture number, in the incoherent case[11], we obtain:

$$\nu_2 = \frac{4SD^2}{(\lambda f)^2} = \frac{4S}{N^2\lambda^2} \qquad [6.44]$$

Approximating the two estimations of the number of degrees of freedom made before and after detection by the photodetector, we obtain the limit case where the two terms ν_1 and ν_2 are equal. This arises in situations where the sensor and lens are able to transmit the same quantity of information, i.e. the two elements are ideally suited. This gives us:

$$\sum_{i,j} \nu(i,j) = \frac{4S}{N^2\lambda^2} \qquad [6.45]$$

In the case where all pixels have an identical noise of variance $\sigma \leq 1$, and considering a sensor of which the whole surface contributes to the signal (with no space set aside for electronic elements – Figure 3.7), and for which, therefore, $S = n_x n_y l_x l_y$, we obtain the following relationship between the number of useful bits in image n and the aperture number of lens N:

$$N^2 = \frac{l_x l_y}{2^{n-2}\lambda^2} \qquad [6.46]$$

Two different regimes are shown in Figure 6.13: below the curves, the lens measures more degrees of freedom than the sensor is able to transmit. The camera is thus limited by the sensor. Above the curves, on the other hand, the sensor transmits more signals than the lens is able to send. These signals are, therefore, not independent, as the lens is not sufficiently powerful, and the sensor is underexploited.

In the highly simplified form used here, information theory indicates what each system is capable or incapable of in an ideal situation. Here, we have considered that the lens, focus and signal conversion process are entirely

11 Note that using this calculation, the incoherent case gives a number of degrees of freedom which is 4 times higher (as the incoherent pupil is the convolution function of the coherent function, and this function extends twice as far in each direction). However, we must take account of the fact that the samples obtained in incoherent optics are not independent, as they form a positively-defined signal, with its origins in another positively-defined signal present before image formation. This point is discussed in greater detail in section 2.6.5.

flawless. However, there are many reasons why pixels from neighboring sites may not be independent; they are often correlated, either during the formation of the image on the sensor, and beyond the limiting optical resolution considered here – focusing errors, aberrations, parasitic reflection, electronic diaphony between pixels, etc. – or during reconstruction (demosaicing, antialiasing, low-pass filtering, etc.) – or during coding (e.g. the resetting of high frequencies when using JPEG). Similarly, the operational lighting conditions we have considered allow us to maintain noise below one gray level, whatever architecture is used, and, more specifically, for any size of photosite. The hypotheses underpinning this reasoning will be considered in greater detail later. Note, finally, that information theory gives us no indications as to the way in which signals should be effectively treated in order to profit fully from the number of available degrees of freedom.

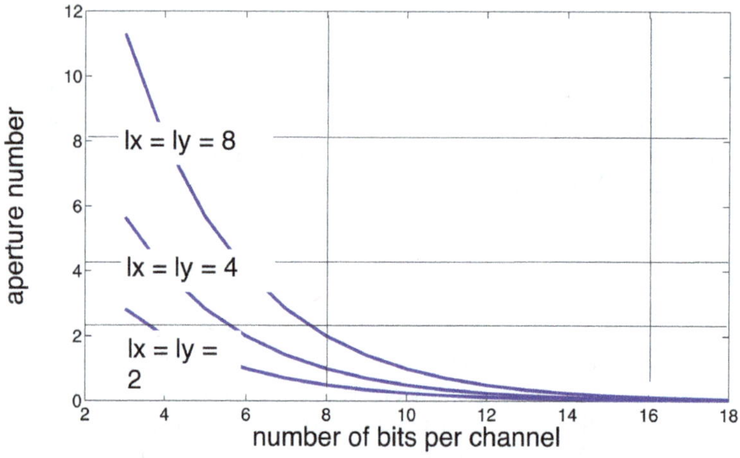

Figure 6.13. *Relationship between the number of bits assigned to each pixel and the f-number of the optical element, in the case where the sensor and lens match perfectly, i.e. the sensor transmits exactly the same number of degrees of freedom recognized by the lens. Above these curves, the sensor is overdimensioned; below the curves, the sensor is underdimensioned. Results are shown for a green wavelength (500 nm) and three dimensions of photosites*

6.3.1.1. *Information theory and coded apertures*

Approaches which take account of the number of degrees of freedom in a system have recently been applied to a new area, due to the use of coded apertures or lens matrices in the context of computational photography, which requires us to measure different quantities to those used in traditional imaging (particularly in terms of depth of field); this requires us to modify equation [6.44] [STE 04, LIA 08].

6.3.1.2. *Color images*

The case of color images is more complex.

From a physical perspective, examining the way in which various wavelengths transmit various types of information, we first need to examine the role of sources, then that of objects in a scene. The degree of independence of the various wavelengths emitted is dependent on the emission type (thermal, fluorescent, etc.). The emission bands are generally large and strongly correlated, with few narrow spectral lines (this point is discussed in section 4.1.2), meaning that the various wavelengths are only weakly independent. This enables us to carry out white balancing using a very small number of measures (see section 5.3).

However, the media through which waves are transmitted (air, glass, water, etc.) and those involved in reflecting light lead to the creation of significant distinctions within spectral bands, increasing the level of diversity between frequencies. The phenomena governing the complexity of the received spectrum are, essentially, those discussed in section 5.1.1, and are highly dependent on the type of scene being photographed; each case needs to be studied individually in order to determine the quantity of information carried by the incident wave. Note that a number of studies have shown that representations using between 8 and 20 channels have proved entirely satisfactory for spectral representation of chromatically complex scenes [TRU 00, RIB 03].

Leaving aside these physical considerations, we will now return to an approach based on subjective quality, which uses the visual trivariance of human observers to quantify the information carried by the luminous flux.

In this approach, spectral continuity is reduced to three channels, R, G and B. For each channel, the calculations made above may be applied, but it is then necessary to take account of interchannel correlation, and of noise on a channel-by-channel basis.

The procedure consists of representing the image in a colorimetric perceptual space (see section 5.2.4) which is as universal as possible, in which we measure the degrees of freedom in each channel, using a procedure similar to that described above. To do this, we may follow the recommendations set out in standard [ISO 06b]. The approach used to determine the number of degrees of freedom is set out in [CAO 10b]; it takes place following white balancing, which corrects issues associated with lighting and with the sensor, and after correction of chromatic distortions due to the sensor (for example, by minimizing the CIELab measure between a known test card and the measured values for the illuminant in question). We then calculate the covariance matrix of the noise in the space sRGB. From this, we determine three specific values of σ_k for $k = 1, 2, 3$. Ratio $2^n/(max(1, \sigma(i,j)))$ from equation [6.42] is then replaced by the number of independent colors available to pixel (i, j), and equation [6.42]:

$$\nu_c(i,j) = \frac{2^{3n}}{\prod_{k=1:3}(\max(1, \sigma_k(i,j)))} \qquad [6.47]$$

Equation [6.44] then needs to be modified to take account of the RGB spectrum. For this, we use three primaries of the sRGB standard, $\lambda_B = 0.460\, nm$, $\lambda_V = 0.540\, nm$ and $\lambda_R = 0.640\, nm$, adding a factor of 2 to the denominator for the green channel and a factor of 4 for the red and blue channels, due to the geometry of the Bayer matrix. The maximum number of degrees of freedom perceived by the photographic objective is then obtained using equation [6.44], which gives:

$$\nu_{2_{RGB}} = \nu_{2_R}\nu_{2_V}\nu_{2_B} = \frac{2S^3}{N^6\lambda_R^2\lambda_V^2\lambda_B^2} \qquad [6.48]$$

For a sensor of size $S = l_x \times l_y$ (expressed in micrometers) and an f-number N, this becomes:

$$\nu_{2_{RGB}} \sim \frac{80\ S^3}{N^6} \qquad [6.49]$$

6.3.1.3. *Unified formulation of the frequency space*

An integral space-time approach, using the Wigner–Ville transformation, provides an elegant expression [LOH 96] of the diversity of descriptions of image f in a space with 2 × 2 dimensions: two dimensions are used for the space variables (x, y), and two dimensions are used for the orientation variables (μ, ν) of the incident rays:

$$W(x, y, \mu, \nu) = \int i(x + \frac{x'}{2}, y + \frac{y'}{2})i^\star(x - \frac{x'}{2}, y - \frac{y'}{2}) \times \exp(-2j\pi(\mu x' + \nu y'))dx'dy' \qquad [6.50]$$

This approach will not be used here, as it involves extensive developments which, to our knowledge, have yet to find a practical application in photography.

6.3.2. *Entropy*

The information theory approach to image quality introduced the number of degrees of freedom as a significant parameter in camera dimensioning. Continuing with this approach, it is possible to provide more precise figures concerning the quantity of information transmitted by the camera. This is done using the notion of entropy.

The digital images produced by a sensor can be represented well in terms of entropy, defined as the average incertitude measure carried by the image [SHA 48]. The notion of information, as defined by Hartley, is expressed as the logarithm of the probability of an event (a certain event – with a probability of 1 – provides no information, while a very unlikely event, if it does occur, provides a lot of information). The entropy of an image is expressed in bits per pixel. The image $f(i,j)$ may be seen as a source of pixels, which are, as a first approximation, taken to be independent[12], and transporting the following average quantity of information per pixel:

$$\mathcal{E} = -\sum_{k=1}^{K} p(k)\log(p(k)) \qquad [6.51]$$

where $p(k)$ expresses the probability of the level of gray k (or the triplet $k = \{R, G, B\}$ in the case of a color image); this probability is estimated by measuring the frequency of occurrence of level k among the K available levels. The results of this approach are well known:

– *a priori*, the entropy $\mathcal{E}$ increases as the richness of the image increases at different levels;

– maximum entropy occurs when all levels of gray are equally occupied, with a value of $\mathcal{E} = n$ bits per sample if the image is coded on $K = 2^n$ bits,

12 This approximation, which considers pixels to be independent, is somewhat crude. More refined models use the difference between successive pixels, or small blocks of pixels (Markov models of order 2 or higher, in one or two dimensions) as the signal. These models are more realistic [MAI 08b], but, on the whole, the conclusions reached in this chapter remain the same.

or $\mathcal{E} = 3n$ bits per sample for a color image, coded using three channels of n bits, each represented in a uniform manner;

– entropy is always positive, and only cancels for a uniform image ($\exists k \colon p(k) = 1$);

– entropy calculations in the image space or the frequency space following a Fourier transformation are equivalent, on the condition that the image is stationary (this is rarely the case) and observed over a domain $\mathcal{XY}$ which is sufficiently large that the product $4\mathcal{XY}B_x B_y$ is much greater than 1; B_x and B_y represent half-bandwidths in terms of x and y.

Experience has shown that, unlike most natural signals, images do not possess a histogram model [MAI 08b], except in certain specific cases: seismic imaging, radar imaging [LOP 08], ultrasound, etc. The use of a Gaussian distribution (often applied to high-volume signals) is not justified.

6.3.2.1. *Mapping*

Images are often transformed by changing coordinates, for example to account for perspective or aberrations. This has an effect on the entropy. Transforming the space variables of $\mathbf{x} = (x, y)$ into $\mathbf{x}' = (x', y')$ using the transformation $\mathbf{x}' = T(\mathbf{x})$, which has an inverse transformation $\mathbf{x} = T'(\mathbf{x}')$ and a Jacobian $J(\mathbf{x}|\mathbf{x}')$, the entropy $\mathcal{E}'(\mathbf{x}')$ of image $i'(x', y')$ is expressed as a function of that image $i(x, y)$, $\mathcal{E}(\mathbf{x})$ using the relationship:

$$\mathcal{E}'(\mathbf{x}') = \mathcal{E}(\mathbf{x}) - E_{\mathbf{x}}\left[\log(|J(\mathbf{x}|\mathbf{x}')|)\right] \qquad [6.52]$$

where $E_{\mathbf{x}}\left[u\right]$ denotes the mathematical expectation of function u.

While image entropy is maintained during translations and rotations (since $J = 1$), perspective projections and distortions, which are not isometric, result in changes to the entropy.

6.3.2.2. *Entropy of the source image behind the camera*

It is harder to define in practice. If the object space is expressed as the wavefront at the input lens, we have a continuous signal with continuous values. Definition [6.51] ceases to be applicable, and its extension from the discrete to the continuous domain poses significant mathematical difficulties (see [RIO 07, pp. 60–63], for example). In these cases, we use a formula which is specific to the continuous case, constructed using the probability density $\pi(x)$, which gives us a quantity known as differential entropy[13]. This allows us to extend entropy properties, on the condition that we remain within continuous spaces and we do not modify the space variables:

$$\mathcal{E} = \int_{\mathbf{x}} \pi(\mathbf{x}) \log \frac{1}{\pi(\mathbf{x})} d\mathbf{x} \qquad [6.53]$$

This formula is similar to the direct expression, but does not constitute a continuous extension of this expression.

6.3.2.3. *Loss of entropy*

Equation [6.53] allows us to measure the loss of information from an image in the course of linear filtering ([MID 60] section 6.4). If the image has a bandwidth $B_x \times B_y$ and if the filter has a transfer function $H(u, v)$ in this band, the differential equation loss to which the image is subjected in passing through this filter may be calculated in the Fourier domain (as the FT is isometric and thus conserves entropy). It is measured using:

$$\Delta\mathcal{E} = \frac{1}{B_x B_y} \iint_{-B_x/2, -B_y/2}^{B_x/2, B_y/2} \log(|H(u, v)|^2) du dv \qquad [6.54]$$

If the logarithm is in base 2, this is expressed in bits, and if B_x and B_y are expressed using the interval $[-1/2, 1/2]$, this is expressed in bit/pixel.

Note that this formula is independent of the image, and only characterizes the filter. It indicates a loss of information affecting a signal of uniform spectrum, i.e. white noise. It thus allows us to compare the effects of different filters on a family of images. Applied to a given image, we need to take account of the frequency content in order to determine the information loss affecting the filter, as, for each frequency, the image cannot lose more information than it actually possesses.

6.3.3. *Information capacity in photography*

Various elements contribute to the capacity of information which may be carried by an image, recorded by a camera, in specific capture conditions; these

---

13 This entropy value cannot be deduced from the entropy defined in [6.51] using the limits of quantization intervals (applying $p(k) \longrightarrow \pi(x)\delta x$), as this operation results in a divergence of the integral $\sum \pi(x)\delta x. \log(\pi(x)\delta x)\delta x$. The divergent term $\sum \log [\pi(x)\delta x] \delta x$ must be removed, hence the name of the quantity.

may be assembled following the approach set out in [CAO 10b]. To do this, we suppose that the incident signal has a very wide spectrum, greater than the camera resolution, and that the aperture, exposure time and sensitivity are able to cover the full image dynamic without reaching saturation.

The number of photosites, N, is the most visible and most widely-known factor. The dynamics of each pixel, expressed as a number of levels of gray n (256 or more for native-format images), or, better, as the number of effective levels n_e taking account not only of the range of radiances covered by the saturated pixel (corresponding to the full well capacity of the photosite, see section 3.2.3), but also of the noise in the whole dynamic following relationship [6.5], is another key factor.

In the absence of other impairments, the maximum information capacity would be given by $\mathcal{C}_{max} = n_e N$; however, these other potential impairments need to be taken into account.

First, each pixel needs to be connected with its dynamics (which may be affected by vignetting). Vignetting is treated in different ways by different cameras. If vignetting is not taken care of by software elements, the pixel will be subject to dynamic loss (according to the law in terms of $\cos^4 \alpha$ presented in section 2.8.4). If, on the other hand, a gain is added to compensate for vignetting, then the noise level will increase (in accordance with the term g_{cv} in equation [7.10]).

For spatially variable errors, such as chromatic and geometric aberrations, geometric transformation equation [6.52] should be applied to each pixel. These errors may also be taken into account in an approximate manner in the term expressing the noise affecting each pixel[14].

These impairments reduce the number of levels of each pixel to $n_f(x, y) \leq n_e$, and the quantity of information carried by the image becomes:

$$\mathcal{C}_f = \sum_{x,y} n_f \leq \mathcal{C}_{max} = n_e N \qquad [6.55]$$

14 Note that a part of these aberrations can be corrected in full by inverting the geometric deformation. This part, therefore, does not result in any information loss. Information is lost in two sets of circumstances: first, when signals are mixed with others and can no longer be separated, and second, when geometric deformation takes the signals outside of the sensor. This second case is well accounted for by the Jacobian; the first case, on the other hand, is treated by an increase in noise in the affected pixels (or channels).

We then need to take account of the convolution-related impairments which affect pixel formation using equation [6.54]. In this case, we will consider lens diffraction and integration by the sensor (which is almost, although not entirely, invariant under translation).

Spatially invariant impairments affect all pixels, and expression [6.54] can, therefore, be applied in each case. The number $n_n(x, y)$ of bits required to code a pixel (x, y), initially having n_f independent levels but affected by the transfer function H of the lens, is therefore expressed as:

$$n_n(x,y) = \frac{1}{B_x B_y} \iint_{-B_x/2,-B_y/2}^{B_x/2,B_y/2} \times \max\left[0, n_f + \log_2(|H(u,v)|^2)\right] dudv \quad [6.56]$$

The terms n_f and $\max[0, n_f + \ldots]$ reflect the fact that a pixel cannot lose more information than the number of bits used to represent it permits. If the filter is perfect ($H(u, v) = 1, \ \forall(u, v)$), then $n_n = n_f$; generally, however, $0 \leq |H| \leq 1$ and $n_n < n_f$.

Let us apply this expression to a Bayer-type sensor. First, it is likely that we will need to differentiate between the dynamics of the R, G and B signals, and use different terms n_f for each, as chromatic impairments have a different effect on each channel. This is easy to carry out, and in some cases it is not necessary. The problem of sampling is more complex. As we have stated, $B_x = B_y = 1$ must be selected so as to allow expression of the integral in bits/pixel. However, the green channel is sampled using 2 pixel steps. In this case, therefore, we only retain a fourth of the spectrum, and equation [6.56] becomes:

$$n_n(x,y,V) = 4 \iint_{-\frac{1}{4},-\frac{1}{4}}^{\frac{1}{4},\frac{1}{4}} \max\left[0, n_f + \log_2(|H(u,v)|^2)\right] dudv \quad [6.57]$$

Similarly, the red and blue channels give:

$$n_n(x,y,R) = n_n(x,y,B) = 4 \iint_{-\frac{1}{8},-\frac{1}{8}}^{\frac{1}{8},\frac{1}{8}} \times \max\left[0, n_f + \log_2(|H(u,v)|^2)\right] dudv \quad [6.58]$$

Function H is the product of two independent terms: diffraction by the lens H_f and integration at the photosite H_s.

$$H(u,v) = H_s(u,v)H_f(u,v) \qquad [6.59]$$

H_s is determined based on the size of the photosite (presumed to be rectangular) alone. Let δx and δy be the dimensions of this photosite:

$$H_s(u,v) = \mathrm{sinc}(2\pi u\delta x).\mathrm{sinc}(2\pi v\delta y) \qquad [6.60]$$

H_f is determined by the focal distance f and the diameter of the diaphragm D of the lens (and hence by the aperture number f/D), and is dependent on the source wavelength λ:

$$H_f(u,v,\lambda) = \left[\frac{2J_1(\pi\rho D/\lambda f)}{\pi\rho D/\lambda f}\right]^2 \quad \text{where } \rho = \sqrt{u^2+v^2} \qquad [6.61]$$

The maximum capacity of information carried by the green channel thus becomes:

$$\begin{aligned}\mathcal{E}_V = 4\textstyle\sum_{x,y}\iint_{-\frac{1}{8},-\frac{1}{8}}^{\frac{1}{8},\frac{1}{8}} \max[\,0, n_f(x,y,V) + 2[\log_2(|H_s(u,v)|\\ + \log_2(|H_f(u,v,\lambda_V)|)]]dudv\end{aligned} \qquad [6.62]$$

Note that the coupling between lens diffraction and pixel dimension is reflected in the limits of the integral [6.62], as we have chosen to express the bandwidth as the inverse of the maximum frequency accepted by the photosite.

If the losses due to linear filtering are low in comparison to the number of effective levels of a pixel, and if the N pixels in an image have the same dynamics and the same noise level, this equation can be reduced to:

$$\begin{aligned}\mathcal{E}_V = Nn_f(Vert) + 8N\iint_{-1/8,-1/8}^{1/8,1/8}[\log_2(|H_s(u,v)|\\ + \log_2(|H_f(u,v,\lambda_V)|)]dudv\end{aligned} \qquad [6.63]$$

Similar formulas may be written for the blue and red channels. The total maximum information capacity carried by the image is:

$$\mathcal{E} = \mathcal{E}_G + \mathcal{E}_R + \mathcal{E}_B \qquad [6.64]$$

6.3.3.1. *Discussion*

Equation [6.63] shows that the information capacity of the image is calculated as the product of the number of pixels and the effective number of available levels, minus the losses due to pixel integration and lens diffraction (as the integral is always negative); see Figure 6.14.

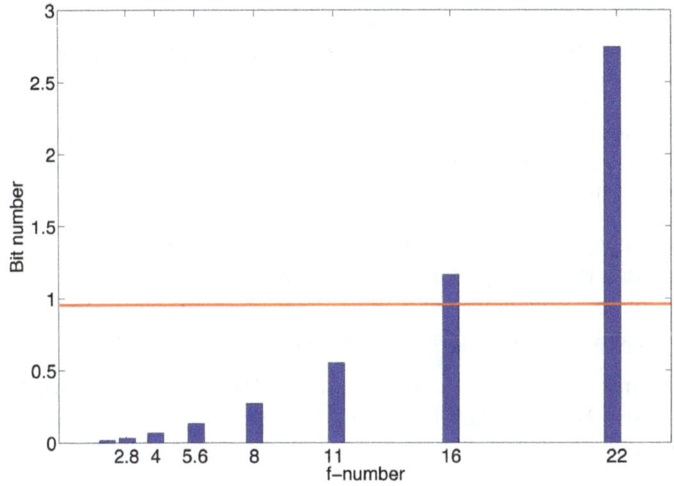

Figure 6.14. *Maximum loss of entropy per pixel due to diffraction as a function of lens aperture (given by term H_f in equation [6.62]), for a detector where the dimension of the photosites is 10 times the wavelength (so approximately 5.5 μm). The horizontal line located at 0.951 bits/pixel indicates the loss of entropy associated with integration by the sensor (given by term H_s). These losses are theoretical, and can only be measured for a signal of uniform spectrum; this never occurs in the case of images*

The consequences of these formulas are listed below:

1) If the dimension of the sensor is fixed, we may wish to increase the number of pixels, N, by reducing the size of each site. This has a number of consequences:

– as N is a factor of the two terms of equation [6.63], the information capacity should increase proportionally;

– however, the noise will also increase, generally following a law proportional to the surface, reducing term n_f in a linear manner;

– the integral over H_s will remain unchanged, as the integrated function and the bounds follow the same law;

– the integral over H_f will increase, as the bounds used for integration over the same function will be more widely spaced.

In all likelihood, the information capacity will initially increase (images will be richer in terms of detail) as the number of pixels increases, up to the point where the noise and blur become too significant.

2) If the number of pixels, the focal distance and aperture are fixed, but we choose to use a larger detector surface, the size of each photosite will increase, hence:

– the number of effective levels will increase, as the available dynamic (fixed by the full well capacity) will increase and the noise (fixed, broadly speaking, by the number of electrons used) will decrease;

– the integral over H_s will increase considerably (fine details will disappear due to the integration over larger photosites), but the total observed field will increase (the image will no longer be the same, as the former version only covered a small part of that which is contained in the new version);

– the integral over H_f will decrease, as H_f will be integrated between much closer bounds (the diffraction figure for each photosite will be negligible).

The scene will, therefore, be different, with fewer fine details but a greater field. Each pixel will be richer in terms of information, and the contrast will be better; the signal will be subject to little noise, and the image quality should appear to be better, as long as vignetting and aberration issues, which will be more significant, are correctly controlled.

3) If we fix the number of pixels and use a larger detector surface, but adapt the focal distance in order to retain the same field:

– the same conclusions as above will be reached in terms of dynamics and noise, and hence in terms of image quality;

– this time, the integral over H_s will remain constant, expressing the fact that integration over the sensor relates to the same details in the image;

– if we succeed in maintaining a constant aperture number, the integral over H_f will decrease, as H_f (which is solely dependent on the numeric aperture) will be integrated between closer bounds; however, lens design in these cases is more complex, as it is hard to obtain identical performances using longer focal distances. The scene will, therefore, be identical in terms of geometry, but of better quality in terms of noise and dynamics.

This discussion illustrates the compromises which need to be made by camera manufacturers. It also helps us to understand the dual aims described at the beginning of Chapter 3: first, the desire to achieve very high resolutions with a fixed sensor size, and second, as VLSI technology progresses, to obtain larger overall surfaces for quasi-constant pixel numbers. The format $24\ mm \times 36\ mm$ is not particularly important in this context in terms of quality optimization, but, over the last three decades, it has come to represent a symbolic objective, connecting digital imaging with the long tradition and prestigious culture of film photography. Moreover, the use of this format allows us to continue to use older, but costly, lenses initially designed for film photography.

The notions presented above have been used in carrying out thorough and objective studies on commercial hardware. In [CAO 10b], for example, the authors determined optimal usage conditions for a number of widely available camera models (body and lens) in specific lighting conditions and with specific exposure times, chosen to reflect typical use, and taking account of particularities in terms of sensitivity, electronic noise, sensor dimensions, numbers of photosites, lens aperture, geometric and chromatic aberrations, and vignetting. The recommendations given in this work allow us to assign scores to the performance of individual cameras, and to determine ISO sensitivity and aperture parameters in order to obtain images with the highest possible information content in a wide range of usage conditions.

6.3.3.1.1. Remark

In older publications, the information theory approach was used to optimize the definition of photographic lenses subjected to specific constraints in order to maximize information transfer from the scene to the captor. Surprisingly, in these cases, for a given optical formulation, the best diaphragm is not that which maximizes the quality of transmitted light, i.e. an all-or-nothing model, but rather a diaphragm which somewhat apodizes the lens, i.e. damps the amplitude slightly in certain zones, in order to reduce the spread of the diffraction figure. The payoff for this reduction is a slight loss of energy [FRI 68]. This result places optimization in terms of energy (in which case the best option would be not to attenuate any frequency at all, leaving the pupil untouched) in conflict with optimization in terms of entropy (in which case, it is best to accept a certain level of energy loss in order to ensure that the impulse response will be as compact as possible).

6.4. What about aesthetics?

The various approaches seen above make use of formal criteria to evaluate the match of the optical signal issued from the photo to the perceptual capabilities of the human visual system. We never tried to evaluate the pleasure or the interest of the observer while examining an image (as for instance one of Figure 6.15). Is it pointless to address these goals which are the major objectives of the photographer?

We will first ignore the documentary interest of the photographic capture, although it concerns the main reason to take a photos, since this interest is tightly attached to the observer personality, culture and context at the time of capture. Obviously we are not able to evaluate this interest from the image alone. The question raised by evaluating this aspect of a photo clearly overruns the only properties of the image and appertains to a broader question: What are we looking for in a piece of information? Judicious elements of answer may be found in [DES 08b], but more relevant from the modelling of the observer than from the modelling of the media (the image) or the channel (the camera).

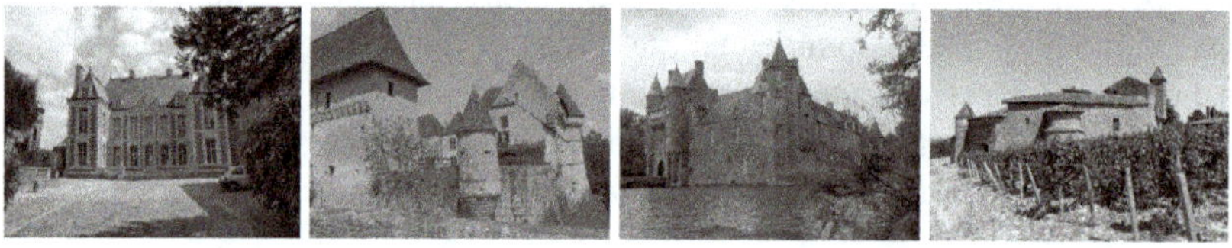

Figure 6.15. *In order to rank images from a purely aesthetical point of view, it is important to ensure that no subjectif effect perturbates the evaluation. In the comparison made here, the images share the same semantics (a castel in a quite similar context which discards emotional biases). Moreover, the elementary technical properties of each photo should be similar: format, resolution, contrast, noise, etc.*

Up to a certain point, aesthetics deserves the same remarks. But during the last 25 centuries, aesthetics received the greatest attention of philosophers who developed very elaborated theories, some likely to be relevant in the field of photography. Aesthetics is the study of beauty, and as such, it concerns all the arts. Since Greek philosophers, the adopted approaches follow more or less two opposite tracks:

– for "objectivists", beauty lies in the objet: its form, its colors, its proportions. Therefore, beauty is universal and should be perceived the same by every observer;

– for “subjectivists”, beauty is a personal experience perceived by a given observer in specific conditions, and under the only influence of the observer perceptual state.

The objective way, as taught by Platon and Aristote for instance, dominated the world of arts for a long time. It has been well formulated in the field of painting during the Quattrocento and the Classical period. However, Romantic period has seen the subjective current be set up[15]. During the 19th and the 20th century, a fruitful dialog established between the two approaches, in the purpose of completely modelling aesthetics. Aside generalist philosophers (Hegel, Schopenhauer, Adorno, Heidegger, ...), many authors specialize themselves on painting and on the relationship between the perceptual work and the spectator [GOM 82, ARN 10]. Understanding the mechanisms of the human visual paths as neurobiology allowed to do at the end of the 20th century, greatly favoured a purely subjectivist interpretation of aesthetics [ZEK 99, LIV 02], which is on top now-days.

6.4.1. *Birkhoff's measure of beauty*

The first mathematical “formules” derived to translate the aesthetic feeling are due to Henry ([HEN 85], quoted in [NEV 95]). Rather complex, they are today forgotten. However they explicitly refer to a least action principle, similar to the one acting in thermodynamics(is beautiful what is easily perceived). Such a principle is a foundation of the more achieved work of the mathematician Birkhoff [BIR 33]. For Birkhoff, an object is more beautiful, if it offers to the observer in a simpler way a more complex reality. Following this idea, beauty–denoted as $\mathcal{B}$–may be expressed by the ratio;

$$\mathcal{B} = \mathcal{O}/\mathcal{C} \qquad [6.65]$$

where $\mathcal{O}$ expresses the masterpiece simplicity (as measured by the order in its features: symmetry, alignment, regularity, ...) and $\mathcal{C}$ its complexity (number and diversity of the features). Unfortunately, at the time of his discovery, Birkhoff only had a limited toolbox to develop his theory. He adapted it in an ad'hoc way on specific examples: vases, frescoes, poems and music. Birkhoff work is nowdays revisited with a renovated scientific equipment.

15 At the same time than photography (was it by chance?).

6.4.2. *Gestalt theory*

Gestalt theory, or psychology of perception [KOE 30], provides the principles allowing to distinguish in the visual flow the important elements which are perceived as indivisible objects[16]. Gestaltism also indicates how these elements should be associated and how they should be priorised. In [MOL 57, BEN 69] Birkoff's ideas are exploited with beauty explicitly expressed from the laws of gestaltism. Line drawing have often been used by gestaltism as mean of demonstration. In recent years, in the domain of image processing, a contrario methods [DES 08, MUS 06] have revisited gestaltism within a probabilistic framework with excellent results.

6.4.3. *Shannon information theory, Kolmogorov Complexity and Computational Complexity theory*

The second tool which could have helped Birkhoff is information theory[17] which could be used to replace terms in formula [6.65] by some more adapted terms. For instance, complexity C could be expressed as the information capacity, product of the number of pixels by the image entropy. Simplicity O is less easy to measure. However, computational complexity theory, as expressed by Kolmogorov provides interesting issues. Kolmogorov complexity is given as the length of the shortest program able to reproduce the image on a universal machine. Unfortunately, it cannot be theoretically alculated but it may be majorated by empirical techniques. For instance, in [RIG 08], the authors make use of lossless compression techniques to estimate the "beauty" of paintings. In other works, lossy compressions are used instead, with almost not noticeable defects.

Of course these quite preliminary studies are only providing orientations for a possible use of aesthetics measure to determine the beauty of a photo. No guarantee exists that so simple approaches will lead to usable results.

6.4.4. *Learning aesthetic by machine*

Aside this analytical approaches when defining aesthetics, a new trend appeared in the last ten years, based on machine learning techniques and taking advantage of the public enthusiasm for social nets. The purpose of

---

16 Early works on gestaltism were contemporary from those of Birkoff and, apparently did not inspired him.

17 But it was created in 1948, after Birkhoff's work.

these studies was to confirm the role of psychophysiological rules of perception in the evaluation of beauty and to propose tools to measure it. Typical rules concern the decay in $1/u^2$ of the Fourier spectrum, the rule of the third concerning the spatial distribution of objects in the scene, the preferred palette of colours, etc. [AMI 14, GRA 10a, PAL 10].

Very large databases exist on the web, collecting photos either from professionals or from more or less skilled amateurs. These collections are often associated with some form of quality assessment obtained from the free judgement of the visitors (ranking for instance on a 5 grade scale). In other cases, judgements come from a jury and may then be more explicit. From these databases and taking advantage of the available ranking, algorithms may be trained to associate some measures made on any image with its computer-guessed beauty. In the early studies, the measures were rather simple: dynamics, contrast, richness of the colour palette, accuracy of contours, spatial distribution of regions, variety of textures, etc. Classification is often made using deep neural networks which need very large training sets but may rather well reproduce a human ranking.

When a new image is proposed to the system, the most significant attributes are extracted and processed by the neural network up to the last layer where a mark is given to the image. Even if the methods are still coarse, they provide today rather good results for instance to sort images depending on their quality level: "amateur" versus "professional" [AYD 15, LI 10, LO 12, LU 15]. It is likely that this domain will receive a greater attention in the close future because of the urgent need for the reliable automatic evaluation of the image quality for the consumer market as well as for the corporate one.

7

Noise in Digital Photography

Noise is one of the intrinsic factors which limits the ability of a photograph to reproduce a scene as accurately as possible.

In *film photography*, the main noise affecting images is encountered during the formation of the latent image (physical absorption of a photon by the sensitive material used). It then becomes apparent during the chemical transformation of this material during development: rather than being limited to the photon impact point, the image spreads to a small, compact local structure. This is known as speckle in a film photograph [KOW 72, p. 95]. Speckle consists of the production of uniform areas of gray by the juxtaposition of very small "atoms" or "aggregates" of varying density, which constitute the active matter of the emulsion (generally silver halide). When the noise is stronger, these aggregates are larger, and the variance of their gray level increases; they thus become visible, even in moderate enlargements. The size of the grains in mass-market products typically varies, from $5\,\mu m$ to $50\,\mu m$ in diameter. This constitutes an increasing function of film sensitivity (measured by the ISO number, see section 4.5), but also of the potential contrast of the film.

In *digital photography*, the reasons for noise are very different, and result in very different visible impairments. Noise does not directly affect image resolution in this case, but rather the gray level value of each pixel. Thus, noise in digital photography has a very similar meaning to noise in other areas of signal processing.

The greater sensitivity of digital sensors makes it necessary to take account of noises which are generally negligible in the context of film photography, except in certain specific scientific domains (such as astronomy).

In this chapter, we will consider the sources of this noise[1], and attempt to model them, notably based on the work of [AGU 12].

7.1. Photon noise

This type of noise has its origins in the fluctuations of the optical signal. These fluctuations are linked to the quantum nature of the photons (bosons) making up the luminous flux. Their emission is naturally random. The variation in the photon count of a constant signal is very low if the signal is strong. For very low luminosities, however, it ceases to be negligible.

Optical fluctuations are transmitted to the digital signal during conversion within the photodetector (see section 3.2.1), where it becomes known as shot, or Schottky, noise.

7.1.1. *Fluctuations in the optical signal*

In conventional photography, fluctuations in the optical signal are always considered to follow a Poisson distribution, and this is verified by experience; however, the theoretical reasons for this distribution are complex [SEI 11].

Three classic situations should be considered when examining the noise affecting photon emission in optoelectronic conversion processes ([SAL 07 pp. 460–470], see also section 2.6):

– coherent imaging;

– thermal, or totally incoherent imaging;

– partially coherent imaging.

7.1.1.1. *Coherent imaging*

In the coherent case[2], the statistics of the photon flux are Poissonian due to the quantum (unpredictable) and independent nature of emission events.

If the average number of incident photons on a photosite during the exposure period τ_i is n_{ph}, then $n_{ph} = C_{ph}\tau_i$, where C_{ph} is the radiance

1 Note that certain impairments which are commonly referred to as "noise" do not, strictly speaking, belong to this category, as they are neither parasitic impairments, nor random errors.

2 See section 2.6 for the distinction between coherent and incoherent lighting.

(expressed here in photons per unit of time). The probability that the flux contains k photons is given by the Poisson distribution:

$$P(k) = \mathcal{P}\text{oisson } (k\,;\,n_{ph}) = \frac{(n_{ph})^k}{k!}e^{-n_{ph}} \qquad [7.1]$$

7.1.1.2. *Incoherent imaging*

In thermal imaging, typical emissions take the form of a black body (Boltzmann equations). The noise affecting a specific mode of the black body verifies a Bose–Einstein statistic ([SAL 07] p. 467), with very strong variance (greater than the square of the mean): $\sigma^2_{ph} = n_{ph} + n^2_{ph} > n^2_{ph}$ and:

$$P(k) = \frac{1}{n_{ph}+1}\left[\frac{n_{ph}}{n_{ph}+1}\right]^k \qquad [7.2]$$

and n_{ph} is linked to the temperature T by Planck's law (equation [4.14]):

$$n_{ph} = \frac{1}{\exp\left(h\nu/kT\right) - 1} \qquad [7.3]$$

In practice, however, light sources (the sun, lamp filaments, etc.) have a large number of emission modes. The number of modes M is determined by the volume V of the resonator (the emission zone) and the bandwidth $\Delta\lambda$ in question:

$$M = \frac{8\pi V \Delta\lambda}{\lambda^4} \qquad [7.4]$$

For a bandwidth of 0.1 μm and a source of 1 mm^3, the number of modes in the visible domain is of the order of $M = 2.10^{10}$.

The variance of the number of photons is then given by [MAN 95a]:

$$\sigma^2_{ph} = n_{ph} + n^2_{ph}/M \qquad [7.5]$$

This value is practically identical to the mean value n_{ph}, as in the Poisson case, and [SAL 78] shows that the statistics of photon numbers are very close to following a Poisson distribution.

In the context of photography, the incoherent case is most interesting, as it applies to scenes illuminated by the sun, flames, incandescent lighting (ordinary light bulbs and flash), etc.

7.1.1.3. *Partially coherent imaging*

Cases of partially coherent imaging include:

– scenes lit by electroluminescent diodes;

– scenes lit by discharge lamps, which emit over very narrow wavelength bands for very short periods;

– microscopy, which involves very small dimensions, and where objects are often illuminated by filtered light sources.

These situations involve relatively complex coherency problems, which are often treated on a case-by-case basis. Diodes are covered in [SEI 11], while [SAL 07], p. 468, concerns "almost coherent" cases. These examinations generally show non-Poisson distributions (Mandel equation), although these distributions are approximately Poissonian at the limits.

Conclusion.– Leaving aside the last specific situation, we see that the photon flux is almost always Poissonian, but may differ somewhat from this distribution in the specific case of microphotography.

Properties of the Poisson distribution.– The Poisson distribution has a mean equal to its parameter n_{ph} and a variance equal to the mean. As this parameter increases, the Poisson distribution tends toward a Gaussian distribution; in practice, this occurs rapidly, for a photon number of the order of 20 (see Figure 7.1). If the number of events is large, we may therefore consider the Poisson and Gaussian distributions to be equivalent.

Note that the sum of independent Poisson distributions is a Poisson distribution, with a parameter equal to the sum of the independent parameters. This is important in photography, where a number of independent sources are often used. If these sources are strong, the central limit theorem may also be used to obtain a Gaussian mixture, which is very similar to a Poisson distribution with a high average.

Another property of the Poisson distribution, of parameter n_{ph}, which is important in the context of photography is that, if we remove photons generated by a Poisson flux using a Bernoulli decision applied to each photon (each photon has a probability π of being removed or a probability $1 - \pi$ of being retained, based on a decision which is independent of the photon arrival order), we obtain a new Poisson photon flux with parameter $(1 - \pi)n_{ph}$. The Poissonian character of the flux is therefore conserved when Bernoulli selection is applied to events. Quantum electromagnetic theory explicitly requires the use of this photon selection mechanism in order to explain all optical propagation phenomena (reflections and absorption) and the phenomenon behind the photoelectric photon–electron transformation [FEY 85]. This justifies the application of the Poissonian nature of the source to the photodetector, notwithstanding the multiple interactions affecting the photon flux.

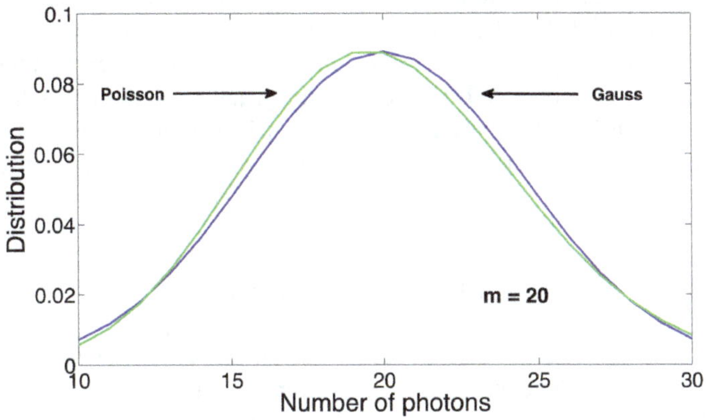

Figure 7.1. *Poisson and Gaussian distributions. For 20 events, the two distributions are very similar, and this similarity increases as the number increases*

Finally, note that we may sometimes wish to move from a signal subjected to Poissonian noise to a signal subjected to Gaussian noise, as many filtering or detection algorithms are optimized based on this hypothesis. This change is possible, under certain hypotheses, using the Anscombe transformation [ANS 48] which replaces the Poisson variable u by a variable v such that $v = 2\sqrt{u + 3/8}$. An approximation of the initial signal can then be obtained using $u_1 = v^2/4 - 3/8$.

7.1.2. *The Poisson hypothesis in practice*

Take equation [7.1]. For lighting conditions corresponding to full sunlight (spectral emission of the order of 1 kW/m^2, around a wavelength λ of $0.5\ \mu m$ – see section 4.1.1, equation [4.6]), the photon flux $I/h\nu = hcI/\lambda$ is of the order of $6{\times}10^{21}$ photons $/m^2$.

For a photosite with sides of $10 \times 10\ \mu m$ and an image obtained in one thousandth of a second, this corresponds to $n_{ph} = 6.10^8$ incident photons.

The fluctuation of the number of photons Δn_{ph} is of the order of $\sqrt{n_{ph}} = 24,000$. Photon noise will thus affect the 15th bit of the binary representation, and will be totally imperceptible using the best sensors currently available.

For a photograph taken by moonlight (10^6 times less luminosity) taken in the same conditions (same sensor, same exposure time), the photon noise will affect the 5th bit, and consequently 8 levels of gray in the case of coding using one byte; this is very noticeable, and considerably higher than the error resulting from high-quality JPEG coding, for instance. Note that in this latter case, the signal is well below the limit established by the "thousand photon rule" (see section 6.1.1); this rule tells us that the noise will be visible to human observers.

7.1.3. *From photon flux to electrical charge*

Photon noise causes fluctuations in the electrical charge (expressed in coulombs) of the photosite receiving the luminous flux, due to the internal photoelectric effect (i.e. the electrons released from the atomic structure do not leave the semiconducting material, but contribute to electrical polarization) ([NEA 11] pp. 480–515). This "photon $\Longrightarrow$ electron" transformation occurs on a one-for-one basis (one photon gives one electron), on the condition that the photon has sufficient energy ($h\nu = \frac{hc}{\lambda}$) in order for the electron to traverse the forbidden zone of the semiconductor. If the energy is insufficient, or if it is not absorbed, the photon traverses the material as a transparent milieu. The excess energy carried by the photon in relation to the forbidden zone becomes the kinetic energy of the electron.

If the electron encounters a hole, then recombination occurs, and the signal is lost. For this reason, the photodetection material is polarized in order to separate holes and electrons.

The conditions required for a photon to encounter an electron are dependent on the doping and thickness of the material. The photon–electron transformation and electron–hole recombination processes are quantum [FEY 85], random and independent, and follow a Bernoulli distribution.

For a photon to be detectable in the visible domain, with $\lambda < 0.8 \times 10^{-6}$ m, $h = 6.62 \times 10^{-34}$ J.s and $c = 3 \times 10^8$ m.s^{-1}, the photodetection material needs to have a forbidden zone energy E_g lower than approximately 2.5×10^{-19} J, i.e. approximately 1.55 eV[3].

The balance of this transformation is determined by three phenomena:

1) Reflections at the point of entry into the semiconductor material, diverting photons away from the photosite. In the absence of optical treatment, these reflections are given by the reflection coefficient, which is a function of the material's index, and, for normal incidence, has a value of:

$$R = \left[\frac{n-1}{n+1}\right]^2 \qquad [7.7]$$

for indexes n of between 3 and 4, as for most photodetectors. The rate of photon diversion for an interaction is therefore between 25% and 36%. For this reason, anti-reflective treatments are needed in order to bring these figures below a threshold of 10% of the whole of the useful bandwidth.

2) Pixel vignetting (see section 2.8.4.1) expresses the loss of photons due to the haze created by the lower position of the photosite and the presence of command electrodes.

3) The internal quantum efficiency ρ_{q_i} of detectors made using new materials is very high, and close to 1 for the visible spectrum, expressing the fact that, in practice, every photon constitutes a source of electric charge.

Reflections, pixel vignetting and internal quantum efficiency are often grouped into a single external quantum efficiency measure, $\rho_{q_e} = (1 - R)\,\rho_{q_i}$, which is therefore generally greater than 0.9 (see Figure 7.2).

In practice, the electrical charge Q of the photosite therefore shares the Poisson properties of the photon flux. The property is now $\rho_{q_e} n_{ph}$. This charge creates a potential V which is measured at the photosite terminals to give the signal value which will be assigned to the pixel. This "charge $\Longrightarrow$ potential" transformation is dependent on the electrical capacitance C of the device (expressed in farads), a value which is solely dependent on the material itself and its geometry, with the relationship $V = Q/C$.

3 Note that the energy of a photon of wavelength λ (in micrometers) may be expressed simply in electron-volts using the formula:

$$E\,(in\,eV) = \frac{1.24}{\lambda\,(in\ \mu m)}. \qquad [7.6]$$

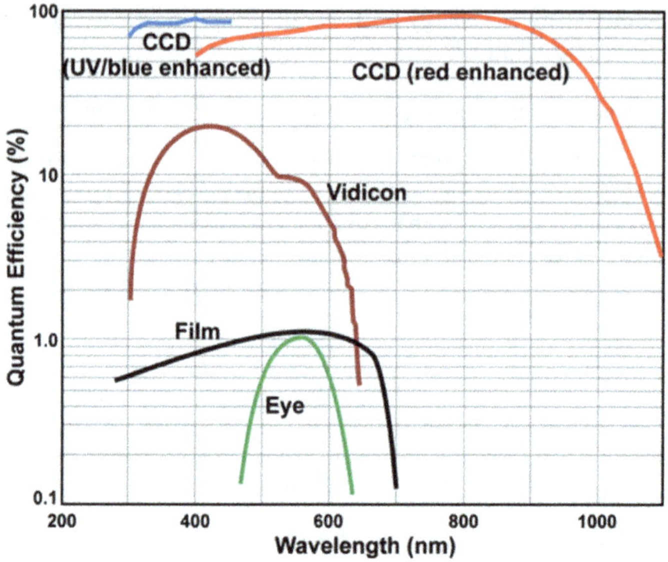

Figure 7.2. *Quantum yield of a solid CCD detector compared to that of the human eye (day vision) and to that of analog detectors (film and video cameras; the visible domain extends from 400 to 800 nm (based on Lincoln Laboratory (MIT) – Advanced Imagers and Silicon Technology)*

In practice, the potential measured at the photosite can be seen to be remarkably linear, a natural consequence of excellent internal quantum efficiency and the absence of charge recombination in the detector[4] as a function of the incident photon flux, until the material saturation point. This linear response is a key difference between solid photodetectors (CMOS and CCD) and film-based receptors, which are subject to nonlinearity, as seen in their H&D curves (see section 4.5).

The potential is therefore also subject to Poissonian fluctuations in the photon flux n_{ph} up to the sensor saturation point. We also need to consider the possibility that the signal may be amplified as a function of the ISO sensitivity determined by the user and the dark current, along with other types of noise which will be discussed below [HEA 94].

4 The deviation of very good CCD sensors from a linear distribution is of a few parts per thousand over 4 or 5 signal decades [HAM 13].

7.2. Electronic noise

7.2.1. *Dark current*

Dark current is the result of thermal noise (Johnson–Nyquist noise) in the sensor, constituted by fluctuations in the "free electron gas" in the sensor. It is a direct function of the temperature of the sensor, and is almost independent of the incoming signal. Mass-market cameras, unlike the scientific devices used in astronomy, for example, do not include a cooling mechanism; in these devices, strong dark currents may be produced in cases with a weak signal and long exposure time. This noise is also dependent on usage conditions (atmospheric temperature); this is something of a rarity in photography.

The dark current for MOS transistors is described in detail in [NEA 11, pp. 455–461], which provides models of spectral noise density under a variety of hypotheses.

The number k of electrons contributing to the dark current can also be shown [THE 96] to follow a Poisson distribution, of which the parameter D_{th} is a function of the sensor temperature and of the electron accumulation period (hence of the exposure time):

$$P(k) = \mathcal{P}\text{oisson } (k\,;\, D_{th}) = \frac{D_{th}^k}{k!} e^{-D_{th}} \qquad [7.8]$$

Parameter D_{th} is determined from the sensitive surface S, the dark current I at the temperature T in question (expressed in Kelvin) and the energy of the forbidden zone E_g:

$$D_{th} = \alpha S I T^{1.5} e^{-E_G/2kT} \qquad [7.9]$$

The dark current for each photosite may be measured during sensor calibration[5], then factored in for each pixel.

Dark current may be reduced by placing an infra-red filter before the sensor in order to limit heating (see section 9.7.2).

5 This is carried out in total darkness, retaining the lens cover. Photographs are then taken in the experimental conditions which we wish to characterize, i.e. exposure time and sensor sensitivity. Note that many cameras leave unexposed pixels in the field during image formation in order to allow noise measurement on an image-by-image basis. These pixels are located around the edge of the photodetector.

7.2.2. *Pixel reading noise*

This type of noise is due to measurement of the charge at a photosite.

The reading noise N_{read} is generally considered to be Gaussian on the basis of theoretical considerations: $P(N_{read}) = \mathcal{N}(0, b_{read})$, and this property is relatively well verified in experimental terms [AGU 12]. The property applies for all pixel values, due to the specific way in which reading circuits operate; the mean output signal of the comparator is always positive, even for an image signal of zero, meaning that it can be subjected to negative fluctuations without becoming negative itself (digitizer offset).

This reading stage is accompanied by an amplification of the charge signal by a factor g (with an associated amplification of photon noise and dark current, which is corrected to a certain extent during the calibration process). This amplification is directly controlled by the user via the choice of ISO sensitivity. The reading noise is thus highly dependent on sensitivity; however, this amplification is linear for all charge values.

7.2.3. *Crosstalk noise*

Crosstalk noise occurs when carriers leave a photosite and are received by a neighboring photosite. This error is particularly common when the distribution of photosites is particularly dense. In order to maintain a high filling rate (see section 3.2.2), we therefore need to reduce the buffer zones between pixels (see Figure 3.2) and the probability of transition increases. Situations involving sensor blooming, where all of the charges are activated, are particularly subject to this phenomenon.

Crosstalk may also result from poor alignment of the microlens with the photosite. It is particularly noticeable at the edge of the sensor field, where the rays received present a strong incline along the axis [AGR 03]. However, this issue can now be easily resolved using placement techniques, by partial metalization of the inter-pixel corridor, or by using non-spherical microlenses to reduce the distance between the lens and the chromatic filter (see Figure 3.8).

Crosstalk noise reduces the photon count assigned to any given pixel, and increases that of the neighboring pixel. In the earliest complementary metal-oxide semiconductor (CMOS) sensors, this phenomenon was particularly problematic (with around 40% of electrons escaping from the site [AGR 03]). A range of solutions have been adopted, involving either material doping or improved circuit design, which present significant reductions in this figure.

The chromatic matrix covering photosites means that crosstalk noise transports energy from one channel to another. The red channel is subject to the greatest electron loss, as the level of penetration is stronger, and the probability of an electron escaping from a red site is higher than that for sites of other colors. The lost electrons are transferred to the green channel, as an R site is surrounded by 4 G sites (see Figure 5.17).

7.2.4. *Reset noise*

In APS CMOS sensors (see section 3.2.3), photosites are recharged, or reset, before exposure. This operation is subject to thermal fluctuations, which need to be taken into account explicitly when assessing the noise affecting a pixel.

In pinned-photodiode devices PPD sensors, each pixel has an electronic component used to measure site charge before and after exposure (see Figure 3.4). This leads to a significant reduction in the thermal noise resulting from the reset procedure.

7.2.5. *Quantization noise*

Quantization noise affects the signal read at each photosite during its conversion into an integer, generally within the interval $[O, \;\; 255]$, but occasionally within a larger range if quantization is carried out using 10, 12 or 14 bits. Quantization takes place after reading, and must take account of all previous stages with their associated sources of noise.

The image signal as a whole does not have a probability distribution which can be modeled (Chapter 1 in [MAI 08a]); the quantization noise K, which is additive, is thus considered to be distributed in a uniform manner across the full signal range. It has a mean value of zero and a variance of $1/12\,\delta$, where $\delta = \frac{\Delta}{n}$ and Δ is the total dynamic of the signal (determined either by the maximum dynamic of the analog–digital converter, or by the maximum number of available converters at the site: the full well capacity) and n is the number of levels of gray. It is generally negligible, particularly in cases of quantization using more than 8 bits.

7.3. Non-uniform noise

7.3.1. *Non-uniformity in detectors*

Dark noise and detection noise are rarely uniform across the whole of the photo-detector (this is particularly true of CMOS sensors). Variations in this noise may be modeled by applying multiplicative factors to the Poisson parameters n_{ph} and D_{ph}. These factors may be determined for each photosite during a calibration process using uniform scenes.

7.3.2. *Salt-and-pepper noise*

Salt-and-pepper noise is a result of imperfections in sensors, either due to photosite faults or to dust on the photodetector.

This type of pulsed noise affects pixels in very precise locations, systematically giving them values of zero. In case of electronic faults in sensors, it may also assign a value of 255 to certain pixels.

Cameras often keep a map of defective photosites in order to apply systematic corrections by interpolation.

In the earliest reflex cameras, the sensor was not protected, and dust-related impairments were a major problem. Efficient systems have since been developed to counteract this issue, often combining mechanical and electronic actions in order to detach and remove dust from the sensitive area. The effects of dust have been considerably reduced, even in relatively inhospitable usage conditions.

7.3.3. *Image reconstruction and compression noise*

Noise also occurs during the final stages of image processing. These stages are not always applied, but may include:

– during demosaicing, used to reconstruct the three components R, G and B from the interlaced Bayer structure (see section 5.4.2.2), the image may need to be interpolated and filtered in order to limit variations to the bandwidth allowed by Shannon's theorem (see equations [5.32] and [5.35]);

– during JPEG or MPEG compression, images are projected onto bases of cosine or wavelet functions, and the coefficients are quantized, resulting in the production of characteristic noise (see section 8.6.1).

7.4. Noise models for image acquisition

We will follow the modeling method presented in [AGU 12], using a simplified model. Image acquisition will be modeled in raw, unprocessed format (see section 8.2), retaining all the dynamics of the original signal, without demosaicing or compression. The salt-and-pepper noise generated by imperfect sensors will be left aside, and we will presume that the image dynamics do not reach the image saturation limit.

The value of the signal over a period of time i is represented by Z_i, which may be written as:

$$Z_i = \Psi \left[(g_{cv}(\rho_{q_e} n_{ph} + D_{th}) + N_{reset}) g_{out} + N_{out} \right] + K \qquad [7.10]$$

where:

– Ψ is a linear function up to the sensor saturation point, where it becomes constant;

– g_{cv} expresses the specific gain of a site, and allows us to represent the non-homogeneous nature of the sensor;

– ρ_{q_e} is the external quantum efficiency (which may also potentially depend on the site due to the possibility of pixel vignetting);

– n_{ph} is the number of incident photons;

– D_{th} is the thermal noise associated with the temperature of the photodetector;

– N_{reset} is the reset noise, which may also include crosstalk noise;

– g_{out} is the electronic gain, which takes account of the ISO sensitivity selected by the user;

– N_{out} expresses the reading noise, with variance b_{read};

– K is the quantization noise, determined by the number of levels of gray n and the total dynamic.

The gain g_{cv} is constant for a charge-coupled device (CCD) sensor. For a CMOS sensor, however, it may vary by site or by site line, reflecting the chosen architecture.

Ideally, the gain g_{out} should be proportional to the ISO sensitivity of the sensor. However, as we have seen (see section 4.5), manufacturers may deviate from this linear relationship within certain limits. Linearity is therefore not always precisely verified experimentally.

The term $\Psi\left[g_{cv}(\rho_{q_e} n_{ph})\right]$ is the image signal, subjected to photon fluctuations of radiance C_{ph} over a period τ_i, read through a photodector with electrical capacitance C: $V_{ph} = \mathcal{P}\text{oisson}\ (., \frac{C_{ph}\tau_i}{C})$.

Taking $N_R = g_{out} N_{reset} + N_{out}$, a Gaussian distribution with mean μ_R and variance σ_R, $g = g_{out} g_{cv}$, and $\omega_i = \rho_{q_e} n_{ph} + D_{th}$, we obtain:

$$Z_i = \Psi\left[g\omega_i + N_R\right] + K \qquad [7.11]$$

In areas which are far from saturation, i.e. for the most general image values, Ψ is linear and of slope 1 (as the output gain g_{out} is expressed separately), the noise expression can thus be simplified, giving:

$$Z_i = g\omega_i + N_R + K \qquad [7.12]$$

This is the sum of a Poisson distribution, a normal distribution and a uniform quantization noise.

Similar equations may be deduced for different architectures, such as those shown in Figures 3.4, 3.5 or 9.4. These equations provide the foundations for integrated noise reduction software.

7.4.1. *Orders of magnitude*

As we have seen, the quantization term is very small and is generally left aside. It is notably considerably lower than reconstruction and compression noise (see section 7.3.3), which were not taken into account in this model as they are not explicitly dependent on the photon flux.

In strong lighting.– In these conditions, photon noise is dominant. However, as we have seen, in the linear zone of the sensor the signal-photon noise ratio is very high in strong lighting, and this noise can therefore be ignored (section 7.1.2), unless the photosite is very small and the exposure time is extremely short. The Poisson aspect of equation [7.11] can thus be replaced by a Gaussian distribution, which covers all noise elements with the exception of quantization noise. This form (Gaussian noise plus uniform quantization noise) serves as the basis for most classic image processing algorithms, which do not include adaptations for specific sensors. These algorithms present explicit solutions with maximum likelihood.

Can this signal then be presumed to be perfect?

A perfect signal cannot be guaranteed, as we need to consider the saturation introduced by Ψ, potentially a major cause of image distortion. Note that this saturation may come from two sources: the dynamic limit of the converter, or the saturation of the site, associated with the maximum carrier number of the photosite (see section 3.2.3.1). While it may seem logical to dimension a converter in order to cover the full capacity of the photosite, this solution is still not perfect, as it requires us to process signals close to the site saturation point, in a zone of significantly reduced linearity (Figure 7.3).

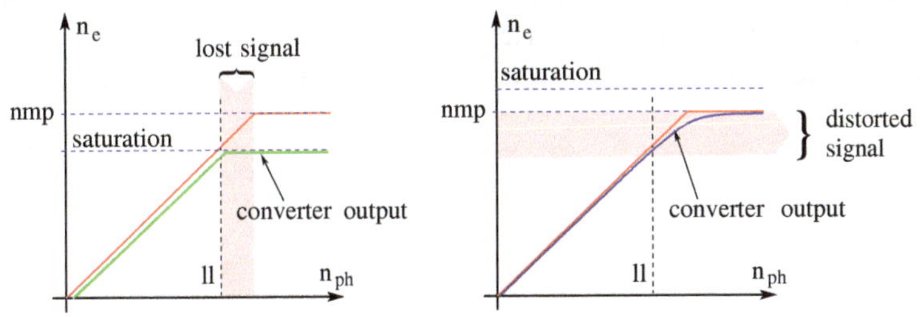

Figure 7.3. *Sensor saturation: number of photons (n_{ph}) converted into electrons (n_e). In red: ideal photon–electron conversion; in blue, conversion in practice; in green, converter output. Two typical cases may be seen. Left: the converter is saturated first. Not all of the photons received by the sensor will be used to make up the image, but linearity will be respected in the converter dynamics (ll= limit of linearity). Right: the converter covers the whole capacity of the photosite (mcn= maximum carrier number), but now needs to take account of a zone of nonlinearity at the point of saturation. For a color version of this figure, see www.iste.co.uk/maitre/pixel.zip*

Many cameras inform users of the presence of these saturation zones and recommend changing the sensitivity (and hence, the gain g_{cv}) or the diaphragm/exposure time, adjusting n_{ph} so as to remain within the linear zone.

If saturation cannot be avoided, the signal is generally considered to be completely unknown in saturated areas. However, [AGU 14b] presents a very interesting approach which estimates the signal at maximum likelihood beyond the saturation point, using a noise probability hypothesis with an equation of the type given in [7.10]. This allows a particularly effective probabilistic extrapolation of the signal. These methods will be applied in sections 10.1.1 and 10.3.1.

Low lighting and short exposure times.– In these conditions, the photon noise, which follows a Poisson distribution, is particularly high, as seen in section 7.1.2, and has major effects on the image. As they require us to use a high ISO level (and thus a high gain g_{out}), the thermal and reset noises, while intrinsically low, are also amplified and, alongside photon noise, contribute significantly to overall noise levels. Precise restoration in these cases is difficult and often requires iterative processing. These treatments will be discussed in section 10.1.1.

Low lighting and long exposure times.– Long exposure times result in a significant reduction in photon noise, but the thermal and reset noises increase to become the dominant sources of noise, whatever the chosen sensitivity level.

8

Image Representation: Coding and Formats

The wide range of formats used to archive and to display images constitutes an obstacle to their use by the general public. While the JPEG format has become widely used by both specialist and general users of photography, and is widely considered to represent a standard, a variety of other formats are available, and often produce better results.

The choice of format is determined by differing and often conflicting aims: reduction of image volume and/or maximum respect of properties (dynamics, chromatic palette, line geometry, quality of solid colors, etc.). Certain formats are best suited for archival purposes, while others are better for transmission, or for interoperability across a range of different treatment or visualization programs.

Images are used in a broad and varied range of applications. Some of these applications require documents to be transmitted at high speeds, whilse others are more concerned with quality, which may be judged according to different domain-specific criteria: representation of nuances or fine details, line sharpness, color reproduction, text readability, capacity for integration into a complex composition, etc.

Nowadays, even the most "ordinary" photograph is made up of at least ten million pixels, each often represented by 3 bytes, one per chromatic channel (red, green or blue). If the signal is not compressed in any way, each image requires several tens of millions of bytes of storage space in both the camera and computer memory. Moreover, hundreds, or even thousands, of photographs may now be taken in the course of an event. Professional and enthusiastic amateur photographers use hundreds of gigabytes of memory for image storage alone. In addition to this archiving cost, image consultation and transfer are also slower and more costly. The variety of available formats described in this chapter have been developed in response to these requirements. A more detailed technical analysis of each format may be found in [BAR 03b, WOO 11], for example.

8.1. "Native" format and metadata

Before considering questions of user access to digital photographs, we must examine the way in which these images are represented for computing purposes.

For a black and white image containing n_l lines of n_p points, where each pixel is coded by a single byte, the simplest method is to record the sequence of $N = n_p n_l$ bytes in a single file. This native format is not completely free from ambiguity, as there are generally multiple ways of constructing a rectangular image from N pixels. To avoid sacrificing levels of gray in order to mark the ends of lines, we must therefore also transmit the values n_p and n_l. These values n_p and n_l constitute the simplest elements of the metadata which needs to accompany an image to ensure readability; many other elements become necessary as we consider different types of images, including the way in which bytes are converted into gray levels (natural code, Gray code, or codes using look-up tables (LUTs)[1]), the way in which color images are represented (by association of 3 image planes or by interweaving the 3 bytes of each pixel along a line), the type of coding used, the number of channels per pixel (for multi-spectrum images) and the number of bytes per pixel (for high dynamic images), the number of dimensions (for medical applications, for example), etc. Nowadays, metadata also includes identification and property marking elements, along with information on capture conditions, and on the treatments applied to images over the course of their lifetime. For photographers, metadata constitutes an obvious location for specifying capture parameters, allowing them to get the best out of their image.

Metadata is sometimes placed in a second (format) file, used alongside the image file. This solution is still used in certain conditions, where a large

1 LUTs allow us to assign a level or color $T(\nu)$, randomly selected from a palette, to a pixel of value ν.

quantity of metadata which does not concern the image itself is required (for example patient information in the context of medical imaging, or mission information in astronomic satellite imaging). Separation of the image and metadata files enables greater confidentiality in the treatment of these files. It also reduces the risk that an error made in handling one of the files will cause significant damage to the other file.

This solution is not generally used in mass-market digital photography, where metadata and image data are combined into a single file, containing information of very different types: some binary, other symbolic. Metadata is usually placed in a header above the image data, using a syntactically-defined format identified by the file extension (*.jpg*, *.tif* or *.bmp*). Reading this metadata provides the keys needed to correctly read and display the image field. It also enables users to make use of information included by the camera manufacturer or by previous users of an image, in order to exploit image properties to the full.

The variety and complexity of available image formats are essentially due to the varying importance of different property types for different groups of users, requiring the inclusion of varying amounts of detail in the metadata.

8.2. RAW (native) format

Saving a signal in the way in which it is recorded by the sensor seems logical, if sufficient care is taken to record the parameters required for future treatment alongside the image. For this reason, more bits than are strictly necessary will be assigned to the signal, the original sensor dynamics will be retained, and image reconstruction elements will be recorded, including the dark current and amplifier gains up to saturation values. The geometry imposed by the objective and by mosaicing must also be specified, along with capture parameters, exposure, aperture, other relevant settings, and the specific colorimetric properties of the chromatic filters, in order to permit subsequent corrections and produce an image of the highest possible quality. This information is placed in the metadata field.

This principle is used in native, or RAW, format. Strictly speaking, it does not constitute a true format, as it is not unique and cannot be shared. However, we will continue to use the term, according to common usage, in cases where it does not cause ambiguity.

Native format is used by camera manufacturers as a way of recording signals immediately from sensor output; this first step needs to be carried out at speed, without considering the operations needed for visualization or transmission to other users. The file obtained is a faithful representation of the measurements received from the sensor (see Figure 8.1). Later treatment, either within the camera or using a separate computer, allow full exploitation of this data. However, the RAW format is highly dependent on manufacturers, and images saved in this way were not initially intended to be conserved in this format outside of the camera production line (whether for embedded or remote processing). The popularity of this "format" is a sign of its success, but it remains subject to a number of faults associated with its intended usage.

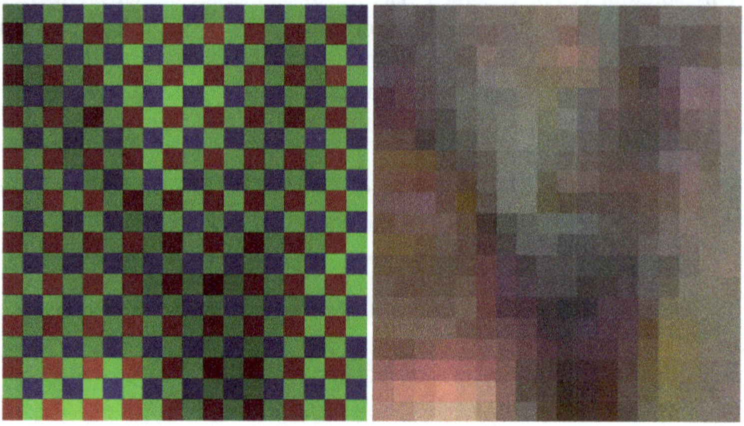

Figure 8.1. *Left: a small section of the color image shown in Figure 8.2 (the top left corner), as recorded by the Bayer matrix and represented in a native (RAW) file, where the three channels, R, G and B, are stacked. For each pixel, only one of the R, G or B values is non-null. Right: the same image, reconstructed by demosaicing. For a color version of this figure, see www.iste.co.uk/maitre/pixel.zip*

An ISO standard does exist (discussed in section 8.2.4), and aims to enable the transmission of these files between users; however, manufacturers continue to use their own versions, which are modified in line with their own specific requirements. Native files are marked by a specific file extension (over one hundred types currently exist[2]), and not by the extension *.raw*, as we might

---

2 For example *.cr2* or *.crw* for Canon, *.nef* or *.nrw* for Nikon, *.orf* for Olympus, *.arw* for Sony, *.ptx* for Pentax, *.srw* for Samsung, *.3fr* for Hasselblad, *.raf* for Fuji, etc. Note that both Leica and Panasonic have adopted a generic *.raw* format alongside their own *.rwl* (Leica) and *.rw2* (Panasonic) extensions. Finally, Adobe's *.dng* extension, which will be discussed in section 8.2.4, is used by certain manufacturers (including Leica) in addition to their own formats.

wish in the case of a shared format. These formats cannot generally be read using ordinary display tools (some examples are even encrypted), but a large number of image processing libraries recognize the most widespread formats, and updates to reading algorithms are made available online as these formats evolve.

There are many reasons why these formats are not universal. First, they reflect the geometry of the photodetector (variable, as seen in section 5.4.2); second, they are tailored to specific architectures in order to take account of the precise noises affecting the signal (dark current, reset current, transfer transistor noise, presence of shared amplifiers, etc.) and provide the correct parameters for noise-reducing filtering equations (such as equation [7.10]). Colorimetric operations are expressed differently as a function of specific chromatic filters, and, finally, each format has a different way of treating known faults, blind pixels, geometric aberrations, vignetting, etc.

The transformation of images from native format to a more widespread format (such as JPEG) occurs during the transfer of the image from the camera memory to a computer, in a process designed by device manufacturers. This operation is also often required for images to be displayed on the camera screen. The solution implemented within the camera or by the host computer using a specific program is often the best available option, as the camera designer is aware of the exact parameters required to best exploit the signal, and knows where to find the necessary parameters in the ancillary data associated with the image. Within the camera, however, the time available for corrections is limited, and more complex correction processes may need to be applied later using a computer. It is also possible for users to override manufacturer settings, which may involve treatments which the user does not require (sharpness reinforcement or color accentuation, for example). In this case, a specific treatment process needs to be applied to the native image, whether using a widely-available free program, DCraw[3], which provides a decoding tool for all commercially-available cameras, or using more general toolkits (openware products such as RawTherapee[4], or commercial products such as PhotoShop, LightRoom, DxO Lab, etc.), giving users access to a wide range of treatments.

---

3 DCraw was developed by Dave Coffin, and includes transformation formulas for almost all native formats. It also forms the basis of a range of other programs and toolkits, and is available from [COF 14].

4 RawTherapee is a specialist program for native file treatment [HOR 14b].

Image processing libraries offer different algorithms to those supplied by manufacturers, and have the potential to produce different reconstructions. These may be better, as more complex algorithms can be used, for example for colorimetrics, noise filtering or demosaicing. Furthermore, these libraries may include treatments for effects which cameras do not always take into account: optical vignetting, pixel vignetting, geometric aberrations etc. In these cases, the reconstruction is better than that obtained using the manufacturer's methods.

8.2.1. *Contents of the RAW format*

The native RAW format needs to include the elements which differentiate different sensors. It generally includes the following elements:

1) in the metadata field:

– a header including the order of bytes (little-endian or big-endian[5]), the name of the image and the address of the raw data;

– information describing the sensor: number of pixels, type of color mastering (using a Bayer matrix or other matrix types, see section 5.4.2.2), the arrangement of photosites[6] and colorimetric profile, expressing the composition of the chromatic filters;

– information describing the image: date, camera type, lens type, exposure time, aperture, focal, selected settings etc., and sometimes geographic position;

– the setting elements required for white balance using sensor measurements or user settings;

– the dynamic of the signal (between 8 and 14 bits) and the parameters of the electronic sequence used to correct this signal (dark current or measures allowing determination of the dark current, bias at zero, amplification factors (or parameters allowing these factors to be calculated using exposure time, ISO sensitivity, and the energy measurement)).

2) in the image field:

---

5 Indicating the order of bytes in a 16-bit word.

6 The exact position of the Bayer matrix is given by a quadruplet indicating the meeting order of colors R, G and B at the start of the first two lines of the image, for example GRGB or RGGB.

– a thumbnail to accompany the image and represent it each time the memory is consulted;

– a reduced JPEG image used to obtain a representation of reasonable size, in certain cases;

– the image data, presented according to sensor geometry (with RGB channels interwoven in a single image plane) (see Figure 8.2).

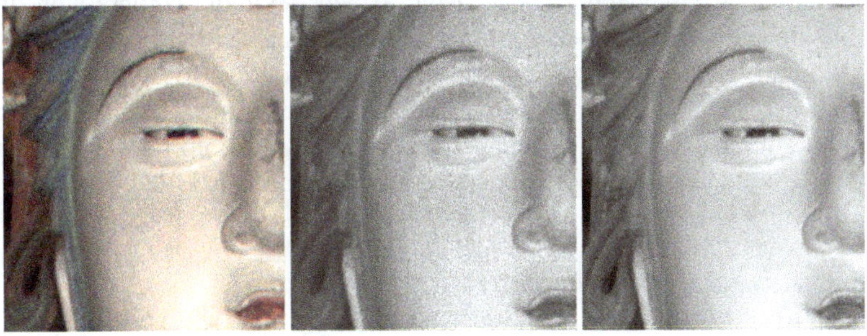

Figure 8.2. *Left: a color image. Center: the corresponding RAW file, represented in grayscale (the R, G and B pixels are interwoven, but have been reduced to gray levels, each coded on one byte). Right: the luminance image obtained from the color image. We see that the geometry of the two images on the left is identical, as the R, G and B channels are layered, but under-sampled. The central image also shows the staggered structure characteristic of the Bayer mask, which is particularly visible on the face, which has low levels of green. This staggered structure is even more evident in Figure 8.1, which is highly magnified. On the lips, we see that the absence of green (which is twice as common in the samples than red or blue) results in the presence of a dark area, while the greenish band in the hair appears more luminous than the hair itself, which is made up of red and blue. The luminance image does not show this structure, as the RGB channels have been interpolated. For a color version of this figure, see www.iste.co.uk/maitre/pixel.zip*

As many metadata elements are shared by all images, typical formats for these elements are used (for example Exif); these are discussed below. However, the native format file also contains sensor-specific information (particularly in relation to the precise parameters governing electronic elements: thermal noise, positions of dead pixels, non-homogeneity in gains, corrections for pixel vignetting) which manufacturers do not always wish to make public. For this reason, the metadata of native files is sometimes encrypted in order to prevent them from being read by software outside of the manufacturer's control.

Finally, note that certain manufacturers use native formats in which image files undergo lossless coding (Lempel-Ziv type).

As we have stated, the native format is bulky, and this represents one of its major drawbacks. The header file takes up a few thousand bytes, generally negligible in relation to the size of the image file. For an image with n pixels, each pixel is represented by a single signal in one of the R, G or B channels, and this signal is generally coded using 2 bytes. Without compression, the whole file will therefore take up around $2n$ bytes. Using a lossless compression algorithm, we can obtain a reduction of the order of 2; a RAW file takes up around n bytes[7]. This compression, if selected by the manufacturer, will be applied automatically. Native image files generally take up between 10 and 50 Mb, around ten times more than the corresponding JPEG file.

8.2.2. *Advantages of the native format*

The quality of the signal captured by a camera is best preserved using the native RAW format. Moreover, the accompanying metadata provides all of the elements required for efficient image processing.

This metadata allows us to carry out all of the operations needed to reconstruct an image signal over 8 or 16 bits for any applications, even those which are demanding in terms of quality, such as printing, publication, photocomposition, video integration, etc. JPEG type output is still possible for everyday uses, but if volume is not an issue, formats with a higher dynamic, which better respect the fine details of images, are often preferable: TIFF, DNG, PNG, etc.

Use of the native format is essential in many advanced applications, such as those described in Chapter 10: high dynamic rendering, panoramas, and increasing resolution and focus.

It is often better to use separate computer software for reconstruction processes. These programs, whether free or commercial, give us the opportunity to carry out post-processing as required by specific applications. Post-processing is more effective if the source signal is used than if the signal has already been processed for general usage.

---

7 A lossless compression rate of 2 is high for an image, but not unusual, as the native signal generally does not use all 16 bits of the dynamic.

8.2.3. *Drawbacks of the native format*

Whilst native formats provide the most faithful rendering of images, they are also subject to a number of faults:

– the volume of these images is often prohibitive; for general use, the storage space required is multiplied by a factor of between 5 and 10;

– the size of the image reduces the speed at which successive photographs can be taken, as the flow rate through the bus connecting the sensor to the memory card is limited;

– the time taken to display or navigate through image databases increase accordingly;

– many widespread visualization programs cannot process native files;

– the lack of standardization in native formats leads to uncertainty regarding the future accessibility of images stored in this way. This is a serious issue, as rapid restructuring by camera manufacturers results in the elimination of major players, who protected their reading and writing software using encrypted coding. Moreover, technological developments have also resulted in the evolution of stored information, and manufacturers may create new formats in order to satisfy requirements not met by their existing format.

8.2.4. *Standardization of native formats*

8.2.4.1. *Native formats via TIFF/EP*

The format standardized by the ISO (ISO 12234-2 [ISO 10]) is known as Tagged Image File Format – Electronic Photography (TIFF/EP) and derives from the TIFF 6.0 format (see section 8.2.4.1), which is more widespread, but under the control of Adobe. It aims to constitute a subset of image formats[8].

Unlike TIFF, TIFF/EP is fully accessible to users. Like TIFF products, it is particularly aimed at applications in graphic imaging, like TIFF products[9].

8 Note that all formats eventually claim to constitute subsets of other formats. This is achieved by introducing new extensions after the format has been defined; these new extensions are rarely accepted by the programs which are supposed to use them. These extensions make it particularly difficult to create a map of representation programs; a number of examples will be considered later in this chapter.

9 Graphic images are differentiated from photographic images by the presence of large areas of solid color, highly saturated colors, very fine details (lines, text etc.) with high contrast levels, and a number of specific properties: floating representation of numbers, transparency or animated attributes, etc.

For this reason, particular attention has been paid to document compression aspects, as these processes produce much higher gains in the context of graphic images than for photographic images. Three types of lossless coding may be used: Huffman coding by areas, Huffman-type coding by blocks, or Lempel–Ziv coding, now that this method is in the public domain (see section 8.4). The only lossy compression method currently available in practice is JPEG/DCT.

In the example, given above, of a native file produced by a Bayer sensor, reconstructed then transcoded into TIFF, coding is carried out in TIFF/EP using 2 bytes per channel and per pixel, i.e. 6 bytes per pixel and $6n$ bytes for the full image (120 Mb for a Mpixel image). This is the largest volume this image may take up. Lossless compression is therefore recommended in order to obtain a more reasonable file size; even using lossless methods, significant levels of compression may be attained, as the dynamic does not generally use the full 16 bits.

The TIFF/EP format includes a metadata file relatively similar to the Exif file, which will be discussed below, but based on the TIFF format metadata file, with the exclusion of a certain number of fields. Most of the Exif fields are included in TIFF/EP, but arranged using a different structure and with slight changes in definition in certain cases.

The TIFF/EP format can also be used for tiled representation of images (see section 8.7). It is therefore suitable for use with very large images.

The ISO standard is not respected by most manufacturers, who continue to use their own formats. However, the formats used are often derived from, and constitute extensions to, the TIFF/EP standard.

The lossless coding aspect of TIFF/EP is notably compatible with the DNG format, proposed by Adobe but available in open access, and accompanied by public processing libraries.

8.2.4.2. *Native format via DNG*

Adobe offers another universal format for transportation of RAW files, independent of the proprietary format used in their creation [ADO 12]. This format, DNG (Digital NeGative) is fairly widely recognized by image processing software, and uses the extension *.dng*. Openware programs are available to convert any RAW image into a DNG file, but few manufacturers use this format, preferring their own native formats. However, DNG is widely used in image processing and exploitation software.

DNG is an extension of TIFF 6.0 and uses the same formatting rules.

DNG also uses a metadata file, using EXIF, TIFF/EP or other formats. It retains the proprietary data of these files, offering mechanisms to enable them to be read by other programs. It also transmits specific information from the capture system, concerning the illuminant, calibration, color conversion matrices, representation formats, noise levels, dead pixels, etc. DNG includes fields for post-acquisition processing (floating representation, transparent pixels, etc.). It supports JPEG compression (with or without loss) and predictive lossless compression (using DEFLATE, see section 8.4). DNG is able to use images represented using RGB or YCrCb (see section 5.2.4).

8.3. Metadata

We have seen that metadata is crucial in allowing optimal usage of digital images, and we have considered two file types used for this purpose: TIFF/EP and Exif files. The Exif format is the most widely used in photography and will be discussed in detail further on. First, let us consider the efforts which have been made to standardize this file type.

8.3.1. *The XMP standard*

An ISO standard (16684 – 1:2012) has been created regarding the description of these data, based on the XMP (*extensible metadata platform*) profile, proposed by Adobe. In theory, this format allows conservation of existing metadata alongside an XMP profile; in practice, however, conversion of all metadata into XMP format is highly recommended. Metadata is grouped into mutually independent "packets", which may potentially be processed by separate applications. Users may create their own packets and develop new applications to use these packets, but care is required to avoid affecting other packets.

The resource description framework (RDF) language, popularized by the semantic web, forms the basis for packet description, and XML (*extensible markup language*) syntax is generally used.

The XMP format is widely used in artistic production, photocomposition, the creation of synthetic images and virtual reality. It has been used in a number of commercial developments, in professional and semi-professional processing tools, and has been progressively introduced into photographic treatment toolkits. However, users are generally unfamiliar with this format and rarely possess the tools required to use it themselves, unlike Exif files, which are much more accessible.

8.3.2. *The Exif metadata format*

Exif (*exchangeable image file format*) was designed as a complete sound and image exchange format for audiovisual use, and made up of two parts: one part compressed data, and the other part metadata. Nowadays, only the metadata file accompanying an image is generally used. For the signal compression aspect, JPEG, TIFF or PNG are generally preferred (but the Exif exchange standard includes these compression modes in a list of recommended algorithms). For metadata, however, elements of the Exif file form the basis for a number of other standards: TIFF, JPEG, PNG, etc.

Created by Japanese manufacturers in 1995, the Exif format is used in camera hardware, and, while it has not been standardized or maintained by a specific organization (the latest version was published in 2010), it still constitutes a de facto standard. Unfortunately, however, it has tended to drift away from initial principles, losing its universal nature.

The metadata part of Exif (now referred to simply as Exif) is broadly inspired by the structure of TIFF files. The file is nested within a JPEG or native image file, and contains both public data for general usage and manufacturer-specific data, which is not intended for use outside of image programs. Unfortunately, while image handling programs generally recognize Exif files within images and use certain data from these files, the file is not always retransmitted in derivative images, or is only partially transmitted, leading to a dilution of Exif information and de facto incompatibility. The Exif file can be read easily using simple character sequence recognition tools (for example using PERL), and a wide range of applications for decoding these files are available online.

8.3.2.1. *Data contained in Exif files*

The data contained within an Exif file relates to several areas:

1) hardware characteristics:

– base unit and lens: manufacturers, models, camera series number;

– software: camera software version, Exif version, FlashPix (tiling program) version where applicable;

– flash, where applicable: manufacturer and model;

– dimension of the field available for user comments.

2) capture conditions:

– date and time;

– camera orientation (landscape or portrait; orientation in relation to magnetic North may also be given), GPS coordinates where available;

– respective position of chromatic channels and component configuration;

– compression type and associated parameters in the case of JPEGs;

– resolution in x and in y with the selected units (mm or inch), number of pixels in x and y;

– focal distance and focusing distance;

– choice of light and distance measurements;

– ISO sensitivity number.

3) internal camera settings:

– color space: CMYK or sRGB;

– white balancing parameters;

– exposure time, aperture, exposure corrections where applicable;

– gain, saturation, contrast, definition and digital zoom controls;

– interoperability index and version;

– selected mode and program, if several options are available, with parameters;

– identification of personalized treatments and parameters.

4) data specific to the treatment program:

– processing date;

– program: make and version;

– functions used, with parameters;

– identification of personalized treatments and parameters.

8.3.2.2. *Limitations of Exif files*

Exif files constitute a valuable tool for rational and automated use of images. However, they are subject to a number of drawbacks:

– Exif files only take account of chromatic components coded on 8 bits, which is not sufficient for modern camera hardware;

– Exif is not designed to provide a full description of RAW files requiring different fields, meaning that manufacturers often have to use alternative formats: TIFF or DNG;

– Exif files are fragile, and easily damaged by rewriting or data modification;

– as Exif is not maintained by an official body, it tends to evolve in a haphazard manner in response to various innovations, which adapt parts of the file without considering compatibility with other applications;

– Exif offers little protection in terms of confidentiality and privacy, as, unless specific countermeasures are taken, users' photograph files contain highly specific information regarding capture conditions, including a unique camera reference code, date and, often, geographic location. A number of programs use Exif content in databases constituted from internet archives, email content or social networking sites, enabling them to identify document sources.

8.4. Lossless compression formats

Lossless compression is a reversible operation which reduces the volume occupied by images, without affecting their quality. Lossless compression algorithms look for duplication within an image file, and use Shannon's information theory to come close to the entropic volume limit [GUI 03a, PRO 03].

Black and white images generally do not present high levels of duplication when examined pixel by pixel. Their entropy[10] is often greater than 6 bits/pixel, meaning that the possible gain is of less than 2 bits/pixel; this is not

10 As we have seen, entropy is a good way of measuring the quantity of information carried by an image, and is expressed, as a first approximation, using equation [6.51]: $\mathcal{E} = -\sum_k p(k) \log p(k)$, where $p(k)$ represents the probability of a gray level k; it is therefore limited by the logarithm of the number of levels of gray used in the image, i.e. 8 bits/pixel in the case of black and white.

sufficient to justify the use of a complex coder. Color images generally include higher levels of duplication, and it is often possible to reduce the size of the image file by half with no loss of information.

It is also possible to look for duplication between pixels, coding pixel groups (run lengths or blocks), exploiting spatial or chromatic correlation properties, which are often strong. The coding required is more complex and the process is longer both in terms of coding and decoding, and a significant amount of memory is required for maintaining code words. The tables used also need to be transmitted or recorded, and the cost involved drastically reduces the benefits obtained from compression, except in the case of very large images.

The algorithms used for lossless compression include general examples, widely used in computer science as a whole, and more specific examples, developed for graphic or imaging purposes.

8.4.1. *General lossless coding algorithms*

These coding techniques make use of duplications between strings of bits, without considering the origin of these strings. General forms are included in computer operating systems [GUI 03a]:

– Huffman coding, which produces variable length code, associates short code words with frequently-recurring messages (in this case, the message is the value of a pixel or set of pixels). The best results are obtained for independent messages. In cases where messages are correlated, the Huffman code only exploits this correlation by grouping messages into super-messages. These dependencies are not modeled, meaning that the method is not particularly efficient for images. However, it is particularly useful for coding graphics and drawings, producing significant gains in association with a low level of complexity. Huffman coding by run-lengths or zones is therefore used in a number of standards (including TIFF);

– arithmetic coding does not separate images into pixels, but codes the bit flux of an image using a single arithmetic representation suitable for use with a variety of signal states. Arithmetic coding is not used in image processing, as there is no satisfactory symbol appearance model;

– Lempel–Ziv coding (codes LZ77 and LZ78) uses a sliding window, which moves across the image, and a dictionary, which retains traces of elementary strings from their first occurrence. Later occurrences are replaced by the address of this first example in the dictionary. LZ is now routinely used to code very large files with high levels of duplication, such as TIFF files;

– hybrid coding types use back references, as in the Lempel–Ziv method, along with shorter representations for more frequent strings, constructed using prefix trees. These codes, used in a number of generic formats such as Deflate and zip, are extremely widespread;

– more specific coding types have been developed for use in imaging, making explicit use of two-dimensional (2D) and hierarchical dependencies in images; details of these coding types are given in [PRO 03]. These methods are used by standards such as JBIG (highly suited to progressive document transmission in telescopy) and coders such as Ebcot, which features in the JPEG 2000 standard and will be discussed further in section 8.6.2.

All of these coding methods are asymptotically very close to the Shannon entropy limit for random sequences. The size of the headers involved means that they are not worth using for short messages, but image files rarely fall into this category. These methods are not particularly efficient for images, as duplication statistics are generally relatively low in a pixel flux representation. This duplication is concealed within the image structure, something which hierarchical and JPEG-type coding methods are designed to exploit.

8.4.2. *Lossless JPEG coding*

Specific algorithms have been developed for image coding, based on the specific two-dimensional (2D) properties of images. They have enjoyed a certain level of success in areas such as medical imaging, where image degradation must be avoided at all costs. They are also widely used in graphics, where scanning images are used in conjunction with text and drawings [GUI 03b].

The most widespread approaches are based on predicting the value of a pixel X at position (i, j) using neighbors for which values have already been transmitted and are known to the receiver, specifically those denoted $A(i, j-1)$, $B(i-1, j-1)$, $C(i-1, j-1)$ and $D(i-1, j+1)$ (Figure 8.3, left). The difference between the predicted and actual values is often very low, and often follows a Gaussian distribution, which may be conveniently coded using the Huffman approach.

The lossless JPEG standard, published in 1993, proposes an algorithm for the prediction of X conditional on the presence of a horizontal or vertical contour. There are eight possible situations in this case, leading to eight different decisions. The remainder is coded using the Huffman method. Compression rates of up to 3 have been obtained in medical imaging using this standard, but results are often lower in natural imaging, and lossless JPEG has not been particularly successful in this area.

A second type of lossless JPEG coder has since been proposed. JPEG-LS is based on a typical prediction algorithm, LOCO-1, which counts configurations using a predictor known as the median edge detector. There are 365 possible configurations, and predictions are made using the hypothesis that the error follows a Laplace distribution. Golomb-Rice coding is then used to compress the transmitted sequence (this type of coding is suitable for messages with a high probability of low values). It is optimal for geometric sequences and performs better than lossless JPEG, with a compression rate of up to 3 for natural images (a comparison of coder performances is given in [GUI 03b]). JPEG-LS has been standardized under the reference ISO-14495-1.

The JPEG 2000 standard concerns another lossless coding method (sometimes noted JPEG 2000_R, with R reflecting the fact that the method is reversible). This method uses biorthogonal 5/3 wavelet coding (Le Gall wavelets with a low-pass filter using 5 rational coefficients and a high-pass filter with 3 rational coefficients). This third form of lossless JPEG coding presents the advantage of being scalable[11], progressive[12] and compatible with the image handling facilities of JPEG 2000 (see section 8.6.2). Unlike JPEG-LS, it uses the same principle for lossy and lossless coding, but it is slower, and the implementation is more complex. In terms of compression, it does not perform better than JPEG-LS.

8.5. Image formats for graphic design

8.5.1. *The PNG format*

The PNG (portable network graphics) format was designed for graphic coding in scanning (rather than vector) mode and for lossless compression. It was developed as an alternative to GIF (graphics interchange format), created

11 Scalability is the ability of a coding method to adapt to the channel or display terminal being used, limiting data transmission in cases where the channel is narrow or the display capacity is small.

12 A coder is said to be progressive if it begins by creating a rough reconstruction of the whole image before refining the quality of the image over successive transmissions, allowing users to "see" the image very quickly.

by CompuServe, which was not free to access for a considerable period of time (the relevant patents are now in the public domain).

The PNG format offers full support for color images coded over 24 bits in RGB (unlike GIF, which only codes images over 8 bits), alongside graphics coded using palettes of 24 or 32 bits. It is also able to handle transparency information associated with images (via channel α, which varies from 0 to 1, where 0 = opaque and 1 = completely transparent). However, unlike GIF, PNG is not able to support animated graphics. For images, PNG conversion must occur after the application of corrections (demosaicing, white balancing, etc.) and improvement processes (filtering, enhancement, etc.). Treatment information and details contained in the Exif file are not transmitted, and are therefore lost.

PNG has been subject to ISO standardization ($ISO15948: 2004$).

PNG files contain a header, for identification purposes, followed by a certain number of chunks, some compulsory, other optional, used in order to read the image. Chunks are identified by a 4-symbol code, and protected by an error protection code. The compulsory elements are:

– the IHDR chunk, including the image width, the number of lines and the number of bits per pixel;

– the PLTE chunk, including the color palette for graphic images;

– the IDAT chunk, containing the image data itself, in one or more sections;

– the IEND chunk, marking the end of the file.

Optional chunks may then be used to provide additional information concerning the image background, color space, white corrections, gamma corrections, the homothety factor for display purposes, and even information relating to stereoscopic viewing. PNG has been progressively extended to include color management operations in the post-production phase, i.e. control of the reproduction sequence from the image to the final product.

PNG is widely used in the graphic arts, enabling transparent and economical handling of graphics or binary texts, with graphics defined using up to 16 bits per channel over 4 channels (images in true color + channel α). The main attractions of this format do not, therefore, reside in its compression capacity. However, PNG images are compressed using a combination of predictive coding (particularly useful for color blocks, which have high levels of duplication) and a Deflate coder. The predictive coder is chosen by the user from a list of five options (with possible values of 0 (no prediction), A, B, $(A+B)/2$, A or B or C depending on the closest value of $A+B-C$ (see Figure 8.3)).

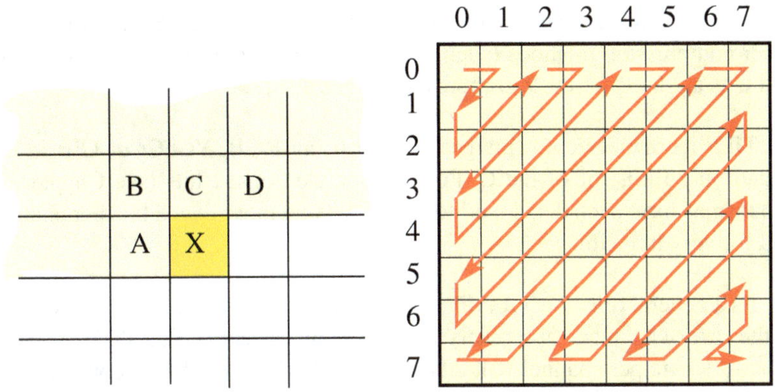

Figure 8.3. *Left: in predictive coding, pixel* X *is coded using neighbors which have already been transmitted:* A, B, C, D. *Right: zigzag scanning order for a DCT window during JPEG coding*

8.5.2. *The TIFF format*

The TIFF format has already been mentioned in connection with the EP extension, which allows it to handle native formats. In this section, we will consider TIFF itself. Owned by Adobe, TIFF was specifically designed for scanned documents used in graphic arts, which evolved progressively from handling binary documents to handling high-definition color documents. Intended as a "container" format for use in conserving documents of varying forms (vector graphics, text, scanned imaged coded with or without loss, etc.), TIFF is an ideal format for archiving and file exchange, guaranteeing the quality of source documents used with other files to create a final document.

TIFF files are made up of blocks of code, identified by tags, where each block specifies data content. Tags are coded on 16 bits. Above 32,000, tags are assigned to companies or organizations for collective requirements. Tags from 65,000 to 65,535 may be specified by users for private applications. There are therefore a number of variations of TIFF, not all of which are fully compatible. The basic version available in 2015, version 6.0, was created over 20 years ago; this version indicates essential points which must be respected in order to ensure inter-system compatibility. TIFF allows very different structures to be contained within a single image: subfiles, bands or tiles, each with their own specific characteristics to enable coexistence.

Algorithms used to read TIFF files must be able to decode lossless Huffman coding by range and JPEG coding by DCT, and, in some cases, other compression methods (such as LZ and wavelet JPEG), as specific flags and tags are used for these compression types.

TIFF recognizes color images created using RGB, YCbCr or CIE la^*b^*, coded on 24 bits, or of the CMYK type, coded on 32 bits (see Chapter 5). Extensions have been developed for images using more than 8 bits per channel and for very large images.

As a general rule, the TIFF file associated with an image is relatively large. In the case of an image of n pixels in sRVB (3 bytes per pixel), a TIFF file using exactly 1 byte per channel will take up $3n$ bytes before lossless compression, and around half this value after compression; however, compression is only carried out if requested by the user, making the method slower.

8.5.3. *The GIF format*

The GIF format is also lossless, and also a commercial product, maintained by CompuServe. While designed for use with graphics, it may also be used to transport images under certain conditions.

GIF is an 8 bit coding method for graphics coded using a color palette, using range-based Lempel-Ziv coding. Black and white images are therefore transmitted with little alteration using a transparent palette (although a certain number of gray levels are often reserved for minimum graphic representation), but with a low compression rate. Color images must be highly quantized (in which case lossless compression is impossible) or split into sections, each with its own palette, limited to 256 colors. Performance in terms of compression is lower in the latter case. Another solution uses the dithering technique, which replaces small uniform blocks of pixels with a mixture of pixels taken from the palette.

8.6. Lossy compression formats

As we have seen, the space gained using lossless image compression is relatively limited. In the best cases, the gain is of factor 2 for a black and white image, factor 3 for a color image or a factor between 5 and 10 for an image coded on 3×16 bits.

However, images can be subjected to minor modifications, via an irreversible coding process, without significant changes to their appearance. This approach exploits the ability of the visual system to tolerate reproduction faults in cases where the signal is sufficiently complex (via the masking effect), i.e. in textured areas and around contours.

This idea formed the basis for the earliest lossy coding approaches, which use the same methods described above: predictive coding or coding by run-lengthes or zones. In this case, however, a certain level of approximation is permitted, allowing us to represent a signal using longer strings. Although these approaches are sophisticated, they do not provide sufficiently high compression rates. As a result, new methods have been developed, which make use of the high levels of spatial duplication across relatively large image blocks, projecting images onto bases of specially-created functions in order to group the energy into a small number of coefficients. These bases include the Fourier transformation (especially the DCT variation) and wavelet transformation [BAR 03b].

For any type of lossy coding, users have access to a parameter α which may be adjusted to specify a desired quality/compression compromise. This compromise is highly approximate, and highly dependent on the image in question. The objective criteria used to judge the effects of compression processes were discussed in detail in Chapter 6. Taking $I(i,j)$ to represent the image before coding and $I'(i,j)$ the image after coding, the most widespread measures are:

1) compression rate:

$$\tau = \frac{number\ of\ bits\ in\ I}{number\ of\ bits\ in\ I'},$$

2) the mean quadratic coding error:

$$e = 1/N \sum_{i,j \in I} (I(i,j) - I'(i,j))^2,$$

3) the signal-to-noise ratio:

$$SNR = 10 \log_{10} \left[\sum_{i,j \in I} (I(i,j) - I'(i,j))^2 / I(i,j)^2 \right],$$

4) and the peak signal-to-noise ratio:

$$PSNR = 10 \log_{10} \left[\sum_{i,j \in I} (I(i,j) - I'(i,j))^2 / 255^2 \right].$$

8.6.1. *JPEG*

In the 1980s, a group of researchers began work on the creation of an effective standard for lossy compression. This goal was attained in the 1990s, and JPEG (joint photographic expert group) format was standardized in 1992, under the name ISO/IEC 10918-1. The standard leaves software developers a considerable amount of freedom, and variations of the coding program exist. The only constraint is that decoding programs must be able to read a prescribed set of test images [GUI 03b].

8.6.1.1. *Principles of JPEG encoding*

JPEG encoding is a six-step process, involving:

1) a color space change to convert the signal, generally presented in the RGB space, into a luminance/chrominance space (generally YCrCb);

2) subsampling of chrominance components using a variable factor, up to 4 in x and y, but usually 2 (this step uses the fact that the eye is less sensitive to fluctuations in color than to fluctuations in gray levels);

3) division of the image into fixed-size blocks: 8×8 pixels for luminance images, 16×16 pixels for chrominance images;

4) transformation of these blocks using an operator which "concentrates" the energy of each block into a small number of coefficients, in this case the DCT (discrete cosine transform);

5) quantization of these coefficients in order to reduce their entropy;

6) transmission of the reordered coefficients by zigzag scanning of the window, followed by Lempel-Ziv coding of the bit chain.

Each of these steps is reversed during the decoding process, with the exception of steps 2 and 5, which involve irreversible information loss.

The direct cosine transform, defined by the matrix in Table 8.1, is a variation of the Fourier transform, which is better suited to real, positive signals, such as images, than the Fourier transform itself [AHM 74]. The

DCT is a 2D, exact and reversible transformation of a scalar signal, which creates a real signal with as many samples as the input signal.

Let $J(i,j)$ be a window of $N \times N$ pixels (where $N = 8$, for example) of the channel being transmitted (the luminance or chrominance channel of the original image). The DCT K_J of J is given by:

$$K_J(k,l) = \frac{2}{N} \sum_{i=0}^{N-1} \sum_{j=0}^{N-1} C(i)C(j)J(i,j) \cos\left[\frac{(2i+1)k\pi}{2N}\right] \times \cos\left[\frac{(2j+1)l\pi}{2N}\right] \quad [8.1]$$

where $C(0) = \sqrt{2}/2$ and $C(i, i \neq 0) = 1$.

The reverse transformation:

$$J(i,j) = \frac{2}{N} \sum_{k=0}^{N-1} \sum_{l=0}^{N-1} C(k)C(l)K_J(k,l) \cos\left[\frac{(2i+1)k\pi}{2N}\right] \times \cos\left[\frac{(2j+1)l\pi}{2N}\right] \quad [8.2]$$

allows us to retrieve the exact value $J(i,j)$, as long as $K_J(k,l)$ has not been subjected to quantization.

Variables k and l are spatial frequencies, as in the case of a Fourier transform. The pair ($k = 0, l = 0$) has a frequency of zero (i.e. the mean value of $J(i,j)$ for the window to within factor N). As k increases, the horizontal frequencies increase. As l increases, so do the vertical frequencies, and all pair values (k,l) where $k \neq 0,\ l \neq 0$ corresponds to a frequency which is inclined in relation to the axes $0x$ and $0y$.

The term $K_J(0,0)$, known as the continuous component, is always much larger than the others. It is treated separately in the DCT, and often transmitted without alterations in order to provide exact, if not precise, information regarding the value of the block.

	0	1	2	3	4	5	6	7
0	1.000	1.000	1.000	1.000	1.000	1.000	1.000	1.000
1	0.981	0.831	0.556	0.195	-0.195	-0.556	-0.831	-0.981
2	0.924	0.383	-0.383	-0.924	-0.924	-0.383	0.383	0.924
3	0.831	-0.195	-0.981	-0.556	0.556	0.981	0.195	-0.831
4	0.707	-0.707	-0.707	0.707	0.707	-0.707	-0.707	0.707
5	0.556	-0.981	0.195	0.831	-0.831	-0.195	0.981	-0.556
6	0.383	-0.924	0.924	-0.383	-0.383	0.924	-0.924	0.383
7	0.195	-0.556	0.831	-0.981	0.981	-0.831	0.556	-0.195

Table 8.1. *DCT filter coefficients for $N = 8$. Lines: i, columns: k (equation [8.1])*

The other terms are subject to quantization, to a greater or lesser extent depending on the desired compression rate. We begin by defining a quantizer $q(k,l)$ and an approximation of coefficient $K_J(k,l)$ using $K'_J(k,l)$ such that:

$$K'_J(k,l) = \lfloor \frac{K_J(k,l) + \lfloor q(k,l)/2 \rfloor}{q(k,l)} \rfloor \qquad [8.3]$$

where $\lfloor \times \rfloor$ denotes the default rounded integer value of x.

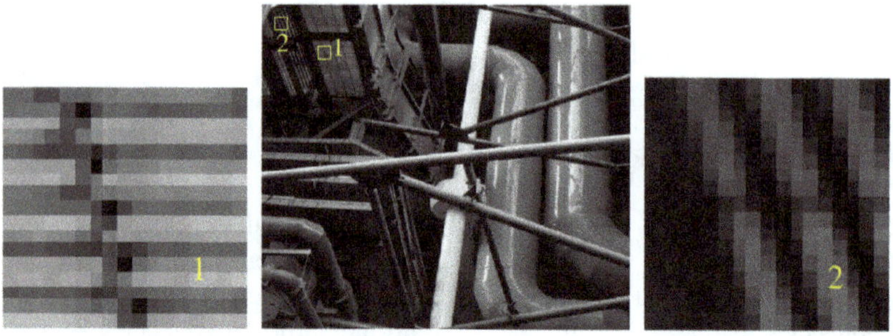

Figure 8.4. *Two thumbnails taken from the central image (top left corner). The DCT transforms of the images, after decimation, are given in Table 8.2 (note that gray levels have been amplified in the display process for ease of reading)*

The set of coefficients $q(k,l)$ thus forms a quantization matrix $\mathcal{Q}$ which is applied to the whole image. This matrix is transmitted with the associated coefficients in order to enable reconstruction of the image. Coders differ in the way in which they use chrominance subsampling and in the choice of quantization matrices.

	0	1	2	3	4	5	6	7		0	1	2	3	4	5	6	7
0	1450	65	0	0	0	0	0	0		980	100	0	0	0	0	0	0
1	0	0	0	0	0	0	0	0		0	0	60	200	0	0	0	0
2	0	0	0	0	0	0	0	0		0	0	0	0	0	0	0	0
3	0	0	0	0	0	0	0	0		0	0	0	0	0	0	0	0
4	0	0	0	0	0	0	0	0		0	0	0	0	0	0	0	0
5	131	0	0	0	0	0	0	0		0	0	0	0	0	0	0	0
6	0	0	0	0	0	0	0	0		0	0	0	0	0	0	0	0
7	0	0	0	0	0	0	0	0		0	0	0	0	0	0	0	0

Table 8.2. *DCT coefficients of thumbnails 1 (left) and 2 (right), after setting all coefficients with an absolute value less than 30 to zero. Note the importance of the term $J(0, 0)$, corresponding to the mean value of the block, and the amplitude of the vertical (in 1) and horizontal (in 2) spatial frequencies corresponding to the periodicities observed in these images*

As a general rule, high frequency terms can be shown to contain very little energy, and these terms are therefore of limited importance in terms of the perceived quality of the image. These terms are set to zero during quantization (see Figure 8.4 and Table 8.1).

The coefficients $K'_J(k, l)$ are then transmitted by zigzag scanning, which places frequencies in increasing 2D order (as far as possible) after the continuous component (see Figure 8.3, right). After a short period, the coefficient string will only contain zeros, which will be compressed effectively using either Liv-Zempel or range-based Huffman techniques.

JPEG images are decompressed by reversing all of the steps described above. Coefficient quantization can highlight faults in individual blocks which are characteristic of the JPEG format. Certain reconstruction algorithms therefore include a low-pass filtering stage which reduces these faults. The subsampling of chromatic components may also require an additional interpolation stage, which may also have damaging effects on the image.

8.6.1.2. *JPEG in practice*

JPEG compression has become the most widely accepted means of transmitting images of reasonable quality and reasonable volume. For color images, a compression rate of 10 is generally considered to cause very little degradation to the image, practically imperceptible to the naked eye, especially for very large image files and when viewed without excessive magnification.

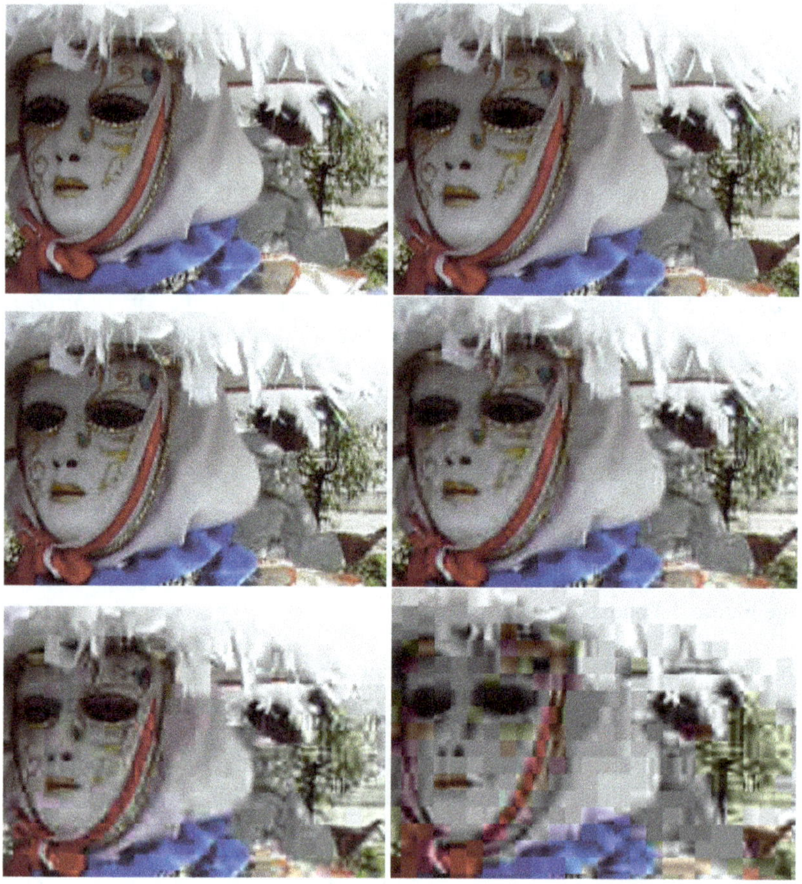

Figure 8.5. *Magnification of an area of an image encoded using JPEG. The top left image is the non-encoded original. The whole image has then been compressed by factors of 5.4 (top right), 12.1 and 20 (central line), and 30 and 40 (bottom line). For a color version of this figure, see www.iste.co.uk/maitre/pixel.zip*

Compression rates of 100 are also widely used, particularly online or when sending images by email. Image quality is highly compromised in this case, and the images in question cease to be suitable for any use except rapid visualization due to the presence of annoying artifacts.

Note that all encoders include a parameter to control compression rates, and this parameter is often unique. However, this indicator is not formally connected to compression gains or to image quality. Applied to two different images, the same index may produce two very different results. For a single image, as the index decreases, the level of degradation and the compression rate increase. This index generally takes a value of 100 in the case of lossless coding, and may be reduced to a minimum value of 1.

Note, however, that lossless JPEG does not use the general principle of the DCT for encoding purposes, as it is based on predictive methods.

JPEG may be used in a progressive manner for transmission purposes: in this case, low frequency terms are transmitted first, and the image is then refined by the transmission of higher frequencies.

A version of the image made up of continuous components alone, 64 times smaller than the original, is often used for display functions or to enable rapid database searches.

8.6.2. *JPEG 2000*

The JPEG 2000 standard is recognized by the ISO under the name ISO/IEC 15444-1, and is a relatively general image exchange standard, offering a variety of functions (archiving, internet and video). In this context, we will focus on the aspects relating to the image format.

JPEG 2000 makes use of the fact that wavelet composition offers better performances than the DCT in terms of image compression: this method provides a better quality image than DCT with the same compression rate, and higher compression rates for the same image quality [BAR 03a].

JPEG 2000 is also scalable.

The format also uses a highly developed arithmetic encoder, EBCOT (embedded block coding with optimized truncation) which makes use of the 2D tree structure of the wavelet decomposition [TAU 00].

Finally, JPEG 2000 gives users the possibility to describe image content at a higher level of the pyramid; this is useful in facilitating intelligent use of image collections, and for optimized transmission over new-generation networks. To do this, the format makes use of the progressive transmission capabilities offered by wavelet scalability, via a codestream transmission structure, and the ability to define regions of interest (ROI) which are transmitted first and with higher quality levels. It explicitly accounts for the possible presence of text or graphics in compound structures. The format is also designed for application to very large images (up to 2^{32} pixels per line), images with a large number of channels (up to 2^{14}) and images with high-dynamic pixels (up to 38 bits per sample).

As with many other standards, JPEG 2000 is open and does not explicitly specify a compression mode, leaving developers the freedom to make improvements over time. The encoder is obliged to produce a binary string which conforms to the standard, but any strategy may be used to do this. Decoders must simply prove their ability to decode the example files included with the standard.

As a transmission standard, JPEG 2000 is highly suited to professional requirements. It has had less success with the general public; its requirements in terms of calculation power and decoding software are higher than those of JPEG, but this is of minor importance compared to the effects of inertia in the general market, where users are not overly concerned with obtaining high quality/flux ratios.

8.6.2.1. *Principles of JPEG 2000*

Like JPEG, JPEG 2000 is able to transform color images created using the RGB space into a luminance/chrominance space (YUV or YC_RC_B). Two types of transformations are available, one reversible, the other irreversible (but more effective in terms of compact representation). The chromatic components are then processed separately.

JPEG 2000 then operates either on the image as a whole, or on tiles, identically-sized sections of the image (with the exception of the edge tiles). These tiles are decomposed hierarchically by a bank of 2D, bi-orthogonal wavelet filters [BAR 03a].

The standard allows the use of two types of filters[13]:

1) Le Gall 5/3 rational coefficient filters, as discussed in section 8.4; these filters only operate using rounded integers, and may be used for lossy or lossless coding, depending on whether all of the non-null coefficients are transmitted (using the EBCOT method, described below),

---

13 Unfortunately, these filters do not possess a simple analytical form, and are expressed either as a list of their coefficients, or using their equation within the polyphase space in which they are defined.

2) Daubechies 9/7 real coefficient filters[14] (see Table 8.3), which can only be used for lossy coding, as calculations are carried out using real numbers with a finite level of precision. These wavelets perform better than rational wavelets in the context of lossy coding.

order of filter	0	± 1	± 2	± 3	± 4
analysis: low-pass filter	0.602949	0.266864	-0.078223	-0.016864	0.026749
analysis: high-pass filter	0.557543	0.295636	-0.028772	-0.045636	0
synthesis: low-pass filter	0.602949	-0.266864	-0.078223	0.016864	0.026749
synthesis: high-pass filter	0.557543	-0.295636	-0.028772	0.045636	0

Table 8.3. *Coefficients of Daubechies 9/7 wavelet filters for image analysis and reconstruction (based on [BAR 03a]). Note that other formulas exist, which differ by a factor $\sqrt{2}$. The analysis and reconstruction filters need to be symmetrical in relation to order 0 to avoid moving image contours*

These filters are bi-orthogonal[15], and defined by a low-pass filter used to calculate images with increasingly low resolutions, starting from the original level of the tile, and a high-pass filter, used to define the contents of details characterizing the specific level in question. These details are contained within the coefficients of the wavelet decomposition: $a_n(i, j)$.

Filtering can be carried out in two ways. The first method, using convolution, is that generally used in signal processing as a whole. The second method, which constitutes a major advantage of the hierarchical wavelet approach, involves the iterative application of two filters, one low-pass, the other high-pass; these filters are identical for all levels of the pyramid. This is known as the lifting process. Using this process (see Figure 8.6), the odd coefficients of a decomposition are expressed using the value of the even coefficients with the application of a corrective term. This corrective term constitutes the specific detail image for the level in question [SWE 96].

The basic wavelet transformation is one-dimensional (1D); it is thus applied once to the lines and once to the columns of the area of interest. This results in the production of 4 images (see Figure 8.7):

– an image with high frequencies in both lines and columns;

14 To be precise, these filters use Cohen-Daubechies-Feauveau wavelets [COH 92].

15 Filters are said to be orthogonal if the scalar product of two filters of different rank is zero. If this particularly demanding constraint cannot be met, it is possible to impose bi-orthogonality, i.e. orthogonality between low-pass and high-pass filters.

– an image with high frequencies in the lines and low frequencies in the columns;

– an image with high frequencies in the columns and low frequencies in the lines;

– an image with low frequencies in both lines and columns.

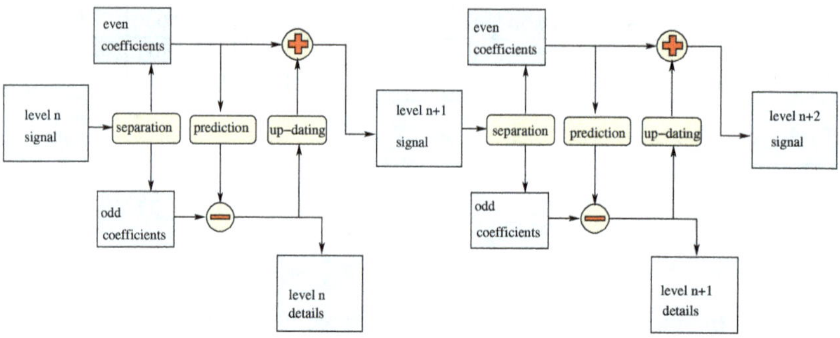

Figure 8.6. *Lifting diagram for 3 levels of n, $n+1$ and $n+2$ wavelet decomposition, showing the separation of odd and even samples. The differences between odd samples and their predictions, made on the basis of the even samples, are known as level n details. The risk of aliasing is taken into account by filtering the detail signal and reinserting it into the even samples, which are then used to form a signal of level $n+1$. The lifting process may then be applied to the signal, and so on, until the required depth is reached*

In a carefully-chosen wavelet base, these coefficients present a number of interesting statistical properties; they are highly concentrated, and can be efficiently modeled using a generalized Gaussian law, allowing efficient quantization:

$$p(a_n) = \alpha \exp\left(-|bx|^d\right) \qquad [8.4]$$

Quantization tables are defined for each resolution level, and calculated in such a way as to provide a wide range of compression levels.

These quantized coefficients, transformed into bit planes and grouped into subbands, are then transmitted to the encoder. As we have seen, the EBCOT coder uses a high-performance contextual encoding process involving a wavelet coefficient tree, predicting each bit using the same bit from the previous resolution level and its immediate neighbors, starting with those bits with the heaviest weighting. The encoder processes subblocks, which are relatively small (with a maximum 64 × 64); the precise size of these subblocks is defined by the user.

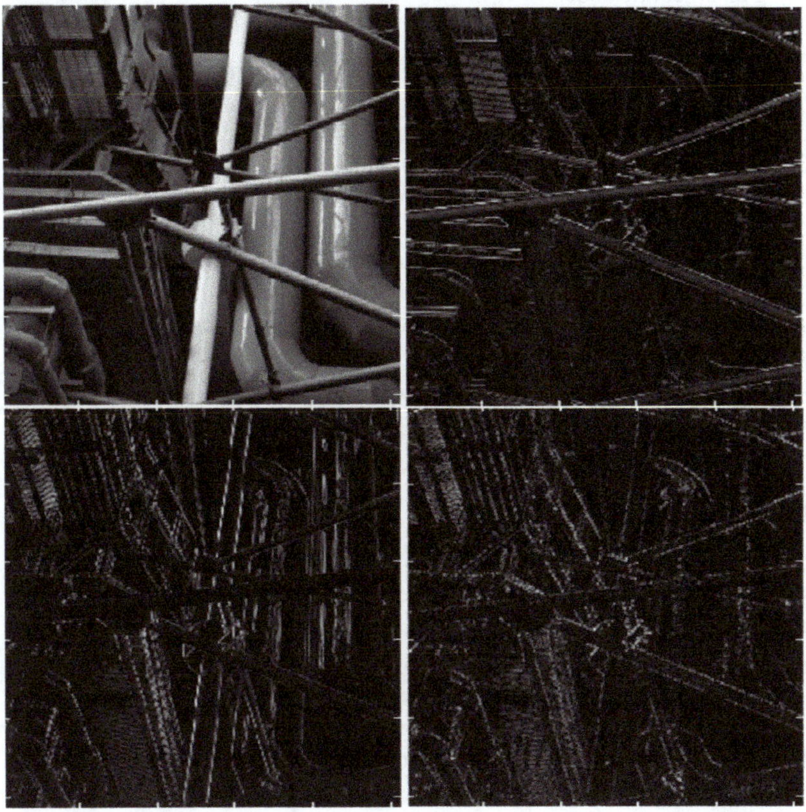

Figure 8.7. *Image coding using JPEG 2000. Top left: level 2 image, with 3 level 1 detail images: top right, detail resulting from a low-pass horizontal filter and a high-pass vertical filter; bottom left: detail resulting from a transposed filter: horizontal high-pass, vertical low-pass; bottom right: result of a low-pass vertical filter and a high-pass horizontal filter. Properly combined, these four images give the original image (using the Daubechies wavelet base)*

The power of the coding method is based on the domain of hierarchical analysis of the wavelet decomposition. We will consider neighborhoods of 6 or 27 connections within a three-dimensional (3D) pyramid of resolution levels. The majority pixels (identified by the bit from the previous level) are not encoded; the others are coded using their sign and deviation from the previous level, if this is significant. Finally, we consider situations where there is no dominant configuration. The resulting tree structure is encoded using an algebraic code.

8.6.2.2. *JPEG 2000: performance*

The compression rates obtained using JPEG 2000 are around 20% higher than those for JPEG for the transmission of fixed images of identical quality. Subjectively speaking, the drawbacks of JPEG 2000 are less problematic than those encountered using JPEG, as the systematic block effect no longer occurs. JPEG is particularly effective for very large images, along with images containing few very high frequencies. JPEG 2000 produces satisfactory results for high compression levels (of the order of 100). The standard is also highly adapted for remote sensing images, due to its ability to process large images using a mode compatible with the tiling approach needed for high-resolution navigation, and to the possibility of specifying zones of interest, which are then subject to greater attention.

8.7. Tiled formats

To complete our discussion of fixed image coding, let us consider the case of very large images. This situation is not yet widely encountered in photography, but with the development of sensors of 50 megapixels and more, these formats may be used by certain devices in the relatively near future.

Tiled formats are used for very large image files (from 10 megapixels up to several gigapixels) and are designed to allow rapid consultation of these images. The images in question are decomposed into tiles (with a default size of 64×64 pixels in FlashPix, the most common form, and 256×256 pixels in the IVUE format). They also use a pyramid shaped, hierarchical binary description of the image, moving from full resolution (several hundred tiles) to a single, coarse image contained in a single tile. A pixel at any level of the pyramid is obtained by averaging the four pixels from the previous level which it is intended to represent. The use of a tree structure allows rapid determination of the level of representation required in response to a user request (for example to display a zone of a given size on a screen of a given size). The use of this tree, along with pre-calculated representation structures, allows very significant gains to be made in terms of image manipulation, both for input/output purposes and for processing, for example when scrolling across a screen.

The payoff for this gain is an increase in the required storage space, as the original image (split into small tiles) needs to be conserved, alongside versions at 1/2, 1/4, 1/8 resolution etc.; this gives a total storage space of $\sum_0^\infty 1/4^n$ times the size of the image, ignoring possible truncation errors, i.e. a 33% increase in memory requirements.

The FlashPix format is able to handle tiles of any type: 8 or 16 bits, 1, 3 or 4 channels, coded using lossy or lossless methods. A description of the Flashpix representation (tile size and number) is given in the associated EXIF file.

Plane representations are replaced by spherical forms in the context of very large datasets, such as those used in astronomy and remote sensing. In this case, a specific tiled representation of the sphere is required. The best-known example, HEALPix (hierarchical equal area isolatitude projection of a sphere) [HEA 14], decomposes the sphere by projection onto 12 squares of a plane, set out in a staggered formation, using different projections according to the position of the zones in question. These squares are oriented at $\pi/2$ in relation to the equator, and centered on either $(0, k\pi/2)$ or $(k\pi/4, \pm\pi/4)$. The resulting squares are then split into 4, following a classic hierarchical approach.

A more intuitive form of tiled projection, hierarchical triangular mesh (HTM) [KUN 01] decomposes the sphere into triangles along large circular planes of the sphere; another format, QuadTree Cube (Q3C) simply uses the quadratic hierarchical structure [KOP 06].

These tiled representations of spherical forms are particularly useful in photographic assemblies and in panoramas with very wide fields (see section 10.1.8).

8.8. Video coding

The domain of video encoding lies outside the scope of this work; however, an increasing number of cameras allow users to record both fixed images and moving sequences. These cameras include video encoding elements which compress the image flux. As in the case of fixed images, these encoders exploit the duplications within each image (intra-image coding), but also duplications between successive images (inter-image coding).

8.8.1. *Video encoding and standardization*

Standardization work has been carried out by the ITU[16], producing recommendations H.120, H.261, H.263 and H.264. The H263 standard was the first to include an inter-image mode; H264 serves as the basis for MPEG-4.

These recommendations have also been used outside of the field of telecommunications, for applications such as archiving, video games and private usage. The moving picture expert group (MPEG) committee, a joint venture by the ISO and the IEC, has worked on creating consensus solutions for the purposes of standardization [NIC 03].

The first recommendation, MPEG-1, concerned low flow rate transmissions for wired transmissions in the period preceding ADSL (using packets of 64 kbit/s, giving a maximum rate of around 1.5 Mbit/s). This very low rate was only suitable for video transmission of very small images (352 $\times$ 288 pixels per image at a rate of 25 image/s). MPEG-1 was widely used in the context of video-CD archiving.

The MPEG-2 standard was designed for television, and specifically for high quality transmission (contribution coding). It is particularly suitable for Hertz transmission, whether via cable or ADSL. The standard allows transmissions of between 2 and 6 Mbit/s for ordinary video and of 15 to 20 Mbit/s for HDTV[17].

MPEG-4 is a multimedia transmission standard, part of which deals specifically with video images. It is particularly concerned with low-rate transmission (around 2 Mbit/s) and uses more elaborate compression mechanisms than those found in MPEG-2; these mechanisms are suitable for use with highly compressed files (although a significant reduction in quality is to be expected). MPEG-4 makes use of the scalability properties included in recommendation H.263. It forms the basis for easy communications with the field of computer science, particularly with regard to copyright protection.

Developments MPEG-7, MPEG-21 and MPEG-x (with x taking a value from A to V) are designed for multimedia applications of video sequences, and define specific functions for archiving, database searches, production

16 ITU = International Telecommunications Union; the section responsible for standardization is known as the ITU-T

17 High definition television = 1920 $\times$ 1080 pixels per image, 50 images per second.

(particularly the integration of a variety of sources), interaction, diffusion and content protection.

8.8.2. *MPEG coding*

8.8.2.1. *Mechanisms used in MPEG*

The video coding standards contained in MPEG-2 and MPEG-4 are currently used in many cameras. Architectures designed for vector-based treatment and integrated into the camera are used to carry out weighty calculation processes, using many of the elements of these standards [NIC 03, PER 02].

We will begin by considering the implementation of MPEG-2. A color image, generally acquired in a RGB space, is converted into a {luminance, chrominance} space, and the chrominance channels are generally subsampled using a factor of 2 for both lines and columns, as in the case of JPEG.

Intra-image coding is carried out by subdividing the image into groups of blocks (GOB), constructed from macroblocks, MB. A macroblock is itself made up of 4 blocks of 8 $\times$ 8 pixels of luminance and, generally, 1 chrominance block (chroma subsampling encoding, $4:2:0$) or 4 blocks (in the case of $4:4:4$ encoding, without subsampling). The binary string is therefore made up of an image layer (I), made up of layers of groups of blocks (GOB), themselves made up of macroblock (MOB) layers. This macroblock layer typically includes the 4 luminance blocks followed by 2 chrominance blocks, one for dominant R, one for dominant B.

These blocks are coded using a discrete cosine transformation (DCT) quantizied, then subjected to entropic encoding (by range length, or, better, arithmetic coding) (Figure 8.8).

There are two possible modes of inter-image encoding:

1) Predictive mode (mode P): one image from the flux is taken as a reference and transmitted with intra-image encoding (image I). The new image (P), to be transmitted later, is compared to I in order to determine the movement affecting each pixel. This movement prediction stage operates using block matching, and determines a movement vector (Δx in x and Δy in y) for each macroblock. The difference between the macroblock in image P and that in image I, offset by $\{\Delta x, \Delta y\}$, constitutes an error macroblock, which is encoded using the DCT. The image is transmitted by sending the address of

the reference macroblock in I, the movement vector $\{\Delta x, \Delta y\}$, and the error block, quantized and entropically compressed.

2) Bidirectional interpolated mode (mode B): using this mode, the image flux is interpolated using both an earlier image I and a later image P to reconstruct intermediate images noted B (see Figure 8.9) using a very small quantity of additional transmission space (only elements essential for positioning the macroblock in B are added, in cases where this cannot be deduced by simple interpolation of that of P). However, this operating mode does result in a delay in image reconstruction; for this reason, it is mostly used for archiving purposes, where real time operation is not essential, or in applications where a very low level of latency will be tolerated.

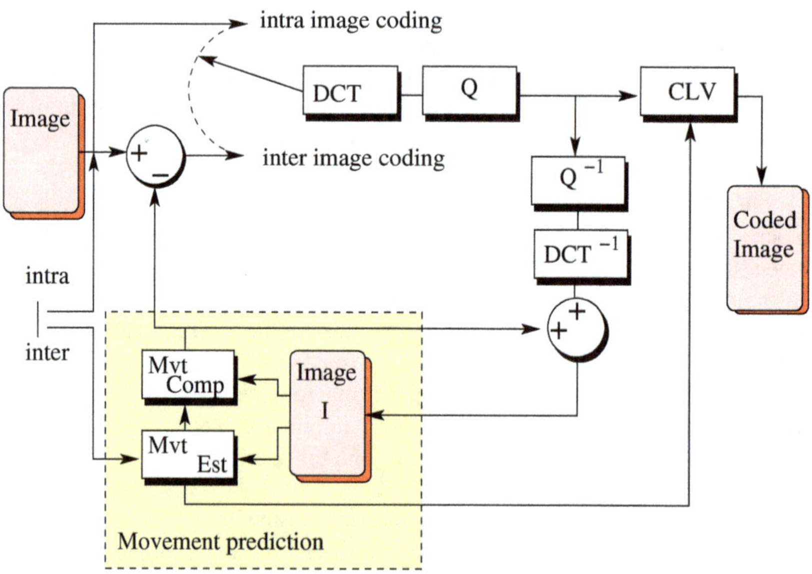

Figure 8.8. *Diagram showing the principle of image encoding in an H.262 flux. Either intra-image or predictive mode may be used. DCT is the cosine transformation of a block of 8 × 8, and DCT^{-1} is the reverse transformation. Q is a quantizer and Q^{-1} is the dequantizer. LVC is a variable-length range coder. The lower left box is the movement predictor. The movement detection stage is followed by movement compensation, used to measure the prediction error, coded in the block using the DCT. The decision to switch to inter- or intra-image mode is based on the position of the image in the flux*

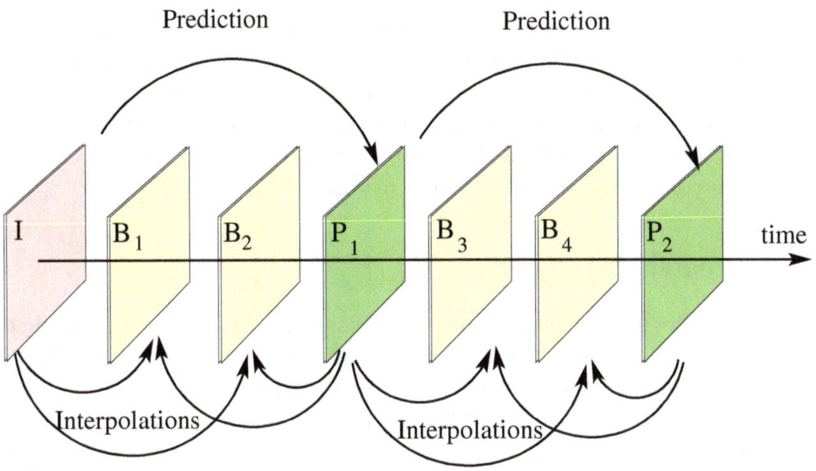

Figure 8.9. *Coding an image flow using MPEG: I is the reference image, coded as an intra-image. P_1 and P_2 are images coded by movement prediction based on I, and on P_1 and I in the case of P_2. B_1 and B_2 are interpolated from I and P_1, and B_3 and B_4 are interpolated from P_1 and P_2*

This coding mode generates significant block effects in the case of high compression rates. A number of improvements have been proposed to eliminate this issue: multiplication of movement vectors for each macroblock, block layering when identifying movement, *a posteriori* filtering, etc.

In MPEG-4, most of the functions of MPEG-2 described above have been improved: better prediction of DCT coefficients, better movement prediction, resynchronization of the data flux, variable length entropic coding, error correction, etc. Part 10 of the standard introduces a new object, the VOP (video object plane), a box surrounding an object of interest. A VOP may be subjected to a specific compression mode different to that used for the rest of the image. A VOP is made up of a number of whole macroblocks, and will be identified in the data flux. For example, a participant in a television program, speaking against a fixed background, may be placed into a VOP with high definition and a high refresh frequency, while the studio background will be coded at a very low rate. VOPs, generally described on an individual basis by users, are not currently used in most cameras (which tend to focus on part 2 of the MPEG-4 recommendations). However, certain cameras now include very similar functions (for example higher-quality coding in a rectangular zone defined by facial recognition).

The use of MPEG-4 coders in cameras and their ability to encode high definition videos (HDTV format[18], in 2015, with 4k format available on certain prototypes for durations of a few seconds) represents a major development in camera technology, as this application currently determines the peak processing power, defines the processor and signal processing architecture, the data bus and the buffer memory.

8.9. Compressed sensing

No chapter on image encoding and compression would be complete without at least a rapid overview of this technique, which, in 2015, shows considerable promise in reducing the volume of image signals beyond the possibilities offered by JPEG 2000. However, compressed sensing is still a long way from constituting a standard or an image forward, and does not yet constitute a reasonable alternative to the standards discussed above [ELD 12].

The basic idea in compressed sensing is the reconstruction of images using a very small number of samples, notably inferior to the number recommended by the Shannon theorem[19] [DON 06, BAR 07, CAN 08]. The proposed approaches consist of applying a number of relevant and complementary measures in order to reconstruct the image using sophisticated techniques which make use of the sparsity of image signals, essentially by using reconstructions of norm $\mathcal{L}_0$ or $\mathcal{L}_1$. Sparsity is the capacity

18 In this context, it is worth re-examining the flow rates required for HDTV video. The commercial considerations associated with mass-market display systems have resulted in the creation of a number of very different HDTV subformats, and it is hard to identify exactly what manufacturers mean by stating that a camera is HDTV-compatible. As we have stated, the production reference value for HDTV resolution is 1,152 lines of 1,920 pixels and 60 (or 30) frames per second [NIC 03]. However, display standards, including HD TNT, HD Ready and HD 1080, are much lower. The 4k format doubles the proposed resolution for both lines (3,840 pixels, hence the term 4k) and columns (2,304 lines). The number of frames is currently used to adapt the number of pixels treated to the power of the processor in question.

19 While compressed coding allows us to reconstruct spatial frequencies beyond the Nyquist limit, it does not invalidate Shannon's theorem, as the reconstruction obtained using compressed sensing uses different hypotheses than those underlying the theorem in question. Shannon may be considered as a specific case of compressed sensing under the hypothesis of signal sparsity across the infinite Fourier bases within the frequency domain $[-B_x, B_x] \times [-B_y, B_y]$. To reconstruct all possible signals of this type, we need to know all of the samples for a step $1/2B_x, 1/2B_y$; in order to obtain a single specific signal, however, it should be possible to measure fewer samples.

of a signal for description in an appropriate base exclusively using a small number of non-null coefficients. Image sparsity cannot be demonstrated, and is not easy to establish. The hypothesis is supported by strong arguments and convincing experimental results, but still needs to be reinforced and defined on a case-by-case basis.

Compressed sensing places image acquisition at the heart of the compression problem; this approach is very different to that used in sensors which focus on regular and dense sampling of whole images, using steps of identical size. However, this "Shannon style" sampling is not necessarily incompatible with all potential developments in compressed sensing. The onboard processor is left to calculate the compressed form in these cases, something which might be carried out directly by sophisticated sensors (although this remains hypothetical in the context of everyday photography).

Presuming that an image can be decomposed using a very small number of coefficients in an appropriate base, we must now consider the best strategy for identifying active coefficients. Candès and Tao have shown [CAN 06] that images should be analyzed using the most "orthogonal" measurement samples available for the selected base functions, so that even a small number of analysis points is guaranteed to demonstrate the effects of the base functions. As we do not know which base functions are active, Candès suggests using a family of independent, random samples [CAN 08]. Elsewhere, Donoho specified conditions for reconstruction in terms of the number of samples required as a function of signal sparsity [DON 09].

In terms of base selection, no single optimal choice currently exists, but there are a certain number of possibilities: Fourier bases, DCT, wavelets (as in the case of JPEG) and wavelet packets used in image processing (curvelets, bandlets, etc.). Reconstruction with a given basis is then relatively straightforward. This process involves cumbersome optimization calculations using a sparsity constraint, for example by searching for the most relevant base function from those available, removing the contributions it explains from the sample, and repeating the process (the classic matching pursuit algorithm [MAL 93]), or by convex optimization.

If an image is principally intended for transmission, reconstruction is not desirable; we therefore need to consider the transmission of random samples. The random character of these samples is not helpful in ensuring efficient transmission, and imposes serious limits on the possible performance of the compressed sensing process, unless we use certain well-mastered quantization and coding forms [GOY 08]. Other types of representation would therefore be useful, and development of these forms is important in order to derive full profit from compressed sensing techniques; however, a lot of work still needs to be done in this area.

Current studies of compressed sensing have yet to produce results of any significance for mass market photography. However, an innovative prototype of a single-pixel camera has been developed, tested [DUA 08] and marketed by InView. This prototype captures images using a matrix of mirrors, individually oriented along both the x and y axes, in order to observe a different aspect of the scene at each instant. All of the signals sent by the mirrors are received by a single sensor, which produces a single pixel for any given instant, made up of a complex mixture of all of these signals. The matrix is then modified following a random law and a new pixel is measured. The flux of successive pixels, measured at a rate of around 1,000 pixels per second, is subjected to a suitable wavelet-based reconstruction process in order to recreate the scene, using around one tenth of the samples previously required. This technology is particularly promising for specific wavelengths for which sensors are extremely expensive (for example infra-red, a main focus of the InView marketing strategy).

Compressed sensing applications are particularly widely used in the field of instrumentation, where they have provided interesting solutions to complex 3D measurement problems in medical imaging (axial tomography, magnetic resonance, etc. [LUS 08]), in microscopy for biological purposes [LEM 12], in holography and for physical instrumentation [DEN 09], etc.

For photographers, compressed sensing is a promising area, but considerable developments are required before these methods are suitable for general use.

9

Elements of Camera Hardware

This chapter will cover most of the components found in camera hardware, with the exception of the lens (discussed in Chapter 2) and the sensor (Chapter 3). We will begin by describing the internal computer, followed by memory, both internal and removable; screens, used for image visualization and for modifying settings; shutters, focus and light measurement mechanisms; image stabilization elements; additional components which may be added to the objective, including rings, filters and lenses; and, finally, batteries.

9.1. Image processors

The processors used in camera hardware have gained in importance over time. The high-treatment capacities required in order to format sensor output signals and to control all of the functions affecting photographic quality have resulted in significant changes to the camera design process. Big-name brands, which have occupied large portions of the market due to their high-quality optic designs and ergonomic hardware elements, found themselves needing to recruit electronics specialists, followed by specialist microcomputing engineers. New actors, mainly from the domain of mass-market audiovisual technology, have also emerged onto the market, often by buying existing brands which were struggling to find the necessary funds to invest in mastering these new technologies. Above and beyond image processing, new functions have been added to cameras, preparing data for storage and printing, using geolocation technology or Internet capabilities, incorporating elements from the field of computer science (such as touch screens) and from the domain of telecommunications.

In the past, the processor market and the optical photography market were two distinct entities. High-quality objective designs were intended to be used over a period of 20 years or more, and used alongside or as an addition to existing products. Microprocessors, on the other hand, evolve rapidly: manufacturers change processor design around once every 3 years, and each new element is intended to replace its predecessors across the whole range. The processor is not designed for a specific camera, and is intended to provide the required performances for a wide variety of purposes in terms of processing speed, input parallelization, memory access, communications facilities, etc. These performances are generally all utilized in high-end professional reflex cameras, which act to showcase the skills and capacities of manufacturers. They are then used to a lesser extent by other models in the range, designed for specific purposes, whether prioritizing video capture, optimizing everyday processing, operating in difficult conditions or ensuring Internet connectivity. In these cases, specific programs are added to the processor, defining the performances and capacities of the camera. These programs are responsible for activating specific processor functions which associate specific sensor measurements with specific settings. The potential of any camera is thus defined by the sensor, processor and software combinations used, all of which are placed at the service of optical components.

9.1.1. *Global architecture and functions*

As we have stated, all of the big players in the camera market maintain their own range of processors: Canon uses the Digic range; Nikon, Expeed; Sony, Bionz; Olympus, TruePic; Panasonic, Venus; Pentax, Prime; and Samsung, DRIMe. However, these products are often developed in collaboration with integration specialists, for example Texas Instruments or Fujitsu, to design cores based on non-proprietary image processing circuits. The processor itself is then completed and integrated by the manufacturer, or co-designed in the context of specific manufacturer/specialist partnerships. For instance, Fujitsu is responsible for the creation and continued development of the Milbeaut architecture, which is included in a wide variety of processors (probably including those used by Leica, Pentax and Sigma) and forms the basis for Nikon's Expeed processors. Texas Instruments worked in partnership with Canon to develop Digic processors, based on their integrated OMAP processor. A smaller number of manufacturers deal with all aspects of component design in-house, including Panasonic, Sony and Samsung. However, the convergence of photography with cellular telephony has led to the integration of increasingly powerful components, with increasingly varied applications outside of the field of optical imaging, and specialized architectures for camera processors are now becoming obsolete[1].

We will not go into detail here regarding the image processors used in camera cores; as we have seen, architectures evolve rapidly, and today's state-of-the-art elements may be obsolete tomorrow. Moreover, hardware configurations are also extremely variable; while most modern cameras include a central processing unit (CPU) associated with a digital signal processor (DSP), some examples also include graphic processors, multiple CPUs or multi-core CPUs with distributed operations. This is particularly true of high-end models, which have higher requirements in terms of processing power. Evidently, the payoff for this increased power is an increase in electrical power requirements, resulting in reduced autonomy. However, this issue lies outside of the processing domain.

Figure 9.1 shows the key elements with requirements in terms of processing power:

– electronic cycle controls, which allow the sensor to acquire images;

– processing the signal recorded by the sensor, transforming it into an image and recording it on the memory card;

– management of signals from a variety of measurement instruments: user settings, photometers, telemeters, battery level, memory state, gyroscope or ground positioning systems (GPS) where applicable; power is also required to alter camera settings, including the diaphragm aperture, focus adjustment and flash;

– communication with the user (program management and menu navigation, image display, memory navigation, etc.) and user interface management (program selection, measurement zone displays, parameters, signal levels, error notifications, etc.).

9.1.2. *The central processing unit*

The CPUs used in modern cameras offer the same functions as general processing units. They are built using classic architectures (for example, TMS320 for the Texas Instruments family, or the ARM-Cortex A5 in the

1 For example, the COACH platform, developed by Zoran and taken up by CSR, is no longer maintained as of early 2014, despite its use in a variety of hardware developed by Kodak, Nikon, Olympus, Pentax, etc.

context of Milbeaut processors), use sophisticated programming languages (for example, a LINUX core in the case of Android platforms) and communication functions using the Internet, BlueTooth or Wifi. While measures are generally taken by manufacturers to prevent users from reprogramming these elements (jailbreaking), a number of open CPUs are available and may be used to host the user's own developments. Moreover, there is extensive evidence of users tampering with camera CPUs, and a number of smaller companies provide users with a means of integrating their own programs into high-end camera hardware.

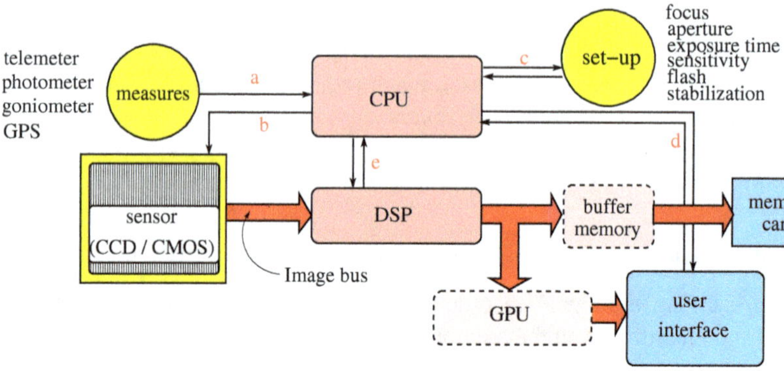

Figure 9.1. *Operating diagram of an image processor. Processing functions are shared between the CPU and the DSP. Signals produced by the sensors (path a) are processed by the CPU and used to adjust optical parameters (path c). The CPU controls sensor command signals (path b) and the parameters of the DSP (path e). It is also responsible for the user interface. The DSP receives the data flux produced by the sensor, formats the data and transmits it to the memory card for storage. It transmits user dialog elements (signal dynamics, saturated zones, etc.) to the CPU, and often relies on a specialized graphics processor (GPU) for display purposes. Image signals travel through a specialized image bus. Certain cameras also include a buffer memory which is quicker to access than the memory card itself*

The CPU is responsible for all high-level camera functions, and is notably responsible for the coordination and sequencing of the functions described above. The CPU also manages the operation of more specialized processors, including the DSP and graphics processing unit (GPU), where applicable.

The CPU takes charge of advanced processing operations, and translates user selections (often expressed in relatively abstract terms: "sport", "backlight", "landscape" , etc.) into precise instructions, tailored to the specific capture conditions: choice of sensitivity, aperture and exposure time, along with the means of measuring focus and aperture, white balance where applicable, and flash use. It also determines focal priorities in cases where sensors produce different results (varying measurement zones in the field, simultaneous use of multiple preferential measurements (e.g. multiple faces), target tracking, etc.). While these decisions are generally made using logical decision trees, a number of other solutions have been recently proposed, using statistical learning techniques alongside databases recorded in the camera itself. This method is notably used in the Canon intelligent scene analysis based on photographic space (ISAPS) system, which controls focus, exposure and white balance on the basis of previously-recorded examples.

The CPU also ensures that other components are configured in accordance with its instructions: focus, diaphragm aperture and capture stabilization.

The CPU is responsible for the creation of metadata files (EXIF file and ICC profile) (see section 8.3.2), archiving the image on the memory card, indexing it to allow users to navigate through previous files, and, where applicable, allowing searches by capture date or theme. Up to a point, it is also involved in managing color printing or transmission to a network, either via a universal serial bus (USB) port or using a wireless communication protocol.

Special effects (black and white or sepia imaging, backlighting, solarization, chromatic effects, etc.) and other non-traditional treatments (panoramas, stereovision and image layering) are also handled by the CPU. We will consider a number of examples of this type of process, with the aim of demonstrating their complexity, something which is rarely apparent to the user:

– Many modern cameras are now able to detect tens of individual faces in a scene, then to select those which seem to be the most important for the purposes of focusing and exposure; these selected faces may also be tracked in burst mode.

– Certain models produced by Canon and intended for family use include a specific setting mode able to detect a child in the center of a scene. This mode includes 17 different configurations, 13 of which are designed for photographing babies; five of these 13 options are intended for sleeping babies, in a variety of lighting conditions. Four configuration options are specifically designed for children in action, and, once again, use different settings depending on the ambient lighting conditions.

– The Fujifilm EXR CRP architecture includes a relatively comprehensive scene identification program, which includes not only classic situations (indoor, outdoor, portrait, landscape, flowers, etc.), but can also detect, recognize and track cats and dogs. This information is located within the proprietary area of the EXIF file (see section 8.3.2) and may be used later in the context of keyword searches.

– In certain configurations, the Olympus TruePic processor applies specific treatments to colors identified as corresponding to human skin and to the sky. These processes operate independently of white balance and are limited to clearly-defined zones in the image, detected and defined prior to processing.

– Canon uses the ISAPS database to offer different strategies for flash management, used to distinguish between situations in which all faces in a scene need to be clearly visible (for example, at a family party) and situations in which unknown faces should be treated as background (for example, at tourist sites). The use of learning databases is becoming increasingly widespread in optimization strategies, for example, in EXPEED 4 in the Nikon D810.

– In portrait mode, the BIONZ processor in Sony α products includes facial processing features which focus precisely on the eyes and mouth of the subject, with reduced focus on skin details; this is intended to replace postprocessing improvement operations generally carried out in a laboratory setting.

9.1.3. *The digital signal processor*

The DSP is an architecture developed specifically for the treatment of image data, using a single instruction multiple data (SIMD)– type pipeline assembly, which repeats the same type of operation multiple times for a very large data flux. It exploits the specific form of calculations involved (using integers, with a reduced number of bits, multiplication by a scalar and addition of successive numbers) to group complex operations into a single operation, applied to long words (128 or 256 bits) in a single clock time (following the VLIW principle: very long instruction word). In cases with higher power requirements, the DSP uses parallelized processing lines, using the multiple instruction multiple data (MIMD) processor architecture, for example using eight parallel treatment paths. The particular structure of images is particularly conducive to this type of parallelization, and processes of this type are easy to transcribe into VLIW instructions.

The DSP collects data output from the photodetector, organizes these data and directs them to various treatment lines. These data can vary considerably depending on the sensor. Note that modern CMOS technology (see Chapter 3) has led to the integration of certain functions in the immediate vicinity of the photosensitive site, including analog/digital conversion, quantization and the removal of certain types of parasitic noise. It has also resulted in the use of specific representation types, in which neighboring pixels are coupled together, either because they share the same electronic component or because they are complementary in terms of representing a particular signal (HDR, stereo or very low-dynamic signals: see Chapter 3). In this case, these processes are applied to data before it reaches the DSP.

The key role of the DSP is to carry out all of the processes required to transform a raw signal into an image for display, generally in JPEG format in sRGB or ICC color mode (see section 5.2.6). Although images are recorded in native (RAW) format, it is essential to give users access to the image in order to monitor the capture process.

A generic example of the series of processes carried out by the DSP is given below:

– The native signal is corrected to compensate for the dark current (dependent on the exposure time and the temperature of the camera). This current is often estimated using dedicated photosites, outside of the useful image field, on the edge of the sensor. The corresponding level is subtracted from all pixels. Bias introduced during the capture process may also be removed in order to exploit the centered character of thermal noise.

– Certain field corrections are applied in order to reduce the effects of noise on very weak signals (spatial means) and to reduce chromatic distortions and vignetting, according to the position of each pixel within the image.

– Initial white balancing may be applied to all signals in order to take account of the spectral response of the sensor and user settings.

– A demosaicing process is then applied in order to create three separate signals, R, G and B.

– Gamma correction may be applied, where necessary, in order to take account of nonlinearity in the sensor.

– Advanced chromatic corrections may also be applied. These may require the use of a perceptual color space, such as Lab and YC_rC_b (see section 5.2.3), and can affect the white balance carried out previously, due to the use of a global optimization strategy (see section 5.3). These corrections are often followed by a process to filter out false colors, which may be due to the residues of chromatic impairments at the edge of the field, in high-contrast images.

– Accentuation algorithms can be used in order to fully utilize the chromatic and dynamic palette, increase contrast and improve sharpness (see section 6.1.2).

Specialized functions may involve operations which are extremely costly in terms of machine time: for example, direct cosine transformation (DCT) is required in order to move from an image signal to a JPEG representation, and for the lossless compression procedures used to limit file volume in RAW format (see section 8.4). Other functions are specifically used for video formats: movement prediction and optical flow (discussed in Chapter 10), audio channel processing, etc.

Modern equipment is able to carry out these functions in a very efficient manner, and bursts of over 10 images per second are now possible (if the memory card permits) for most high-quality cameras of 20 megapixels and above. Note that the use of buffer memory allows us to avoid the bottleneck effect limiting the flow of data into an external memory card. However, this memory, in RAM or SDRAM format, as it supports high flow rates (typically from 500 megabits/second to 1 gigabit/second) has a capacity limited to a few images.

The performance of modern DSPs is not limited by the treatment of fixed images, but rather by video processing, much more costly in terms of processing rates. The reference element used in this context is "4k" video, i.e. 4,000 pixels per line in 16/9 format; this format has twice the resolution, in both lines and columns, than the TVHD standard. Coded using the MPEG 4 standard (ISO H264/AVC), it requires the treatment of 30 images per second using full frame resolution (30 ffps), or 60 images per second in interlaced mode (60 ifps), see section 8.8.2. As of 2015, this still represents a challenge for many manufacturers, who continue to use TVHD format (1,920 pixels per line).

9.1.4. *The graphics processing unit*

In cameras with a graphics processing unit, this element is responsible for displaying images on the camera screen, as well as on the viewfinder, in the case of many hybrid systems where this component is also electronic.

The GPU allows the resolution to be adapted to camera screens, which are generally very small in relation to the size of the image (one megapixel rather than 20). It is responsible for zoom and scrolling functions, and for image rotation (for portrait or landscape visualization). It applies white balancing or special effects in real time, and allows the use of digital zoom to interpolate image pixels between measurement points, for example using bilinear or bicubic interpolation. The GPU also allows the modification of color tables, indicating saturated or blocked zones. It calculates and applies focus histograms and zones, along with all of the graphic or alpha-numeric indications required for user dialog. GPUs are also partly responsible for touch-screen functions in cameras which include this technology.

9.2. Memory

9.2.1. *Volatile memory*

Cameras generally include a volatile memory, allowing temporary storage of images immediately following capture, before archiving in permanent storage memory. Volatile memory is able to store a small number of images in native (RAW) format, and the capacity of this element defines the maximum limit for camera use in burst mode. These memory elements generally use dynamic random access memory (RAM), or DRAM. They are generally synchronized with the image bus (SDRAM = synchronized dynamic random access memory). Input and output, controlled by the bus clock, can be carried out in parallel, serving several DSP processing banks simultaneously in order to speed up operations.

Live memory (which consists of switches, or transistor-capacitance pairs) requires a constant power supply, and data need to be refreshed during the processing period. It is generally programmed to empty spontaneously into the archive memory via the transfer bus as soon as this bus becomes available. The transfer process may also occur when the camera is switched off, as mechanisms are in place to ensure that a power supply will be available, as long as the memory is charged.

Flow rates, both in terms of reading (from the photodetector or the DSP) and writing (to the DSP) can be in excess of 1 gigabyte per second. When writing to a memory card, the flow rate is determined by the reception capacity of the card. For older memory cards, this rate may be up to 100 times lower than that of the SDRAM; in these cases, the buffer action of the SDRAM is particularly valuable.

9.2.2. *Archival memory cards*

The images captured by digital cameras were initially stored on magnetic disks, known as microdrives; nowadays, almost all cameras use removable solid-state memory based on flash technology. Flash memory is used for rewritable recording, and does not require a power supply in order to retain information; it is fast and robust, able to withstand thousands of rewriting cycles and suitable for storing recorded information for a period of several years.

Flash memory is a form of electrically erasable programmable read-only memory (EEPROM).

These elements are made up of floating gate metal oxide semiconductor (MOS) transistors (mainly NAND ports, in the context of photographic applications). The writing process is carried out by hot electron injection: the potential of the control gate is raised to a positive value of 12 V, like the drain, while the source has a potential of 0 V (see Figure 9.2). Electrons are trapped by the floating gate, and remain there until a new voltage is applied. The removal process involves electron emissions using the field effect or Fowler–Nordheim tunnel effect: the control gate is set at 0 V, the source is open and the drain is set at 12 V. Unlike the flash memory used in CPUs, the memory used in removable cards cannot generally be addressed using bits or bytes, but by blocks of bits (often 512 blocks of bytes). The longer the blocks, the higher the potential density of the memory (as each individually addressable block uses its own electronic elements); this makes the writing process quicker for long files, and reduces the cost of the memory.

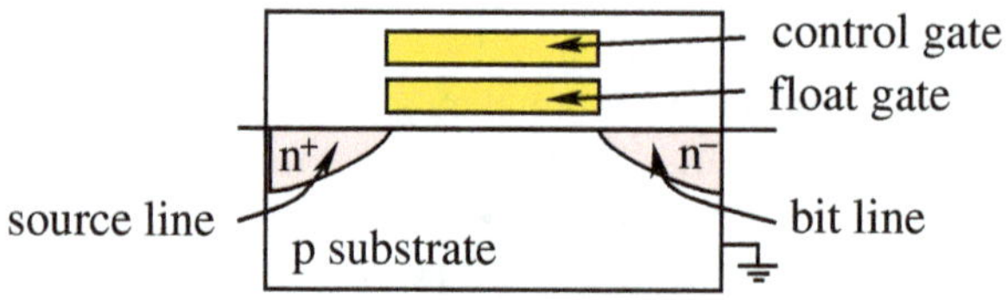

Figure 9.2. *Flash memory: structural diagram of a basic registry*

A large number of different memory formats have been developed, but the number of competing standards has decreased over time, and a smaller number of possibilities are now used. However, each standard may take a variety of forms, both as a result of technological improvements and due to adaptations being made for new markets. Memory technology is currently undergoing rapid change, as in all areas of high-technology information processing. It is essentially controlled by consortia made up of electronic component manufacturers, computer equipment providers and camera manufacturers. The memory used in digital cameras is strongly influenced by product developments in the context of cellular telephony (lighter products, offering considerable flexibility in terms of addressing), and, more recently, in the context of video technology (very high-capacity products with very high flow rates, designed to take full advantage of the formatted structure of data).

Modern cameras often include multiple card slots to allow the use of different memory types and the transfer of images from one card to another, either to facilitate transfer to a computer, or for organizational purposes.

9.2.2.1. *CompactFlash memory*

CompactFlash is one of the oldest archival memory types, created in 1994. While its continued existence has been threatened on a number of occasions by competition from new, better products, CompactFlash has evolved over time and is now one of the most widespread formats used in photography. It is generally referred to by the acronym CF.

CompactFlash units are relatively large (in relation to more recent formats), measuring 42.8 mm × 36.4 mm × either 3.3 or 5 mm, depending on the version, CF-I or CF-II. The shortest side includes 50 very narrow, and therefore fragile, male pins.

The exchange protocol used by CF is compatible with the specifications of the PCMCIA[2] standard, and can be recognized directly using any port of this type, even if the CF card only uses 50 of the 68 pins included in the port. The unit exchanges 16 or 32 bit data with the host port. The exchange interface is of the UDMA[3] type. This interface is characterized by a maximum theoretical flow rate d: UDMA d, where d has a value between 0 and 7 for flow rates between 16.7 and 167 Mb/s. A parallel exchange protocol (PATA) was initially used, and then replaced by a series protocol (SATA[4]). CF is powered in the camera, using either a 3 or 5 V supply. More recent versions are able to withstand hotswapping.

The capacity of CF units is constantly increasing. Initially limited to 100 Gb by the chosen addressing mode, the CF5.0 specification, launched in 2010, increased the addressing mode to 48 bits, giving a theoretical maximum capacity of 144 petabytes. The first 512 Gb CF cards were presented in early 2014, and are still very expensive. These cards can be used to store almost

2 The Personal Computer Memory Card International Association (PCMCIA) standard is a memory standard for personal computers.

3 UDMA = ultra direct memory access.

4 ATA = advanced technology attachment.

10,000 photos, in native format, using the best sensors available. As we might expect, as high-capacity cards become available, the price of less advanced cards has plummeted. Note, however, that the price per Gb of archive space is now determined by writing speed, rather than global capacity.

The transfer speed of CF represented a significant weakness for a long time, with a theoretical limit of 167 Mb/s in CF-II, as we have seen[5]. A new specification, CFast 2.0, was released in late 2012, and allows much higher flow rates (with a current theoretical limit of 600 Mb/s), rendering CF compatible with high-definition video. 160 Mb/s cards are already available, and 200 Mb/s cards, suitable for in-line recording of 4 k video, should be released in the near future.

Cards are identified by their capacity (in Gb) and flow rate (in kb/s), expressed using the formula nX, where X has a value of around 150 kb/s (giving a flow rate of $n150$ kb/s). Thus, a $500X$ card gives a flow rate of 75 Mb/s for reading purposes (this value is generally lower when writing, raising certain doubts as to the performance of the cards).

In spite of their size and the fragility of the pins, CF remains one of the preferred forms of photographic storage, particularly in a professional context, due to its high capacity and the robust nature of the cards, which makes them particularly reliable.

In 2011, the CompactFlash consortium proposed a successor to CF, in the form of the XQD card. These cards claim to offer flow rates of 1 Gb/s and a capacity of 2 teraoctets. Version 2.0 uses the PCI Express 3.0 bus standard, which is the latest form of rapid serial link bus connecting camera hardware and memory cards. A certain number of cards of this type are now available, in 2016, (principally for video use), and some top-end cameras with storage capacity of 128 Gbytes include an XQD host port (physically incompatible with CF).

At one point, microdrives were developed using the physical and electronic format of CF cards, offering higher capacities than those permitted by the CF units available at the time (up to 8 Gb); however, these are now obsolete.

5 This limit of 167 Mb/s remains barely achievable for CF cards, but is lower than we might wish for future applications (particularly video); manufacturers were, therefore, hesitant to include CF ports in their products. This issue has been resolved by the more recent CFast 2.0 specification, which offers some guarantee as to the durability of the standard.

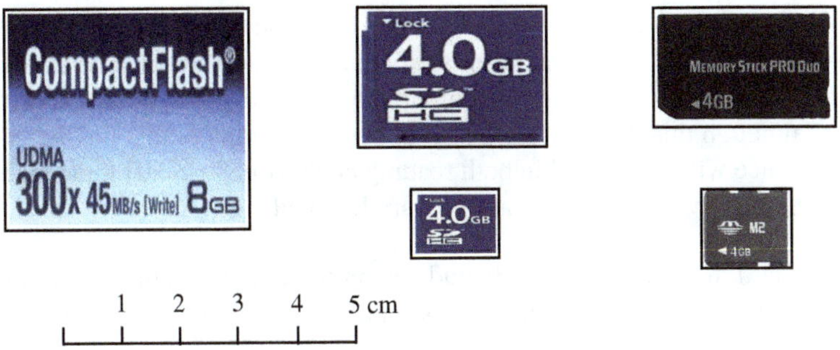

Figure 9.3. *The three most common types of flash memory. Left: CompactFlash. Center: SD, in standard and microformats. Right: Memory Stick, in duo and standard formats*

9.2.2.2. *SD memory*

The first secure digital (SD) memory cards were released in 1999. This standard has taken a number of different forms in the context of photography, both in terms of capacity and unit dimensions. Three sizes of unit are used: "normal", "mini" (extremely rare) and "micro". Capacity is specified in the product name, from "SDSC" (standard capacity, now rare) and "SDHC" (high capacity) up to "SDXC" (extended capacity).

The standard format measures 32 mm × 24 mm × 3.1 mm, and microformat is 15 mm × 11 mm × 1 mm. A passive adapter may be used in order to connect a microcard to a standard port.

SDHC cards, in version 2.0, have a capacity which is limited to 32 Gb and a maximum flow rate of 25 Mb/s. These cards use the FAT32 file system.

SDXC cards have a capacity limited to 2 Tb, and 1 Tb memory cards became available in 2016. The theoretical maximum flow rate for SDXC is 312 Mb/s using a UHS-II (ultra high speed) transfer bus.

Versions 3 and 4 SDXC cards use the exFAT file format, owned by Microsoft; this format is recognized by both PCs and MACs, but is difficult to access using Linux.

As for CF cards, SD memory units are characterized by their capacity and flow rate; however, different labeling conventions are used. Cards belong to classes, C2, C4, ..., C10, where Cn indicates that the reading and writing speed is greater than or equal to $n \times 2$ Mb/s. Above C10, and particularly for video applications, card names are in the format U-n, indicating a minimum flow rate of $n \times 10$ Mb/s in continuous mode. Values of n are indicated on the cards themselves inside a large C or U symbol. An X value is also given on certain cards, although this is less precise as it offers no guarantees that the indicated performance will be reached in both reading and writing[6]. In 2014, the highest commercially available specifications were U-3 and 1,677X.

Standard format SD cards include a physical lock feature, preventing writing and deletion of files (but not copying). Logical commands can be used for permanent read-only conversion.

SD cards can also be password protected, and data can only be accessed (read or written) once the password has been verified. A significant part of the space on these cards is used to trace and manage reading and writing rights (digital right management (DRM)). The content protection for recordable media (CPRM) property management program uses C2 (Cryptomeria Cypher) coding. These two facilities are not often used by cameras. They become more interesting in the context of multimedia applications where cards are inserted into cell phones.

9.2.2.3. *The Sony Memory stick*

The Memory Stick (MS) is a proprietary format, launched in 1998. It takes a variety of forms, both in terms of physical appearance and archival performance. The Memory Stick Micro is used in the context of digital photography and is included in many compact cameras; the Memory Stick Pro HG (MMPro-HG) is used for video recording purposes.

In its current incarnation, the Memory Stick Micro (M2) measures 15 mm × 12 mm × 1.2 mm, and has a theoretical maximum capacity of 32 Gb. 32 Gb versions are currently available. Extensions (XC-Micro and HG-Micro) have increased this limit to 2 Tb by changing the maximum address length. The maximum flow rate, initially set at 20 Mb/s, has been increased to 60 Mb/s by the use of an 8-bit parallel port.

6 Conversion between the two types of flow rate is difficult. A single card may give three different values, creating confusion as to its actual performance: for example, U-3, 1,677X and 280 Mb/s. U-3 indicates a video flow rate of 30 Mb/s in continuous mode, while 1,677X denotes a maximum image transfer capacity in burst mode of 250 Mb/s. 280 Mb/s (1,867X) signifies the maximum transfer rate to an external port when the card is emptied.

As for SD memory, recent Memory Sticks use an exFAT file system, replacing the original FAT32 format. The Memory Stick also offers property management functions via Sony's proprietary Magic Gate data encryption program.

The Memory Stick PRO-HG DUO HX was the fastest removable unit on the market when it was released in 2011. Measuring 20 mm × 31 mm × 1.6 mm, it offers an input/output capacity of 50 Mb/s. However, no more advanced models have been produced since this date.

9.2.2.4. *Other formats*

A wide variety of other formats exist, some very recent, others on the point of obsolescence. Certain examples will be given below.

XD-Picture memory was developed in 2002 by Olympus and Fujifilm, but has not been used in new hardware since 2011, and should be considered obsolete. Slower than the competition, XD-Picture is also less open, as the description is not publicly available; the standard never achieved any success in the camera memory market.

SxS memory, developed by Sony, was designed for video recording, particularly in the context of professional applications, with a focus on total capacity (1 h of video) and the flow rate (limited to 100 Mb/s for the moment). Panasonic's P2 memory was developed for the same market, but using an SD-compatible housing and offering slightly weaker performance.

9.3. Screens

9.3.1. *Two screen types*

All modern digital cameras include at least one display screen, and sometimes two.

The largest screens cover most of the back of the camera unit, and is intended for viewing at distances from a few centimeters to a few tens of centimeters. These screens are used for the user interface (menus and presentation of recorded images) and for viewfinding purposes; they also allow visualization of images and are used to display parameters for calibration. In most of the cases, this screen constitutes the principal means of interacting with the user.

A second screen may be used to replace the matte focusing unit in reflex cameras, allowing precise targeting with the naked eye. Certain setting elements also use this screen, allowing parameters to be adjusted without needing to move. This digital eyepiece represents an ideal tool for image construction in the same way as its optical predecessor.

Large screens were initially used in compact cameras and cell phones for display and viewfinding purposes, but, until recently, only served as an interface element in reflex cameras.

In compact cameras and telephones, these screens generally use a separate optical pathway and sensor to those used to form the recorded image, offering no guarantee that the observed and recorded images will be identical. Observation conditions may not be ideal (particularly in full sunlight), and different gains may be applied to these images, resulting in significant differences in comparison to the recorded image, both in terms of framing and photometry. However, the ease of use of these devices, particularly for hand-held photography, has made them extremely popular; for this reason, reflex cameras, which do not technically require screens of this type, now include these elements. They offer photographers the ability to take photographs from previously inaccessible positions – over the heads of a crowd, or at ground level, for example – and multiple captures when using a tripod in studio conditions.

The introduction of the additional functions offered by these larger screens in reflex cameras (in addition to the initial interface function) was relatively difficult, as the optical pathway of a reflex camera uses a field mirror which blocks the path to the sensor, and modifications were required in order to enable live view mode. Moreover, the focus and light measurement mechanisms also needed to be modified to allow this type of operation. Despite these difficulties, almost all cameras, even high-end models, now include screens of this type, which the photographer may or may not decide to use. The quality of these screens continues to progress.

Viewfinder screens present a different set of issues. These elements are generally not included in compact or low-end cameras, and are still (in 2015) not widely used in reflex cameras, where traditional optical viewfinders, using prisms, are often preferred. They are, however, widespread in the hybrid camera market, popular with experienced amateur photographers. These cameras are lighter and often smaller, due to the absence of the mirror and prism elements; the digital screen is located behind the viewfinder. The use of these screens also allows simplifications in terms of the electronics required to display setting parameters, as one of the optical pathways is removed. Finally, they offer a number of new and attractive possibilities: visual control of white balance, saturation or subexposure of the sensor, color-by-color exposure of focusing, fine focusing using a ring in manual mode, etc. Electronic displays offer a constant and reliable level of visual comfort, whatever the lighting level of the scene; while certain issues still need to be resolved, these elements are likely to be universally adopted by camera manufacturers as display technology progresses.

9.3.2. *Performance*

Digital screen technology is currently progressing rapidly from a technical standpoint [CHE 12]; it is, therefore, pointless to discuss specific performance details, as these will rapidly cease to be relevant.

Whatever the technology used, camera screens now cover most of the back of the host unit. The size of these screens (expressed via the diagonal, in inches) has progressively increased from 2 to 5 inches; some more recent models even include 7-inch screens. Screen resolution is an important criterion, and is expressed in megapixels (between 1 and 3; note that the image recorded by the sensor must, therefore, be resampled[7]). The energy emission level is also an important factor to allow visualization in difficult conditions (i.e. in full sunlight or under a projector lamp). These levels are generally in the hundreds of candelas per square meter. Screen performance can be improved using an antireflection coating, but this leads to a reduction in the possible angle of observation. Another important factor is contrast, expressed as the ratio between the luminous flux from a pure white image (255,255,255) and that from a black image (0,0,0). This quantity varies between 100:1 and 5000:1, and clearly expresses differences in the quality of the selected screen. The image refreshment rate is generally compatible with video frame rates, at 30 or 60 images per second. Movement, therefore, appears fluid in live view mode, free from jerking effects, except in the case of rapid movement, where trails may appear.

Note that these screens often include touch-screen capacities, allowing users to select functions from a menu or to select zones for focusing.

Viewfinder screens are subject to different constraints. The size of the screen is reduced to around 1 cm (1/2 inch, in screen measurement terms),

7 A variable step resampling process is needed in this case, as the display screen generally offers a zoom function.

which creates integration difficulties relating to the provision of sufficient resolution (still of the order of 1 megapixel). However, the energy, reflection and observation angle constraints are much simpler in this case.

9.3.3. *Choice of technology*

The small screens used in camera units belong to one of the two large families, using either LCD (liquid crystal) or LED/OLED (light-emitting diode) technology [CHE 12]. Plasma screens, another widespread form of display technology, are not used for small displays, and quantum dot displays (QDDs), used in certain graphic tablets, have yet to be integrated into cameras.

9.3.3.1. *Liquid crystal displays*

LCD technology has long been used in screen construction [KRA 12]. Initially developed for black-and-white displays in computers and telephones, color LCD screens were first used in compact cameras in the early 1990s. LCD technology was used exclusively for camera screens until the emergence of OLEDs in the 2010s.

Liquid crystal displays filter light polarization using a birefringent medium, controlled by an electrical command mechanism. The element is passive and must be used with a light source, modulated either in transmission or reflection mode. The first mode is generally used in camera hardware, with a few individual LEDs or a small LED matrix serving as a light source.

The birefringent medium is made up of liquid crystals in the nematic phase (i.e. in a situation between solid and liquid states), hence the name of the technology. Liquid crystals are highly anisotropic structures, in both geometric and electrical terms. When subjected to an electrical field, these crystals take on specific directions in this field, creating birefringence effects (creation of an extraordinary ray), which leads to variations in polarization (see section 9.7.4.1). An LCD cell is made up of a layer of liquid crystals positioned between two transparent electrodes. At least one of these electrodes must have a matrix structure, allowing access to zones corresponding to individual pixels. In more recent configurations, one of the electrodes is replaced by a thin film transistor (TFT) matrix. Two polarizers, generally crossed, ensure that the image will be completely black (or white, depending on the selected element) in the absence of a signal. Applying a voltage to a pixel causes the wave to rotate. The amplitude of the voltage determines the scale of the rotation and the transmitted fraction, rendering the pixel more or less luminous.

Color images are created by stacking three cells per pixel, and applying a chromatic mask to the inside of the “sandwich”. This mask is often of the “strip” type (see Figure 5.17, right), but staggered profiles may also be used. The R, G and B masks are generally separated by black zones in order to avoid color bleeding; these zones are particularly characteristic of LCD screens.

The earliest LCD screens used twisted nematic (TN) technology, which is simple, but relatively inefficient, and does not offer particularly good performances (low contrast and number of colors limited to 3×64). Its replacement by TFT technology and intensive use of command transistors have resulted in a significant increase in modulation quality and display rates.

Recent work has focused on higher levels of cell integration (several million pixels on a 5-inch screen, for example), improved refreshment rates and on reducing one of the most noticeable issues with LCD technology, the inability of crystals to completely block light. This results in impaired black levels and a reduction in contrast. Vertical alignment techniques (VA: whether MVA, multi-domain vertical alignment, or PVA, patterned vertical alignment) improve the light blocking capacities of nematic molecules, producing a noticeable reduction in this effect. Blocking levels of up to 0.15 cd/m^2 have been recorded, sufficient to allow contrast of around 1,000:1; unfortunately, these levels are not generally achieved at the time of writing.

9.3.3.2. *Light-emitting diode matrices: OLED*

Nowadays, LCDs face stiff competition from organic light-emitting diodes, OLEDs [MA 12a, TEM 14], used in viewfinders and screens in the form of active matrices (AMOLED). The first cameras and telephones using OLED screens appeared between 2008 and 2010, and demand for these products continues to increase, in spite of the increased cost.

OLEDs are organic semi-conductors, placed in layers between electrodes, with one emitting and one conducting layer. As the material is electroluminescent, no lighting is required. A voltage is applied to the anode, in order to create an electron current from the cathode to the anode. Electrostatic forces result in the assembly of electrons and holes, which then recombine, emitting a photon. This is then observed on the screen. The wavelength emitted is a function of the gap between the molecular orbitals of the conducting layer and the emitting layer; the materials used in constructing these elements are chosen so that this wavelength will fall within the visible spectrum. Organometallic chelate compounds (such as Alq_3, with formula $Al(C_9H_6NO)_3$, which emits greens particularly well) or fluorescent colorings (which allow us to control the wavelength) are, therefore, used in combination with materials responsible for hole transportation, such as tri-phenylamine. Doping materials are placed between the metallic cathode (aluminum or calcium) and the electroluminescent layer in order to increase the gain of the process.

The anode is constructed using a material which facilitates hole injection, while remaining highly transparent (such as indium tin oxide, ITO, as used in TFT-LCDs and photodetectors). This layer constitutes the entry window for the cell [MA 12b].

Two competing technologies may be used for the three RGB channels[8]. In the first case, single-layer cells are stacked, as in color photodetectors. In the second case, three layers are stacked within the thickness of the matrix, using an additive synthesis process. Note that in the first case, a Bayer matrix is not strictly necessary, and different pixel distributions may be chosen, for example leaving more space for blue pixels, which have a shorter lifespan (an increase in number results in a decreased workload for each pixel).

OLEDs present very good optical characteristics, with a high blocking rate (the theoretical rate is extremely high, $\geq$ 1,000,000:1; in practice, in cameras, a rate of between 3000:1 and 5000:1 is generally obtained), resulting in deep black levels and good chromatic rendering, covering the whole sRGB diagram (see Figure 5.8). They have high emissivity, making the screens easy to see even in full sunlight (almost 1,000 cd/m^2). OLEDs also have a very short response time ($\leq$ 1 ms), compatible with the refreshment rate; they also have the potential to be integrated with emission zones reduced to single molecules, and can be used with a wide variety of supports (notably flexible materials).

Unfortunately, the life expectancy of OLEDs is still too low (a few thousand hours in the case of blue wavelengths), and they require large amounts of power in order to display light-colored images.

Originally limited to viewfinders, OLEDS are now beginning to be used in display screens, notably due to the increase in image quality.

8 The discovery of blue-emitting LEDs, in 1990, represented a major step forward in the lighting domain, and earned Akasaki, Hamano and Nakamura the 2014 Nobel Prize for physics.

9.4. The shutter

9.4.1. *Mechanical shutters*

Shutters were originally mechanical, later becoming electromechanical (i.e. the mechanical shutter was controlled and synchronized electronically). The shutter is located either in the center of the lens assembly, or just in front of the image plane (in this case, it is known as a focal plane shutter). The shutter is used to control exposure time, with durations ranging from several seconds or even hours down to $1/10{,}000^{th}$ s, in some cases.

A focal plane shutter is made up of two mobile plates, which move separately (in opposite directions) in the case of long exposure times, or simultaneously for very short exposure times. Shutters located within the lens assembly are made up of plates which open and close radially, in the same way as diaphragms.

These mechanical diaphragms have different effects on an image. Focal-plane shutters expose the top and bottom of an image at different times, potentially creating a distortion if the scene includes rapid movement. A central shutter, placed over the diaphragm, may modify the pulse response (*bokeh* effect, see section 2.1) and cause additional vignetting (see section 2.8.4).

Mechanical shutters are still commonly used in digital cameras. The earliest digital sensors (particularly CCDs) collected photons in photosites during the whole of the exposure period, and this period, therefore, needed to be limited to the precise instant of image capture in order to avoid sensor dazzling.

9.4.2. *Electronic shutters*

As we have seen, the newer CMOS-based sensors use a different process, as each photosite is subject to a reset operation before image capture (see section 3.2.3.3). This allows the use of a purely electronic shutter function. Electronic control is particularly useful for very short exposure times, and is a requirement for certain effects which can only be obtained in digital mode:

– burst effects (more than 10 images per second);

– bracketing effects, i.e. taking several images in succession using slightly different parameters: aperture, exposure time, focus, etc.;

– flash effects, particularly when a very short flash is used with a longer exposure time. In this case, we obtain a front-curtain effect if the flash is triggered at the start of the exposure period (giving the impression of acceleration for mobile subjects), or a rear-curtain effect if the flash is triggered at the end of the exposure period.

The use of an electronic shutter which are included in a number of cameras, is also essential for video functions, replacing the rotating shutters traditionally used in video cameras. Electronic shutters also have the potential to be used in solving a number of difficult problems, for example in the acquisition of high-dynamic images (HDR, see section 10.3.1) or even more advanced operations, such as flutter-shutter capture (see section 10.3.4).

9.4.2.1. *Rolling shutters*

As we have seen, the pixel reset function dumps the charges present in the drain, and occurs in parallel with a differential measurement (correlated double sampling) controlled by the clock distributed across all sites. This electronic element behaves in the same way as a shutter, and may be sufficient in itself, without recourse to a mechanical shutter. However, charges are dumped in succession (line-by-line), and, as they are not stored on-site, the precise instant of measurement is dependent on position within the image. In this case, the shutter, therefore, operates line-by-line, and is known as a rolling shutter, or RS.

Mechanical shutters are no longer used in most compact cameras or telephones, and exposure is now controlled electronically.

However, rolling shutters present certain drawbacks for very short capture periods. As the instant of exposure is slightly different from one line to another, images of rapid movement are subject to distortion. Successive lines are measured with a delay of between a few and a few tens of microseconds; cumulated from top to bottom of an image, this leads to a difference of several milliseconds, which is non-negligible when photographing rapidly-moving phenomena. The resulting effect is similar to that obtained using a mechanical shutter.

Top-of-the-range cameras currently still include a mechanical shutter in order to preserve compatibility with older lens assemblies. A central mechanical shutter is particularly valuable in reducing rolling effects, as the photons recorded by each cell all come from the same exposure; electronic control is, therefore, used in addition to an electromechanical shutter mechanism.

9.4.2.2. *Global shutter*

Global shutters (GS) require the use of a memory at pixel level; in the simplest assemblies, this prevents the use of correlated double sampling [MEY 15]. However, this technique is useful in guaranteeing measurement quality. Relatively complex designs have, therefore, been proposed, involving the use of eight transistors per pixel (8T-CMOS), some of which may be shared (Figure 9.4). Noise models (see equation [7.10]) also need to be modified to take account of the new components (eight transistors and two capacitors: see Figure 9.4). Finally, backlit assemblies present a further difficulty for these types of architecture. Given the complexity, this type of assembly justifies the use of stacked technology (see section 3.2.4).

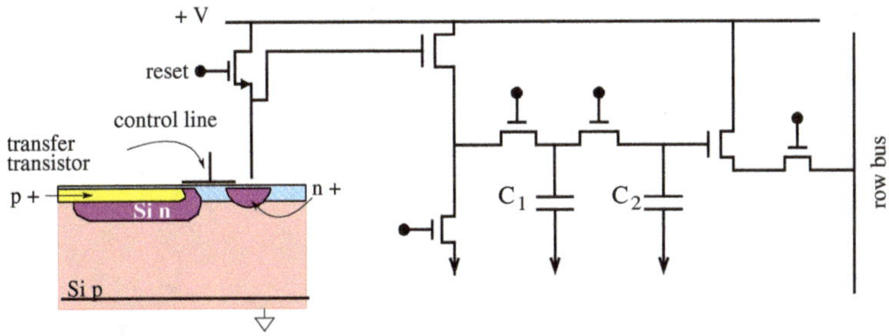

Figure 9.4. *Command electronics for a CMOS photodetector pixel used for global shuttering [MEY 15]. The two additional capacitances C_1 and C_2 are used to maintain both the measured signal and the level of recharge needed for correlated double sampling (CDS) measurement. This design, using eight transistors and two capacitances (denoted by 8T), is similar to the equivalent rolling shutter assembly (4T) shown in Figure 3.4, right*

9.5. Measuring focus

For many years, focus was determined by physical measurement of the space between the object and the camera by the photographer. This distance was then applied to the lens assembly to calibrate capture, using graduations on the objective ring (the higher objective expressed in meters or feet, as shown in the photograph in Figure 2.7). The photographer then examined the image using a matte screen, compensating for errors as required.

Later, analog cameras were fitted with rangefinders, giving users access to more precise tools than those offered by simple visualization on a matte screen. These systems, known as stigmometers or, more commonly, split image focusing mechanisms, operated on the principle of matching images obtained using different optical pathways. Using this technique, two split images were presented side-by-side in the viewfinder field, with a lateral shift determined by the focal impairment. The optical pathway of the viewfinder isolates two small portions of the field, preferably taken from the areas furthest from the optical axis. Straight prisms are used in a top-to-bottom assembly so that the images provided are stacked with the main image, provided by the rest of the field, in the viewfinder field. This gives three images, which match perfectly if the image is in focus, but will be separated if the object is out of focus. The focal impairment may also be amplified using microprisms. The photographer then modifies the flange (lens-sensor distance) until these three images combine to form a single, well-defined image (see Figure 9.5).

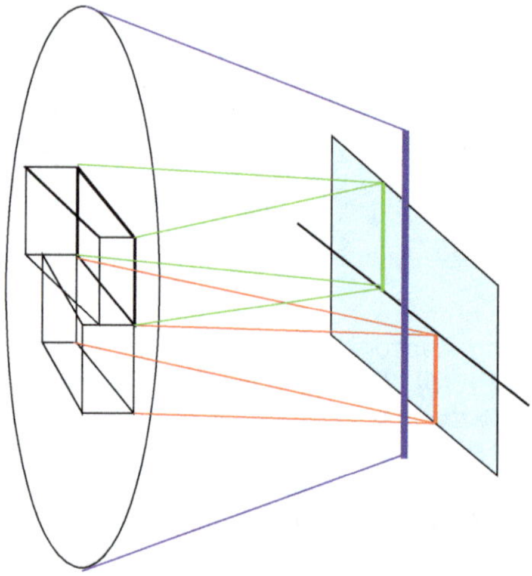

Figure 9.5. *Principle of split-image focusing (stigmometer) as used in analog film cameras. The two right prisms are placed top to bottom, and use a very small part of the viewfinder field (highly magnified in the diagram). They produce two images, which take different positions in the image produced by the rest of the optical pathway in the case of focal errors*

In other cases, the distance is determined by measuring the travel time of an acoustic or an emitted and modulated infrared wave; the echo from the object is then detected and the angle is measured in order to determine the position of the target by triangulation. These two techniques are known as active telemetry, as a wave must be emitted. They are particularly useful in systems operating with very low lighting levels, but are less precise than split image focusing techniques. Using these systems, the value determined by the telemeter is often transmitted to a motor, which moves the lens assembly to the recommended distance. These systems were first used in film photography, and proved to be particularly useful in automatic capture systems (triggered by a particular event in a scene, whether in the context of surveillance, sports or nature photography).

With the development of solid-state sensors, telemeters became more or less universally used, and were subject to considerable improvements in terms of performance, to the point where user input was rarely required. In these cases, an electric motor is used in association with the lens assembly, and coupled with the telemeter. The focal technique is based on two different principles: first, attempting to obtain maximum contrast, and second (and most commonly), phase detection. However, ultrasound or infrared systems have also been used, often in a supporting role in addition to the main telemeter, to provide assistance in difficult situations (such as low lighting levels and reflecting or plain surfaces).

9.5.1. *Maximum contrast detection*

The basic idea behind this technique is to find the focal settings which give the highest level of detail in a portion of the image. Contrast telemetry, therefore, uses a small part of an image (chosen by the user, or fixed in the center of the field) and measures a function reflecting this quality. A second measurement is then carried out with a different focal distance. By comparing these measures, we see whether the focus should be adjusted further in this direction, or in the opposite direction. An optimum for the chosen function is obtained after a certain number of iterations. Three different functions may be used, which are all increasing functions of the level of detail in the image:

– dynamics;

– contrast;

– variance.

Let $i(x, y)$ be the clear image, presumed to be taken from a scene entirely located at a single distance p from the camera (see Figure 9.6).

When the image is not in focus, if the difference in relation to the precise focal plane is δ, the focal error is expressed in the approximation of the geometric optic by a pulsed response: circle$(2\rho/\varepsilon)$ and the blurred image is given by:

$$i'(x,y) = i(x,y) * \text{circle}(2\rho/\varepsilon) \qquad [9.1]$$

where ε is obtained using formula [2.1].

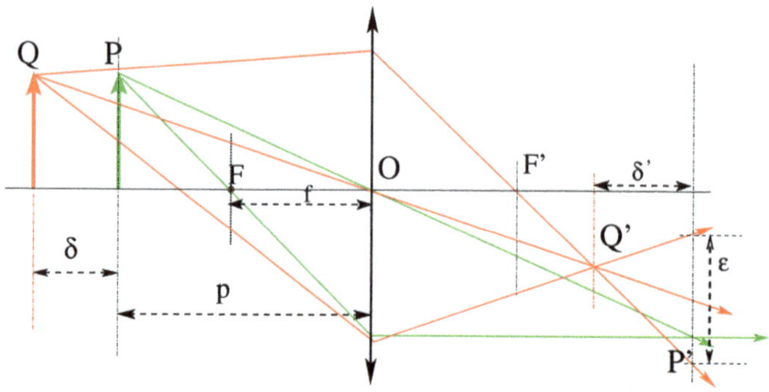

Figure 9.6. *Construction of a blurred image: point P is in the focal plane, and focuses on point P'. If we move in direction Q, by δ, the image is formed at Q', at a distance $\delta' = \Gamma\delta$ (where Γ is the longitudinal magnification, see equation [1.3]). The blur of Q' in the image plane of P' will be of diameter $\varepsilon = D\delta'/p = \delta' f/Np$, where N is the aperture number of the lens assembly*

The study domain for variations in i may be, for example, a line of 64 or 128 sensors, with a size of 1 pixel.

We immediately see that dynamic $\Delta(i')$ must be lower than $\Delta(i)$, as $i'_{max} \leq i_{max}$ and $i'_{min} \geq i_{min}$, due to the use of a low-pass filter which reduces the maximum and increases the minimum. Moreover, if $d' < d$, then $\Delta(i') < \Delta(i)$, and the process will lead to correct focus, whatever the focal error, as $\Delta(i)$ is a decreasing function of d.

Contrast measurement is rather more complex. Contrast evolves in the following way:

$$C(i) = \frac{i_{max} - i_{min}}{i_{max} + i_{min}} \qquad [9.2]$$

When we change the focus, the maximum luminosity is reduced by η and the minimum is increased by η' (where η and η' are positive): $i'_{max} = i_{max} - \eta$, $i'_{min} = i_{min} + \eta$. The contrast becomes:

$$C'(i) = \frac{i_{max} - \eta - i_{min} - \eta'}{i_{max} - \eta + i_{min} + \eta'} \qquad [9.3]$$

and:

$$C(i) - C'(i) = \frac{2\eta i_{min} + 2\eta' i_{max}}{(i_{max} - \eta + i_{min} + \eta')(i_{max} + i_{min})} \qquad [9.4]$$

This is always positive, leading to the same conclusion obtained using dynamics; however, the command law obtained in this way is independent of the mean level of the image, something which does not occur when using dynamics.

For a variance measurement, $v(i) = <(i- <i>)^2> = <i^2> - <i>^2$. The mean value $<i>$ is hardly affected by the focal error due to the conservation of energy, whether the image is clear or blurred: $<i>^2 \sim <i'>^2$. However, $<i^2>$ is affected by the focal error. Using the Fourier space, let I be the Fourier transform of $i(x, y)$ $(I = \mathcal{F}(i))$. Parseval's theorem tells us that $<i^2> = <I^2>$. In this case, from equation [9.1], we have:

$$I' = I.(J_1(\rho)/\rho)^2 \qquad [9.5]$$

with J_1 the Bessel function of the first kind (equation [2.42]), and

$$<I'>^2 \;\leq\; <I>^2 <(J_1(\rho)/\rho)^4> \;<\; <I>^2 \qquad [9.6]$$

Once again, the quantity decreases, and the process will finish at the point of optimum focus. The calculations involved when using variance are more substantial, but a greater number of measurements will be used before a decision is reached; this results in a better rendering of textures.

Using the maximum contrast method, two measurements are required in a reference point, to which the second measurement is compared in order to determine whether or not we have moved closer to the optimum. From thesc two values, we are able to determine the strategy to follow: continue moving in the same direction, or reverse the direction of movement.

9.5.2. *Phase detection*

Phase detection focus operates in a very similar way to split image focusing, comparing pairs of small images obtained using different optical pathways. However, the obtained images are not presented using a prism image, but rather using a beam splitter, extracting two smaller beams from the main beam and directing them toward two dedicated telemetry sensors for comparison (see Figure 9.7).

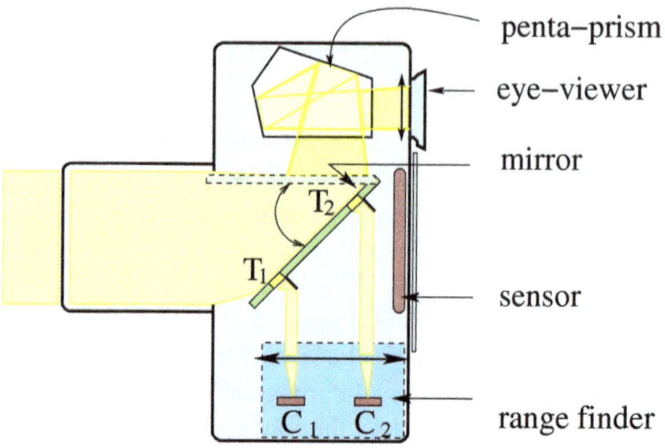

Figure 9.7. *Simplified diagram of telemetric measurement in a digital reflex camera by phase detection. The semi-reflective zones T_1 and T_2 of the mirror direct light back toward photodetector bars, where signals from the edges of the target field are then analyzed*

Phase matching occurs when the two images coincide exactly for a correctly calibrated position. The respective positions of the maximums on both sensors are used to deduce the sign and amplitude of the movement to apply, combined to give a single measurement. These systems, therefore, have the potential for faster operation, and are particularly useful when focusing on a moving object. However, phase detection is less precise than contrast detection, particularly for low apertures, as the angular difference between the beams is not sufficient for there to be a significant variation, notably as a function of the focal distance.

9.5.3. *Focusing on multiple targets*

Focusing is relatively easy when the target is a single object. However, different issues arise when several objects of interest are present in the same scene.

9.5.3.1. *Simultaneous focus on two objects*

The most conventional case of multiple focus involves focusing on two main objects. From Figure 9.6 and equations [2.8]. We show that if ε is very low, then the focal distance required for optimal clarity of two points situated at distances q_1 and q_2 from the photographer is the geometric mean of the distances:

$$p^2 \approx q_1 q_2 \qquad [9.7]$$

and not the arithmetic mean, $(q_1 + q_2)/2$, as we might think.

9.5.3.2. *Average focus*

Modern cameras include complex telemetry systems which determine the distance q_i of a significant number of points in a field, typically from 5 to 20 (see Figure 9.8). The conversion of these measurements, $\{q_i,\ i = 1, \ldots, n\}$, into a single focal recommendation generally depends on the manufacturer, and targets are weighted differently depending on their position in the field.

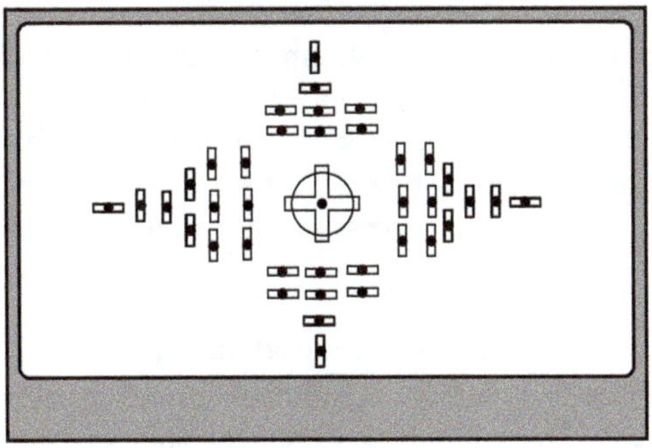

Figure 9.8. *Focal measurement zones, as they appear to the user in the viewfinder. By default, measurement is carried out along the two central perpendicular lines. Users may displace the active zone along any of the available horizontal or vertical lines, in order to focus on a site away from the center of the lens. In the case of automatic measurement across a wide field, the measurement taken from the center may be corrected to take account of the presence of contrasting targets in lateral zones*

There are a number of possible solutions to this problem.

One option is to select the distance p which minimizes the greatest blur from all of the targets. In this case, the calculation carried out in section 9.5.3.1 is applied to the nearest and furthest targets in the scene. Hence:

$$p = \sqrt{\max_i(q_i) . \min_i(q_i)} \qquad [9.8]$$

Another solution is to minimize the variance of the blur, so that we effectively take account of all measurements and not only the two extreme values. Using equation [2.8], the blur ε_i of a point q_i can be expressed when focusing in p:

$$\varepsilon_i = \frac{2f^2(q_i - p)}{Np^2(q_i - f)}$$

but the mean square error: $E = 1/n \sum_i \varepsilon_i^2$ does not allow precise expression of the value p which minimizes this error. In these conditions, we may use iterative resolutions or empirical formulas, for instance only taking account of objects in front of the central target, and only if they are close to the optical center.

9.5.4. *Telemeter configuration and geometry*

For both types of distance measurement, the simplest sensors are small unidimensional bars of a few tens of pixels. In both modes, measurement is, therefore, dependent on there being sufficient contrast in the direction of the sensor. Contrast due to vertical lines is statistically more common in images, so a horizontal bar will be used (in cases with a single bar), placed in the center of the field.

Performance may be improved by the addition of other sensors, parallel or perpendicular to the first sensor, or even of cross-shaped sensors. Modern systems now use several tens of measurement points (see Figure 9.8). This raises questions considering the management of potentially different responses. Most systems prioritize the central sensor on principle, with other sensors used to accelerate the focusing process in cases of low contrast around the first sensor. Other strategies may also be used, as discussed in section 9.5.3: use of a weighted mean to take account of the position of objects around the central point, majority detection, compromise between extrema, hyperfocal detection, etc.

Most systems which go beyond the basic level allow users to select from a range of different options: concentrating on a specific object in the center of an image, a large area or the scene as a whole. Advanced systems allow users to select the exact position of the focal zone.

Phase-matching assemblies are well suited to use in reflex assemblies, as the beams can be derived using the field mirror (see Figure 9.7). They are harder to integrate into hybrid or compact cameras, which, therefore, mostly use contrast measurement. Nevertheless, if focusing is carried out while the field mirror is in place, it becomes difficult to refocus in the context of burst photography of a mobile object.

If contrast measurement is carried out without beam splitting, pixels in the image plane need to be dedicated to this distance measurement. This means that the image for these pixels needs to be reconstructed using neighboring pixels (this does not have a noticeable negative effect on the image, as it only involves a few tens of pixels, from a total of several million). In the most up-to-date sensors, pixels are able to fulfill two functions successively, i.e. telemetry and imaging, and to switch between the two at a frequency of around 100 hertz. This idea has been generalized in certain sensors, where all pixels (2 by 2 or 4 by 4) are used in phase or contrast measurement.

9.5.5. *Mechanics of the autofocus system*

Focus commands are transmitted to the lens assembly. To ensure compatibility with manual adjustment, it acts via the intermediary of a helicoidal guide ring, used for both rotation and translation. Generally, all of the lenses in the assembly are shifted in a single movement. In more complex assemblies, it may act separately on the central system configuration, transmitting different movement instructions to subsets of lenses, divided into groups. In very large lens assemblies, the camera itself moves (as it is much lighter than the lens assembly), and the movement is, therefore, purely linear.

Movement is carried out by micromotors placed around the lenses. These motors constitute a specific form of piezoelectric motor, known as ultrasonic motors. Current models use lead zirconate titanate (PZT), a perovskite ceramic, to create progressive acoustic surface waves via a piezoelectric effect, leading to controlled dilation of very fine fins at a frequency of a few megahertzs. These fins, which are active in alternation, "catch" and move the rotor associated with the assembly, as shown in Figure 9.9. Note that these motors present a number of advantages: they allow very long driving periods, are bidirectional and do not create vibration, as the fins move by less than 1 μm at a time.

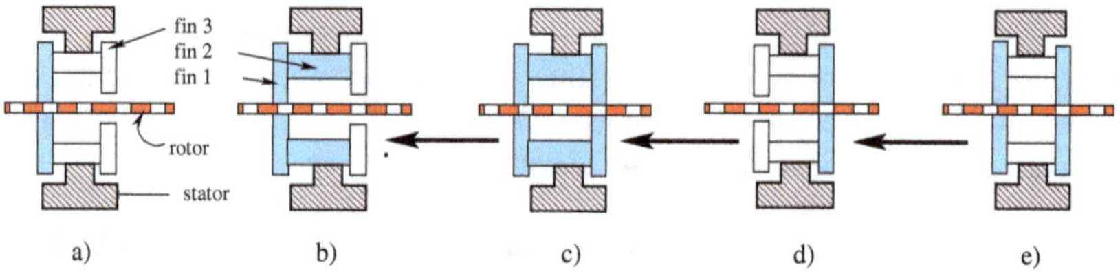

Figure 9.9. *Operation of an ultrasonic autofocus motor. The terms "rotor" and "stator" are used by analogy with electric motors, despite the very different operating mode of acoustic wave motors. The movable ring (rotor) is the central structure. The stator is made up of three piezoelectric elements, which are either passive (shown in white) or active (dark) following a regular cycle, allowing the rotor to be gripped and moved in the course of a cycle. These elements dilate when active, producing gripping (stators 1 and 3) and movement (stator 2) effects. Carried by a progressive acoustic wave, the cycle repeats at very high rates without requiring any form of movement elsewhere than in the rotor*

9.5.6. *Autofocus in practice*

Considerable progress has been made in the domain of focus systems in recent years, both for contrast and phase detection methods. With the assistance of ultra-fast motors, very precise focus can now be achieved in very short periods, i.e. a fraction of a second. This progress is due to the development of focusing strategies and the multiplication of measurement points, helping to remove ambiguity in difficult cases; further advances have been made due to improvements in lens assemblies, which are now better suited to this type of automation (with fewer lenses and lighter lens weights).

Recent developments have included the possibility of connecting autofocus to a mobile object, i.e. tracking a target over time, and the possibility of triggering capture when an object reaches a predefined position.

However, certain situations are problematic for autofocus:

– highly uniform zones: cloudless sky, water surfaces, bare walls, fog, etc.;

– zones with moving reflections (waves, mirrors, etc.);

– zones with close repeating structures;

– zones of very fast movement (humming bird or insect wings, turning machines, etc.);

– very dark zones.

In practice, a situation known as "pumping" is relatively widespread in autofocus systems. This occurs when the focus switches continually between two relatively distant positions without finding an optimal position.

Note that while camera manufacturers aim to create systems with the best possible focus and the greatest depth of field, photographers may have different requirements. Excessively sharp focus in a portrait or on a flower, for example, may highlight details which the photographer may wish to conceal. Similarly, excessive depth of field can impair artistic composition, highlighting secondary elements that we would prefer to leave in the background. Photographers, therefore, need to modify the optical aperture or make use of movement in order to focus the observer's attention on the desired elements (see Figure 9.10, left).

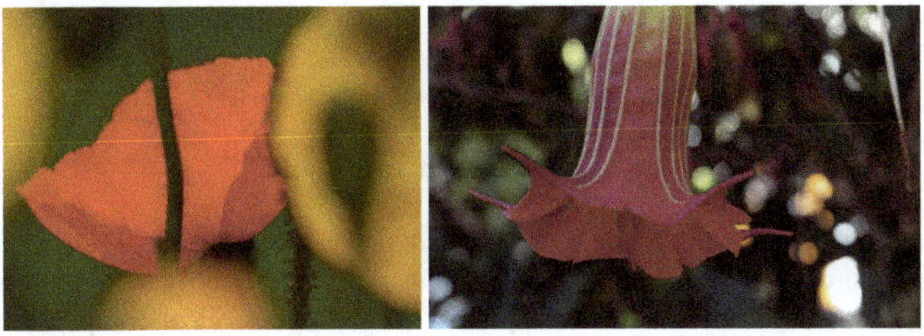

Figure 9.10. *Blurring is an important aspect of the aesthetic properties of a photograph. Left: spatial discrimination of objects exclusively due to focus. The choice of a suitable diaphragm is essential in softening contours. Right: the lights in the background are modulated using the optical aperture, giving a classic bokeh effect. For a color version of this figure, see www.iste.co.uk/maitre/pixel.zip*

As the quality of blurring is important, photographers use specific forms of diaphragm to produce harmonious graduation. In this case, diaphragms with multiple plates, which are almost circular, are better than diaphragms with five or six plates for the creation of soft blurring, which is not too sudden or angular. Different shapes of diaphragm may also be used, such as right angles and stars, in which case source points in the background appear through the blur in the precise shape of the diaphragm image (creating bokeh). Bokeh is generally produced against dark backgrounds (distant light sources at night, reflected sunlight; see Figure 9.10, right). Note that catadioptric lens assemblies[9], in which the optical elements are bent using ring-shaped mirrors, produce a very distinctive form of ring-shaped bokeh.

9.6. Stabilization

All modern cameras include an image stabilization system, used to limit the effects of user movement in long shots taken without a tripod.

Traditionally, in hand-held photography, without stabilization, and using full-format cameras (i.e. with 24 mm $\times$ 36 mm sensors)[10], in order to avoid the risk of movement, the exposure speed should generally not be below $1/f$ (f expressed in millimeters, giving an exposure speed in seconds). This is known as the inverse focal length rule. Thus, a photograph taken with a 50 mm objective may be taken in up to $1/50$ s, while a photo taken with a 300 mm distance lens should be exposed for $3/1000$ s at most[11].

Stabilization reduces the need to adhere to this rule.

Lens assembly stabilization is carried out using a motion sensor. The signals produced by these sensors are sent to the processor, which identifies the most suitable correction to apply, by projecting the ideal compensation onto the mobile elements available for stabilization, acting either on the lens assembly (lens-based stabilization) or on the sensor itself (body-based stabilization).

9.6.1. *Motion sensors*

Two types of sensors may be used in cameras: mechanical sensors, made up of accelerometers and gyrometers, and optical sensors, which analyze the image flow over time. Linear accelerometers are simple and lightweight components, traditionally composed of flyweights which create an electrical

9 These lens assemblies offer the advantage of a very long focal length, with reduced weight; see Figure 2.10.

10 For smaller sensors, the rule is adapted using the conversion factor (see section 2.3.2); thus, for a sensor of 16 mm $\times$ 24 mm and a 50 mm objective, the minimum exposure speed is 1/75 s (factor = 1.5).

11 This formula is extremely conservative, and experienced, careful photographers may use significantly longer exposure times.

current by a variety of effects (Hall, piezoelectric or piezoresistive). These components have now been highly miniaturized and integrated into micro-electro-mechanical systems (MEMS) [IEE 14]. Linear accelerometers obtained using photolithography are often designed as two very fine combs, made of pure silicon, which interlock face-to-face. During acceleration, the capacitance between the combs changes, providing the measurement (see Figure 9.11). Gyrometers, which identify angular rotations, generally measure disturbances to the movement of a rotating or vibrating body (measuring the Coriolis force, which leads to a precession of movement around the rotation axis). The physical effects used in MEMS structures are very fine, and principally result from various types of waves resonating in piezoelectric cavities, of ring, disk or bell form; unlike earlier mechanisms, these techniques do not require rotation of the sensor.

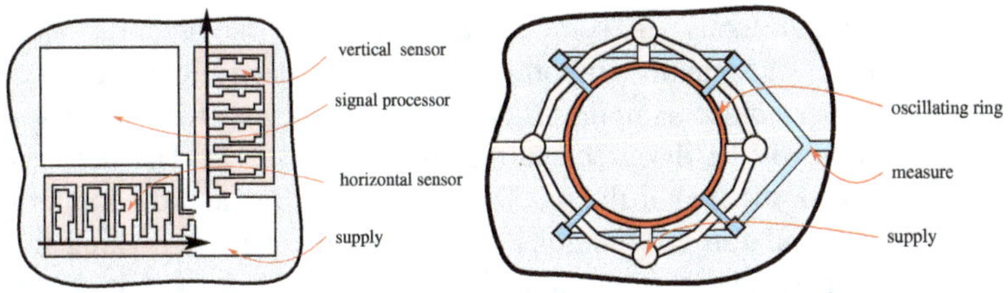

Figure 9.11. *Left: two linear accelerometers using MEMS technology, made from interlinking combs, integrated into the same housing. These devices are used to determine translation movement in the plane of the host circuit. Right: a gyroscope, consisting of a vibrating ring (red, center). A circuit (shown in pink) is used to maintain vibration of the ring, while another circuit (in blue) measures deformation of the ring due to rotational movement. These two types of captors are produced from raw silicon. For a color version of this figure, see www.iste.co.uk/maitre/pixel.zip*

Two accelerometers need to be used in a camera in order to determine the level of translation in the image plane. A third, the length of the optical axis, is rarely solicited. The three gyrometers are often integrated into a single circuit. The combination of the accelerometer and gyrometer gives us an inertial platform. Magnetometers may also be used to determine movement in absolute terms. The influence of the guidance system and games markets, the car manufacturing industry and telecommunications has led to miniaturization of these systems and a considerable reduction in cost. However, many cameras only use linear displacement accelerometers. Measurements obtained using mechanical sensors have the advantage of insensitivity to the movement of objects in a scene, unlike optical sensors. Mechanical sensors also have a much higher capacity than optical sensors to measure very large movements. If movement sensors are used in an open loop, the measured movement value must be used to deduce the effect to apply to the image sensor or correcting lens in order to correct the effects of this movement on the image. To do this, we need to take account of information concerning optical settings (focal length and distance). For this reason, it may be useful to use a closed loop mode, locking the mobile pieces into the control loop; however, this method requires the use of a mechanism within the image to measure stability. In practice, this mechanism carries out a function very similar to one required by the sensor, taken from the image, which will be discussed in the following section.

Optical flux analyzers implement similar functions and mechanisms to those used for autofocus (in some cases, these are one and the same). Measurements of the spatio-temporal gradient are determined across the whole field, at very close sampling times, using block-matching techniques. These operations will be discussed further in section 10.1.6. The denser the measurement field, the higher the precision of calculation, and the higher the charge on the calculator. Calculations may be accelerated in tracking mode by maintaining parameters using predictive models (ARMA models, Kalman filters or particle filters), either directly using measurements at each individual point, or on the basis of global models obtained by analyzing all of the points as a whole:

– if the speed field is relatively constant across the image, the movement is a bi-dimensional translation; this is the first fault to be corrected, and is generally the most significant error;

– if the speed field presents rotational symmetry along with linear and radial growth, this demonstrates the presence of rotation around the optical axis (see Figure 9.12);

– if the symmetry of the speed field is axial, we have rotational movement in a downward or upward direction (tilt).

Optical sensors express the movement detected in an image in terms directly compatible with the compensation to apply to the image. These calculations are made considerably easier by the presence of the DSP and by the use of multiple parallel processing lines.

Note that different movement measurements are obtained from optical and mechanical motion sensors:

– if the whole scene is scrolling, an optical sensor will integrate the movement of the scene into that of the camera, something which the user may not necessarily wish to happen;

– if a camera is accelerating in a moving vehicle, an accelerometer will combine the movement of the camera with that of the camera, which, once again, may be contrary to the wishes of the photographer.

To take account of this type of major movement by carriers, certain manufacturers have introduced a “drive” mode which specifically compensates for significant, uniform movement. This mode should not be confused with stabilized mode, and will not be discussed in detail here.

Note that gyrometers (which provide tiny temporal variations in the angular orientation) should not be confused with gyroscopes (which measure the orientation of the sensor in relation to the Earth: North-South, East-West, azimuth-nadir), which are included in certain high-end products and provide additional details regarding the position of a scene, adding the three orientation angles to the GPS coordinates (the full equations involved in this stabilization are obtained using the calibration equations [2.47]).

9.6.2. *Compensating for movement*

Once the corrections needed to compensate for movement have been identified, we need to apply these changes to the actuators involved in compensation. As we have seen, two options exist in these cases: moving the sensor, or moving the lens.

Body-based stabilization is an economical solution, as it allows the use of any lens assembly, even those not adapted for stabilization. Furthermore, the sensor is lightweight and easy to move, giving it the ability to respond rapidly to a stabilization order. The actuator command laws are deduced directly from the measurements obtained by the sensor, a fact which simplifies the calculations required and reduces compensation errors. For motion compensation purposes, the sensor is placed onto a mobile unit, which moves across ball bearings using a piezoelectric motor (similar to those discussed in section 9.5) or a coil/electromagnet assembly. In reflex cameras, antidust systems may also be used for stabilization, as they apply vibrations to the sensor in the same way as the stabilization. However, body-(or sensor-)based stabilization does have its drawbacks. First, it may lead to excessively large sensor movements in cameras with a long focal length (several tens of pixels). Next, to compensate for complex errors (for example, dips and raises), complex movements are required, while the sensor must be kept in a very specific position in order to ensure capture quality.

If we act instead on optical elements, an additional connection must be established between the body and the lens assembly, and the camera in question can only be used with lens assemblies designed to be compatible with the specific stabilizer model. To limit the weight of the mobile element, a small group of lenses is generally moved, selected to be as light as possible. For assemblies with a long focal length, acting on the lens requires smaller movements than those required when moving the sensor. Crossed electromagnets are often used as actuators in order to generate this compensation movement. The compensation law requires us to take account of the image formation sequence as a whole (requiring calculations using the setting parameters). On a more positive note, lens position is less critical than that of the sensor in body-based stabilization. Note that in this case, it is not possible to compensate for rotation around the optical axis, but it is relatively easy to correct rotation around the horizontal axis. Unlike body-based stabilization, lens-based stabilization is beneficial for all camera functions: focus, light measurement, framing, etc. This is particularly beneficial when using long focal lengths.

Nowadays, the two types of stabilization system offer relatively similar performance levels. Note that lens-based stabilization was developed very early on for use in high-end film cameras[12]. It is also used in binoculars, microscopes and telescopes, and therefore has a longer and more impressive pedigree. Body-based stabilization was developed for solid-state sensors, at a time when stabilized lenses were already available. This solution is becoming increasingly popular. Newer propositions include hybrid systems, combining the two stabilization modes using a strategy determined by individual manufacturers; in these approaches, sensor stabilization is generally used to compensate for vibrations, which require rapid response times and significant lens movement [ATK 08]. This form of hybrid stabilization could become the dominant technology in the field if the command laws were to be made public, allowing interoperability between products.

It is also important to consider the types of movement compensated by an image stabilizer. First, motion compensation is particularly suited to correcting errors resulting from hand-held photography, i.e. erratic

12 For film cameras, image stabilization is only possible using lens-based techniques.

movements, generally vibrations, of low amplitude, affecting the sensor at an approximate rate of under 50 Hz. The main components of this motion are generally horizontal and vertical translations. Entry-level cameras and cell phones generally include stabilization functions for this type of error, and are also often able to compensate for rotation around the target axis. These three correction types are included in the three-axis stabilization approach (see Figure 9.12).

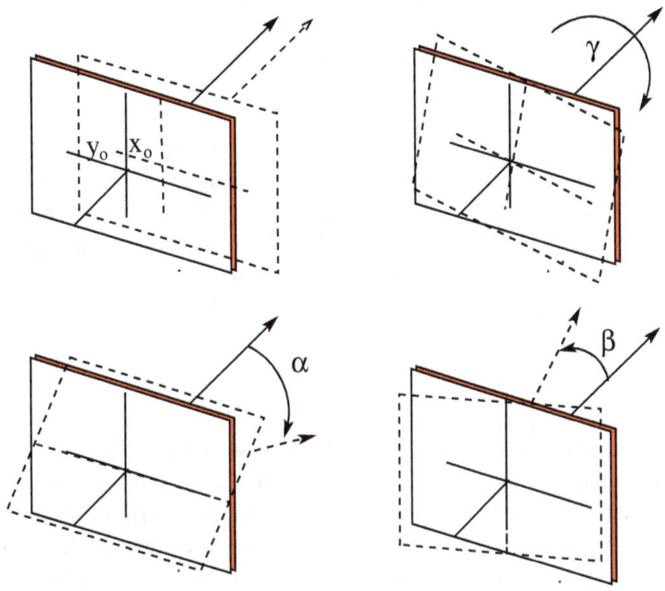

Figure 9.12. *Top: stabilization in response to three movements: translation, x_o and y_o (left) and rotation γ around the optical axis (right). Bottom: additional stabilization, rotating around the horizontal axis: dipped (α) and raised (β), giving a total of five parameters to control (the notation used is taken from equations [2.47]–[2.49])*

More advanced systems offer five-axis stabilization, which compensates for upward and downward tilt in addition to the three other movement types. Note that the distortions resulting from tilt (unlike those due to translation and rotation) are not isometric, and result in spatially variable blurring, which are difficult to correct by deconvolution[13].

Stabilization is remarkably effective in terms of allowing longer exposure times. It is usually expressed in diaphragms, and can reach four or five diaphragms in the best cases; this results in a reduction of minimum exposure speed by a factor of 16 or 32. It is now possible to take photographs of fixed objects with exposure times of several tenths of a second without using a tripod, something which was barely possible 20 years ago. A more appropriate form of the inverse focal length rule, discussed at the beginning of this section, for modern apparatus would be 10 times the inverse of the focal length.

The possibility for confusion between movement in a scene and movement of the camera, mentioned above, requires further consideration. For this reason, it is best to avoid using the stabilizer if the camera is placed on a tripod[14]. Another, related, issue is the high-power consumption of a stabilized camera, as stabilization is active from the moment the camera is switched on, whether or not a photograph is being taken.

9.6.3. *Video stabilization*

The specific context of video photography, which concerns an increasing number of cameras, raises different issues to those involved in stabilizing fixed images. In this case, the main aim is not to reduce blurring during capture, but to keep the scene stable in spite of operator movements. Images must, therefore, be globally readjusted in relation to each other, while respecting the progressive movement of the frame imposed by the operator. Images are readjusted to each other using the motion detection methods (three or five axes) described earlier, but on the basis of 25 or 30 images per second, rather than the exposure times used in still photography. The available processing time is, therefore, longer, but the work required is more complex, as the movement accepted for an image n must be compatible with that accepted for images

13 Many cameras also offer an antishake mode, designed to compensate for sudden and significant movements. This mode is generally based on a simple change in sensitivity, with priority given to shutter speed; this allows significant reductions in exposure time, but the shaking motion itself is not analyzed or corrected.

14 Certain cameras automatically pass into non-stabilized mode when a tripod is used. Others offer a specific "tripod" mode, where only vibrations due to the movement of the mirror are taken into consideration.

$n-1$, $n-2$, etc. The edge of fields is particularly problematic, as images are offset in relation to each other. The pixels from image n which have left the field due to parasitic movement, but should still be present as they were seen in $n-1$ and $n-2$, therefore, need to be temporally extrapolated. This specific treatment is relatively similar to that carried out by the P lines in MPEG coding (see Figure 8.9). These operations may be carried out in real time in the camera using motion prediction. Offline treatment also remains possible, notably for replacing extrapolations with interpolations, and linear prediction with non-local approaches, which are costly in terms of processing time but are highly effective.

Another problem affecting video images taken with a CMOS sensor with an electronic sensor resides in the fact that the capture instant is not the same at the top and bottom of the image, as the CMOS dumping process is sequential (see section 3.2.3 and the discussion in section 9.4). If the camera is subject to non-vertical movement, then vertical lines will be distorted. This effect should also be corrected during the readjustment process.

9.7. Additions to the lens assembly: supplementary lenses and filters

Let us now consider the accessories used to improve or modify the properties of the lens assembly, as discussed in Chapter 2. We will begin by discussing additional lenses, used to modify the focal distance of the objective, followed by filters, which affect the photometry of the image. To conclude this section, we will then consider polarizers. All of these filters are placed in front of the lens assembly (between the assembly and the object being photographed), with rare exceptions being placed between the lens assembly and the body of the camera.

9.7.1. *Focal length adjustment*

9.7.1.1. *Close-up lenses*

Photographers regularly make use of additional lenses, fixed onto the lens assemblies, known as close-up filters or close-up lenses. These often consist of a single convergent lens, or an achromatic doublet, and always take the form of a lightweight element which is added to the objective in the same way as a filter. Close-up lenses operate by bringing the infinite plane in to a short distance from the camera, in order to allow more significant magnification than those obtained with the lens assembly alone. Used for excessive zooming, close-up lenses can cause aberrations, limiting their usefulness. They are most effective when used with long focal length assemblies, increasing the magnification still further. For this reason, close-up lenses are often particularly useful in outdoor photography.

Close-up lenses are defined by their optical power, expressed in diopters[15]. Most close-up lenses have a power of between 1 δ and 10 δ.

9.7.1.2. *Teleconverters*

A teleconverter is a secondary lens assembly which is placed between the camera body and the lens, and fulfills the same role as a close-up lens. 2$\times$ teleconverters are the most widespread. 3$\times$ teleconverters also exist, alongside lower multipliers, for example 1.3 and 1.5. Teleconverters are relatively complex optical systems (up to 10 lenses) and are relatively large (particularly in comparison with close-up lenses), with a thickness of several centimeters. They operate in the same way as a divergent lens. 2$\times$ teleconverters generally produce fewer aberrations than close-up lenses, and for this reason they are often used for photographs taken at short or very short distances, where close-up lenses are often unsuitable.

As with close-up lenses, teleconverters magnify the core of an image in the sensor, reducing the energy received by the sensor; higher exposure settings are, therefore, required (a 2$\times$ teleconverter requires an exposure time 4 times that of the lens assembly alone for the same aperture settings).

9.7.1.3. *Focal length dividers*

A system symmetrical to the one described above may be used to replace a divergent assembly with a convergent assembly, placed between the lens assembly and the camera body, in order to reduce the size of the image. The advantages of this technique are not immediately apparent, but it can prove beneficial. These devices constitute a second type of teleconverter, sometimes referred to as a focal length divider (in contrast to the multipliers discussed above)[16]. This type of teleconverter produces a smaller image of the subject and covers a wider area of the scene. The received energy is, therefore, concentrated, allowing a reduction in exposure time (or lens closure; see

15 The diopter is a unit of optical vergence, represented by the letter δ. It is the inverse of a length (dimension $[L^{-1}]$, and more specifically the inverse of the focal distance). A lens with a focal distance of 0.1 m has a vergence of 10 δ.

16 These mechanisms are also known by a number of proprietary trade names, for example the MetaBones SpeedTurbo and the Sony LensBooster.

Figure 9.13). These elements are sometimes presented as a means of increasing the aperture of the lens assembly (this is not true, and not physically possible, as it would require the collection of light from outside of the diaphragm). However, as they reduce the focal distance, for a constant aperture, they produce an effect identical to that obtained by increasing the aperture number, as $N = f/D$.

Focal length dividers were first introduced for use with digital cameras; as we clearly see from Figure 9.13, they can only work if the lens assembly is overdimensioned in relation to the sensor, otherwise an unacceptable level of vignetting is produced. Moreover, they can only be used if the flange focal distance (FFD, the distance between the sensor and the last lens) can be significantly reduced, as the equivalent focal distance is shorter. This is very hard to do using a reflex camera due to the presence of the mirror tilting mechanism. Focal length dividers are, therefore, only used in non-reflex cameras with changeable lens assemblies [CAL 12].

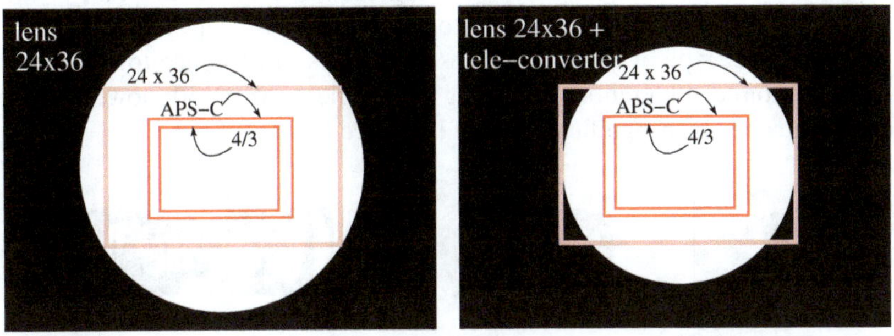

Figure 9.13. *Effect of a focal length divider (lens booster). Left: the field covered by a lens assembly for use with a 24×36 sensor (the white circle), shown for three different formats of commercial sensor: 24×36, APS-C and 4/3. Right: the same lens assembly, with the addition of a focal length divider. The 24×36 format is now subject to high levels of vignetting. The two others are still in full light and benefit from the effects of focal length reduction. In practice, focal length dividers are used in this final configuration*

9.7.1.4. *Extension tubes and bellows*

Extension tubes are optically passive elements, placed between the camera body and the objective, which are used to increase the flange. These elements are essential in macrophotography, allowing increasingly strong enlargements G as the length increases, following relationship [1.3]: $G = -(f + p')/f$. In order to conserve camera settings, these tubes also need to be able to transmit sensor measurements to the actuators, making them complex and costly.

Camera bellows function using the same principle. One of the rings is placed on a mobile unit, and the length of the element can then be adjusted to suit the desired enlargement size. Tubes generally have a length of a few centimeters, but bellows can cover tens of centimeters, allowing significant levels of enlargement. However, this results in a very shallow depth of field and requires long exposure times. Both extension tubes and bellows are generally used with mid- and short-focal length assemblies with high apertures.

Other specific types of ring include inversion rings, which allow lens assemblies to be mounted in the wrong direction: in these cases, the input lens is situated nearest the sensor, while the original fixture mounting is closest to the object. For mid-focal length lens assemblies (between 30 and 85 mm), this creates very significant enlargements (as the nodal plane of the sensor is pushed back), comparable to those obtained using long extension tubes. Another type of ring allows lens assemblies to be joined together in a top-to-bottom configuration, producing enlargements of 20–50; however, this is accompanied by a significant loss of light (see Figure 9.14).

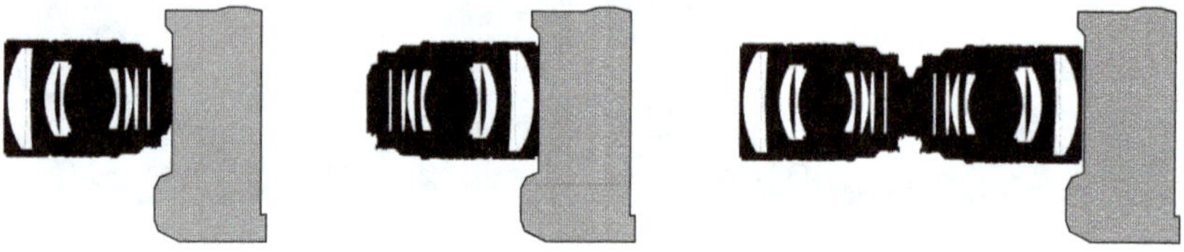

a b c

Figure 9.14. *Three camera lens assemblies: left, standard assembly, allowing focal lengths from tens of centimeters to infinity; center: inverse assembly, producing significant enlargements for nearby objects; right: two assemblies, placed top to bottom in order to produce extremely high enlargement levels*

9.7.2. *Infra-red filters*

We will only provide a brief description of infra-red (IR) filters here. These filters are generally placed on the input end of the lens assembly. They fulfill two key roles:

– as their name indicates, they suppress IR radiation (with a wave length greater than approximately 780 nm), to which the silicon used in sensors is particularly sensitive (see Figure 5.22); this radiation should not contribute to image formation, as it is not visible to the naked eye;

– they protect the external face of the lens assembly from mechanical shock and dirt (water, dust, etc.).

This second property is often the most important, as modern solid-state sensors are well protected against IR radiation by treatments applied to the photodetector itself.

IR filters are made from either glass with a high level of absorption in the IR spectrum, or by applying thin layers of metallic oxides to thin glass plates, which act as interference filters. Interference filters have sharper cut-off lines, but their effectiveness is dependent on the angle of the rays, and they, therefore, offer very limited protection for lenses with a strongly-inclined capture angle (such as fisheye lenses).

Commercial filters differ in the sharpness of their cut-off limits, both above the red spectrum and below the blue spectrum. Excessively powerful treatments can result in a blue or, in the other direction, pink tinge in the resulting images.

Besides their performance in rejecting IR radiation, the quality of IR filters is determined by their mechanical hardness, their transparency across the visible spectrum and the antireflection treatment used.

Note that it is also possible to use filters to remove all light except for IR in order to create special effects, or to select thermal radiation in night-time photography polluted by parasitic noise. As we have seen, this is only possible if the IR filter on the sensor is also removed.

9.7.3. *Attenuation filters*

Neutral filters attenuate the beam following a wavelength law which is generally uniform[17]. These filters are characterized by their optical density, i.e. the inverse of the common logarithm of their transparency. Thus, a filter with a density of $OD = 1$ attenuates light with a ratio of 10, which may be offset either by multiplying the diameter of the diaphragm by around 3.16, or by multiplying the exposure time by 10. Filters are available with a density of

17 Note that the attenuation law is not uniform in terms of wave numbers, $\sigma = 1/\lambda$, another way in which uniformity might be perceived.

between 1 and 10. Attenuations are also sometimes described using the number of equivalent diaphragms, ND. This is the base 2 logarithm of the $OD: ND = OD/\log_{10}(2) = 3.32OD$.

The interest of using a neutral filter, which reduces the amount of light in an image, lies first in the possibility of photographing scenes with very high luminosity, such as those encountered in industrial or scientific photography (eclipses, volcanoes, furnaces, etc.). It also allows more "everyday" photographs to be taken using very long exposure times, allowing the creation of special effects, for example for a water course, the sea or a moving crowd. Finally, neutral filters are very useful when using wide lens apertures in scenes where we wish to reduce the depth of field.

Certain neutral filters may only cover half of the optical field, in order to adapt to contrasting lighting conditions; in this way, the sky and ground are treated in different ways. This local treatment may also affect spectral content, for example increasing the levels of blue in the sky.

These filters are made using glass plates or gels. They are screwed onto the lens assembly, or held in place by a filter holder structure. Note that the use of two mobile polarizers, which can be moved in relation to each other, or, better, two polarizers followed by a quarter wave plate, constitutes an excellent variable attenuator, for reasons which will be explained in the following section.

Note that high attenuator densities render the photosensitive cells, which determine exposure time, and autofocus systems less efficient.

9.7.4. *Polarizing filters*

Polarizing filters are often used to produce specific effects on an image, for example damping reflections, reinforcing contrasts and tints, or modifying the appearance of certain shiny objects.

These effects can be explained by physical phenomena relating to electromagnetic propagation.

9.7.4.1. *Physical considerations*

Equation [1.1], as seen at the beginning of this book

$$A(\mathbf{x}, t) = \sum_i a_i \cos(2\pi\nu_i t - \mathbf{k}_i.\mathbf{x} + \phi_i)$$

is written in a specific scalar form for use in optics, which only takes account of the amplitude of the electromagnetic field, without considering its orientation. A fuller, vectorial form provides a more complete description of the electromagnetic field, describing the two components of the field in a plane normal to the propagation vector **k**. In air, a homogeneous medium (and the medium for almost all photographs), the electromagnetic field consists of two vectors, the electric field **E** and the magnetic field **B**, which are in phase but orthogonal to each other, in a plane orthogonal to the propagation vector **k**, and tangential to the wave front. Choosing a fixed direction for this plane, **E** can be subjected to unique decomposition, for example into two components $E_{||}$ and $E_{\perp}$.

$$E_{||}(\mathbf{x},t) = a_i \cos(2\pi\nu_i t - \mathbf{k}_i\mathbf{x} + \phi_i)$$
$$E_{\perp}(\mathbf{x},t) = a'_i \cos(2\pi\nu_i t - \mathbf{k}_i\mathbf{x} + \phi'_i)$$

The relationship between $E_{||}$ and $E_{\perp}$ defines the polarization of the wave:

– if the two components are unrelated, the light is not polarized;

– if $E_{||}$ and $E_{\perp}$ are in phase ($\phi_i = \phi'_i$), the polarization is said to be linear. In one direction, the component of **E**, vectorial sum of $E_{||}$ and $E_{\perp}$, is maximized; in an orthogonal direction, its component is always null;

– if $E_{||}$ and $E_{\perp}$ are dephased, the vector **E**, sum of these two components, turns around vector **k**. In this case, the polarization is said to be elliptical. If $||E_{||}|| = ||E_{\perp}||$, we have circular polarization, and the wave will have the same energy whatever the direction used for analysis.

9.7.4.1.1. Changing polarization:

It is possible to move from a non-polarized or elliptically polarized wave to a linear polarized wave by using a polarizing filter, reducing the energy of the incident beam by a factor of around 2. A quarter wave plate is used to change a linear polarized wave to a circularly polarized wave. To transform a polarized wave (of any type) into a non-polarized wave, a random diffuser is placed in the optical pathway [PER 94].

Polarizing filters are generally made up of microstructured materials, such as Polaroid, which will be described later. Waves which are polarized in the direction of alignment are stopped, while those which are perpendicular to the direction of alignment are transmitted with almost no attenuation.

Quarter wave plates are made of a birefringent material, presenting two different indices dependent on the direction of polarization. Placed at 45° to a polarized wave, they decompose this wave into two orthogonal components[18].

Natural light is generally not polarized. However, the state of polarization can vary in relation to the medium through which it travels [FLE 62], as we will see below.

In photographic terms, non-polarized light and circularly-polarized light are indiscernible, and the terms are often used synonymously even in specialist publications.

9.7.4.1.2. Polarization and photography

If the incident light received by the sensor is selected on the basis of polarization, it is possible to increase or dampen contrast in an image. The simplest approach consists of placing a linear polarizer in front of the lens assembly. If the whole of the scene is made of non-polarized light, then the resulting image will only be damped (theoretically by a factor of 2, but sometimes slightly more in practice, as the glass used to support the filters absorb and reflect a small quantity of light in addition to the effects of the polarizer). Given a sufficiently long exposure time, however, the image will be identical to the original, whatever the direction of the polarizer. This is also true in cases of circularly-polarized light, although this situation rarely occurs in natural conditions.

If the light is polarized for certain objects, the resulting intensity will vary based on the direction of the filter, which may be placed parallel or perpendicular to the direction of polarization, from complete removal to full lighting. In the perpendicular case, all of the light from the polarized objects will be transmitted, unlike that of the rest of the scene, and these objects will be highlighted, appearing twice as luminous as the scene itself. In the parallel case, light from the polarized objects will be completely blocked, and these objects will appear dark. For an angle α between the filter and the incident polarization, the resulting intensity $||\mathbf{E}_s||$ is obtained as a function of the incident intensity $||\mathbf{E}_e||$ using:

$$||\mathbf{E}_s|| = ||\mathbf{E}_e|| \ \cos^2 \alpha \qquad [9.9]$$

18 The "ordinary" and "extraordinary" rays. The pathway followed by the extraordinary ray is offset in relation to the ordinary ray, and is subject to a very brief delay. The thickness of a quarter wave plate is calculated so that this delay corresponds to a quarter of the period, but this property will not be used here.

This is Malus' law [PER 94]. It is, therefore, possible to highlight or conceal certain zones of an image simply by turning the polarizer around the optical axis. To identify the zones concerned by this treatment, we need to consider the causes of polarization.

9.7.4.2. *Nature and polarization*

9.7.4.2.1. Specular reflection

The Snell–Descartes law [JAC 62] states that if a wave is reflected from a dielectric of index n at an angle of i in relation to the normal plane, the reflected wave follows an angle which is symmetrical to i in relation to this normal plane. Polarization which is parallel to the reflecting surface reflects with an amplitude of $E_{||}$ (see Figure 9.15, left) and the orthogonal component reflects with an amplitude $E_{\perp}$. For a non-polarized incident wave, ratio R varies as a function of angle i following the relationship:

$$R = \frac{E_{\perp}^2}{E_{||}^2} = \frac{(\sqrt{(1-\sin^2 i)(n^2-\sin^2 i)} - \sin^2 i)^2}{(\sqrt{(1-\sin^2 i)(n^2-\sin^2 i)} + \sin^2 i)^2} \qquad [9.10]$$

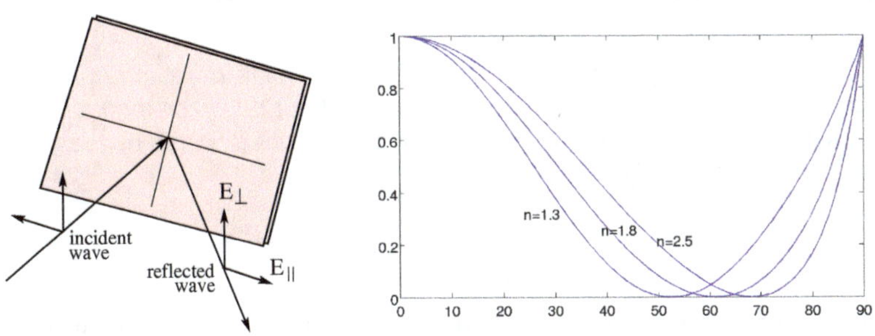

Figure 9.15. *Left: reflection of a circularly-polarized wave on a plane and notation of emergent waves. Right: relationship between the energy of the emergent wave, polarized in parallel to the reflection plane, and that of the wave polarized in parallel to the plane, as a function of the angle of the incident wave and the normal of the reflection plane. The reflecting medium is a dielectric of index n, using three values of n. The incident wave is taken to be non-polarized. When the relationship cancels out, the emergent wave is completely polarized; this occurs at Brewster's angle*

This function is illustrated on the right-hand side of Figure 9.15. We see that the ratio cancels out for an angle at which the emergent wave is fully polarized. This is Brewster's angle. At normal incidence ($i = 0$) and grazing incidence ($i = 90°$, but this is not particularly interesting, as both waves are null in this case) the wave remains circular, with equal components. In natural imaging, it is very hard to remove all reflection; however, it is possible to achieve a significant reduction (see Figure 9.16).

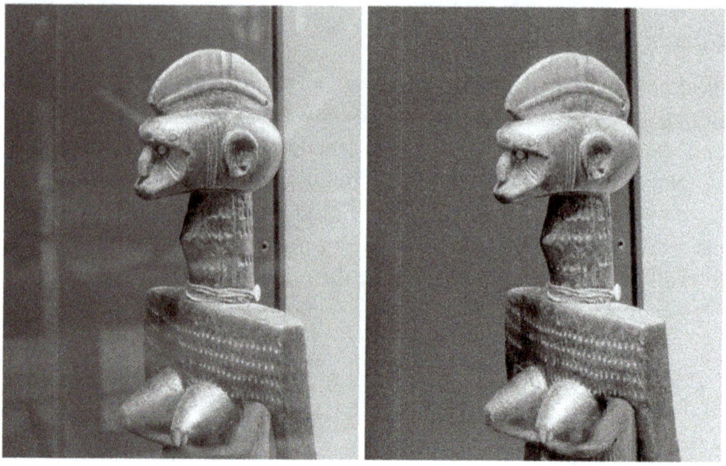

Figure 9.16. *Photograph taken through a glass display case, showing the effects of a linear polarizer. Left: image taken without polarizer; right: image taken using a polarizer oriented so as to reduce the light reflected by the glass (Musée des Arts Premiers, Paris)*

9.7.4.2.2. Microstructured materials

Light, like all electromagnetic waves, reacts in a very specific way to periodic structures with a period close to that of the wavelength, and its response is strongly linked to polarization. In the field of optics, these dimensions are less than 1 μm, making them barely perceptible for observers; however, the polarization effects are still significant. This effect concerns materials such as insect exoskeletons, hummingbird feathers, certain organic structures, pearlized surfaces and varnishes, and also certain biological tissue types (collagen fibers). These effects may apply to surfaces or volumes, and may affect certain specific wavelengths, but not others.

Polaroid, invented by E.H. Land, is an artificial structured material made of a sheet of polyvinyl alcohol polymer, stretched in the manufacturing process in order to create aligned, linear chains of molecules. These chains are then treated with iodine to make them conduct, and take on strongly dichroic behavior: waves which are polarized in the direction of alignment are absorbed, while those polarized perpendicular to the chains are transmitted with almost no attenuation. Polaroid polarizers are widely used in photography.

9.7.4.2.3. Blue skies

The fact that the light emitted by a blue sky is polarized is generally explained by Rayleigh diffraction resulting from particles in the upper atmosphere ([BOR 70, p. 655]). These particles are typically smaller than $\lambda/10$ in size. Solar radiation traversing these particles creates diffused radiation which is naturally polarized. Short wavelengths (i.e. blue) are diffused most (which gives the sky its blue color). Seen by an observer on the Earth, the degree of polarization π depends on the angle γ between the direction of the sun and the direction of the target point:

$$\pi = \frac{\sin^2 \gamma}{1 + \cos^2 \gamma} \qquad [9.11]$$

The degree of polarization is, therefore, null in the direction of the sun, and reaches a maximum at 90°. In physical terms, matters become much more complicated if we take account of multiple diffusions of the same ray, as short wavelengths are more susceptible to this type of diffusion, which removes polarization. The same effect can also result from diffusion by large particles (observed close to the horizon), which obey Mie diffraction laws; once again, this leads to greater desaturation of short wavelengths. The longest wavelengths (producing white light) may be absorbed by a carefully oriented polarizer, while blue wavelengths will represent the majority of those transmitted; however, these wavelengths remain rare, producing deep blue shades in the image.

9.7.4.2.4. Birefringent materials

Finally, polarization may also result from materials with rotary powers, which are said to be optically active. These materials naturally appear differently depending on the incident polarization. This is encountered in crystals (such as quartz) and certain liquids (such as benzene) ([FLE 62], Chapter 12).

9.7.4.3. *Polarizer use*

The use of linear polarizers to improve images is particularly simple. The best orientation is determined by direct observation of results on the monitor screen. However, this simple filter may have undesirable effects on the operation of a camera, as the sensors (both for autofocus and exposure) often operate by deriving part of the luminous flux traveling through the lens assembly using semi-reflective plates (see Figure 9.5). If the light is polarized, then the response from these captors will be wrong (as a direct function of Malus' law, cited above). Many cameras also use an antialiasing filter (see Figure 3.10) which only operates correctly with non-polarized or circularly-polarized light.

The use of a more complex assembly for linear polarization is, therefore, recommended. This assembly consists of a linear polarizer followed by a quarter wave plate, placed at an angle of 45° in order to re-establish full polarization. These devices are generally known as circular polarizers. They only operate in one direction (with the linear polarizer in front of the quarter wave plate) and cannot be used in series for damping purposes, something which is possible with simple linear polarizers, unless the filter is mounted upside down; this is generally not possible due to the fixation mechanisms used.

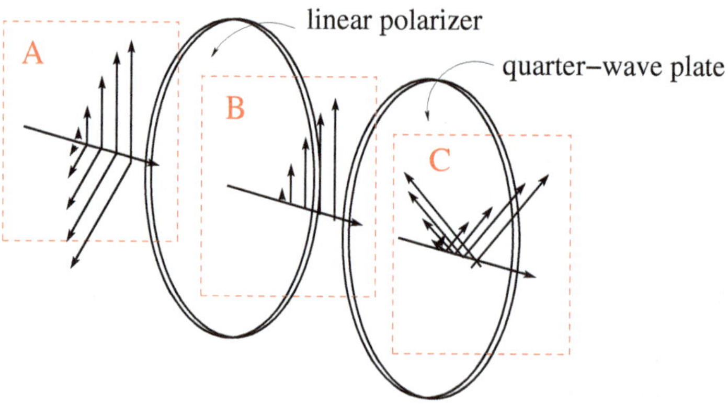

Figure 9.17. *Polarization using a linear polarizer followed by a quarter wave plate, allowing reconstruction of a non-polarized wave (C), from the incident light (A) of which only a linearly polarized component (B) has been transmitted by the linear polarizer. Assemblies of this type allow the optical components of the camera to function correctly, while removing one of the polarizations of the scene. The two plates should be moved as a single unit during rotations in order to find the angle of absorption*

In photography, polarizing filters are used to process natural scenes. In a scientific context, these filters may be used in a more complex way, by controlling both the polarization of the light source and the orientation of the analysis filter. This is known as polarimetric imaging. The process consists of a series of exposures, and by analyzing the images produced, it is possible to obtain a variety of information:

– the nature of the surfaces under examination, which may retain or rotate the polarization (metallic or dielectric reflections) or destroy the polarization (rough surfaces, absorbent or diffusing materials);

– the orientation of surfaces relative to the retention or rotation of polarization.

9.7.5. *Chromatic filters*

In this section, we will consider filters which modify the overall appearance of a scene, and not chromatic selection filters, which are placed over the sensor and were examined in section 5.4.

The most important filters of this type are color correction filters. We have already seen (Chapter 4) the way in which the color temperature of a scene is defined, particularly in the case of artificial lighting. The spectral content of this light may be different from that desired for a photograph, either because the light is too cold (too much blue), or too warm (too much red). This problem may be corrected by the introduction of a filter, which acts on the incident spectrum so that it meets the photographer's requirements.

Let T_i be the color temperature of the incident source and T_o the desired color temperature. Wien's formula (which is a very good approximation of Planck's law for the temperatures ordinarily used in photography) provides an expression of the spectral luminance corresponding to these values (equation [4.17]):

$$\mathcal{L}(\lambda, T_i) = 2\pi hc^2\lambda^{-5}e^{\frac{-hc}{\lambda k T_i}} \qquad [9.12]$$

where h is Planck's constant, c is the speed of light and k is the Boltzmann constant.

The filter used to pass from T_i to T_o thus has a spectral transmittance $\mathcal{T}(\lambda, T_i, T_o)$:

$$\mathcal{T}(\lambda, T_i, T_o) = \alpha e^{\frac{-hc}{k\lambda}\left[\frac{1}{T_o} - \frac{1}{T_i}\right]} \qquad [9.13]$$

where α is an arbitrary constant which ensures that $\mathcal{T}(\lambda, T_i, T_o)$ is less than 1 across the whole spectrum. Filters are often described using their optical density (the normal log of their transmittance): $\mathcal{D}(\lambda, T_i, T_o) = -\log_{10}[\mathcal{T}(\lambda, T_i, T_o)]$. Hence:

$$\mathcal{D}(\lambda, T_i, T_o) = \frac{\beta}{\lambda}\left[\frac{1}{T_o} - \frac{1}{T_i}\right] + \mathcal{D}_l \quad [9.14]$$

The spectral density of the filter is therefore linear in terms of the wave number $\sigma = 1/\lambda$, and is a function of the inverse of the temperature [BUK 12]. Each temperature is associated with a source characteristic, expressed in reciprocal megaKelvins[19], equal to $10^6/T$, and the filter is defined by its value in reciprocal megaKelvins. Thus, a 133 reciprocal megaKelvin filter converts light from an incandescent lamp ($T_i = 3,000\ K$, i.e. 133 MK^{-1}) to the equivalent of lightly softened sunlight ($T_o = 5,000\ K$, i.e. 200 MK^{-1}).

9.7.6. *Colored filters*

In addition to temperature correction filters, colored filters are often widely used, essentially in the context of black and white photography, in order to create specific effects by heightening certain tints. They are often used for this purpose by professional photographers. Notable examples include:

– red filters: generally used for dramatic or "fantasy" effects, absorbing blues and greens. They can be used to increase contrast in clouds, waves or whirlpools, for example. They are also essential for daytime IR photography;

– orange filters: often used for portraits, as they can produce a range of flesh tones, eliminating zones of redness. They are also useful when photographing natural environments with a high mineral content, such as deserts and mountains;

– yellow filters are used for landscape photography in cases of low light, as they harden blue tones and increase contrast;

– green filters are generally used to highlight vegetation, and are suitable for use in landscape photography. They also give a tanned appearance to flesh tones.

---

19 Reciprocal megaKelvin is the recommended name for an old unit of measurement, which is still widely used, known as the mired (microreciprocal degree).

9.7.7. *Special effect filters*

We will not go into detail concerning the wide range of special effect filters available here; we will simply describe a number of broad categories.

Soft filters have an attenuating effect on microcontrast. They are often made from a very fine diffusing structure (microbubbles in glass, very fine matte effects, etc.), used to reduce the level of fine detail in a scene, generally in order to reinforce a steamy or misty effect. These filters are widely used in portrait and nude photography, where they help to reduce the appearance of skin imperfections; they may also be used in landscape photography, and are widely used in advertising. Note that soft filters do not necessarily have to cover the whole field; they may only cover half of the space, the center of the field or only the edges of the field. Another type of soft filter produces a graduated effect across the field.

Directional filters consist of very fine, transparent structures, oriented either in a linear manner (one or two dimensions, crossed or in a star configuration) or radially. They contribute to a modification of the pulse response, and introduce a highly visible diffraction figure, resulting in the presence of stars or irisations (bokeh, as discussed in section 9.5). They are particularly useful when photographing scenes with strong light sources (e.g. in night photography).

Even more dramatic effects may be obtained using binary masks, which only allow light to pass through specific apertures: hearts, spirals, stars, clouds, silhouettes, etc. In mathematical terms, the effects of these masks may be analyzed using equations [2.35], [2.36] and [2.37], but they are only used in response to specific requirements, deforming the pulse response using a particular motif and distorting fine details in an image to give a particular aesthetic effect.

Note that these filters act in exactly the same way as coded apertures, which will be discussed below.

9.8. Power cells

Modern cameras all require a source of current in order to power integrated electronic components. While these components themselves often have very limited requirements, the calculation power involved, along with a number of newer functions (stabilization, GPS, display, Internet connectivity, etc.), which are active even when a photograph is not being taken, is highly dependent on the power supply. Camera battery life, often expressed as a number of photographs, is not a particularly reliable or useful indicator. The practical battery life of a camera is often a major usage constraint in cases of intensive use, and a number of different options may be envisaged: use of mains power in a studio context, extension grips or, simply, carrying spare batteries.

Considerable changes are currently taking place in the domain of rechargeable batteries, and the field may be very different in the future from what we currently know [CHA 14]. However, for lightweight, portable devices such as cameras, the broad principles are unlikely to change, and are discussed below.

9.8.1. *Batteries*

The first cameras used either zinc-carbon (Zn-C) cells (zinc anode[20], carbon cathode, with an ammonium or zinc chloride electrolyte) or, better, alkaline cells ($Zn\text{-}MnO_2$: zinc anode, manganese dioxide cathode and potassium hydroxide electrolyte, KOH) which have a slightly longer life expectancy and a higher charge density. These cells, and their numerous variations, operate by a process of oxido-reduction, producing voltages of between 1.5 and 1.7 V. Taking the form of cylinders or disks, these cells are traditionally used in series and conditioned in parallelepipedic casings, giving a total voltage of up to 9 V. Batteries of this type are not generally rechargeable, but are cheap and widely available.

9.8.2. *Rechargeable Ni-Cd batteries*

For many years, nickel-cadmium (Ni-Cd) rechargeable batteries were used as an energy source in most cameras. This technology has been progressively replaced by Ni-MH (nickel-metal hydride) batteries, due to the toxicity of cadmium, a heavy metal which is now closely monitored and which was made illegal for mass-market usage in Europe in 2006.

Ni-Cd batteries use cadmium electrodes on one side and nickel oxide and hydroxide (NiOOH) electrodes on the other side. They deliver a relatively low

20 The terms “anode” and “cathode” are used here in place of negative and positive electrode, following standard usage; however, strictly speaking, the terms are only equivalent in discharge mode [HOR 14a].

voltage (1.2 V) and need to be placed in series or used with amplification methods to power camera processors. They present reasonable performance levels (energy-weight ratio of 40– 60 Wh/kg, charge/discharge yield of around 80%, a life expectancy of a few years, a cycle number of over 1,000 and a spontaneous discharge rate of around 20% per year).

Ni-Cd batteries are widely believed to be subject to "memory effects", a generic term which covers two specific faults: first, a behavior which causes a battery, operating regularly between two precise charge and discharge points (as in the case of batteries integrated into satellites, which recharge during a specific phase of the circadian cycle) to lose the ability to function beyond this limit, and second, the formation of cadmium hydroxide crystals, reducing the usable surface of the anode, if the battery remains charged over a long period. In both cases, the result is the same: a significant loss in battery capacity. The first of these faults (that which may properly referred to as a memory effect) is still the subject of debate, and no irrefutable scientific explanation exists. The second fault, on the other hand, is known to be real. This fault does not produce a memory effect, but produces natural evolutions within the battery; furthermore, this change is reversible via the application of a suitable recharge cycle (reconditioning). However, most chargers are unable to carry out this process.

It is best to charge and discharge Ni-Cd batteries on a regular basis in order to avoid cadmium oxide crystallization.

Ni-MH (nickel-metal hydride) batteries were developed to replace cadmium at the anode once the ban became effective. These batteries use lanthane and nickel hydrides (of type $LaNi_5$), in a potassium hydroxide electrolyte (KOH). The voltage of individual elements is the same as that obtained from a Ni-Cd battery, but the energy/weight ratio has increased to 100 Wh/kg. However, the discharge/charge ratio using this technology is only 66%, and the discharge rate is in excess of 30% per year.

Different manufacturers have issued very different recommendations concerning the best way of conserving Ni-MH batteries in order to attain the best possible capacity and life expectancy.

9.8.3. *Lithium-ion batteries*

The use of lithium for energy accumulation purposes is a relatively recent development in commercial terms (1991), although there was widespread awareness of its possibilities well before this date; this late development is due to the difficulties involved in conditioning these products.

Lithium-ion batteries operate via the exchange of lithium ions between a composite anode (for example, a carbon or graphite matrix) into which lithium is injected in a reversible manner, and a cathode made of a metallic oxide (such as cobalt oxide and manganese oxide) [YOS 09, HOR 14a]. Ions are exchanged within an aprotic solvent electrolyte, i.e. an electrolyte which will not release a H^+ ion (lithium salt in an organic solvent). Rechargeable lithium batteries differ from their non-rechargeable cousins at the anode: disposable cells use metallic lithium directly, meaning that the cycle is irreversible.

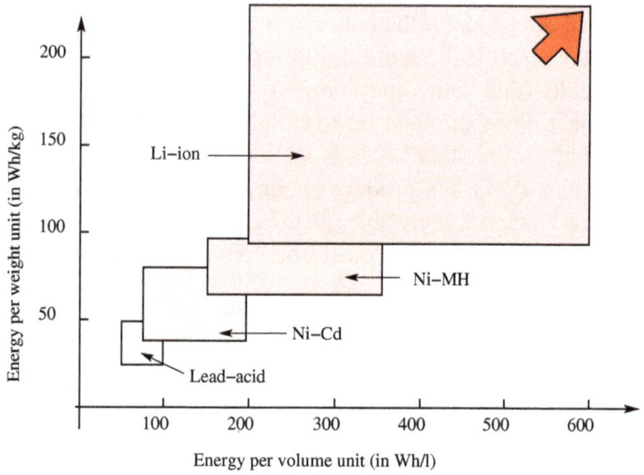

Figure 9.18. *The performances of rechargeable battery types used in photography are shown here using two variables: the energy/weight ratio and the energy/volume ratio, which constitutes an important criterion in miniaturized systems. Li-ion batteries currently offer the best performance from both perspectives (based on [HOR 14a])*

Rechargeable Li-ion batteries offer a number of remarkable properties (see Figure 9.18). They provide high voltages of around 3.6 V, considerably higher to those obtained using different technologies; this voltage is directly compatible with integrated electronic circuits, and the batteries do not need to be placed in series. The energy/weight ratio is more than double that of Ni-Cd batteries, from 100 to 250 Wh/kg. The charge-discharge rate is greater than 80% and may even exceed 90%, and the spontaneous discharge rate is very low (less than 10% per year). Li-ion batteries are not subject to memory effects, removing the need for an initial learning cycle and for specific maintenance operations. Note, however, that for optimal storage, these batteries should be kept at around 40% charge, and at a temperature of around 15°C.

However, Li-ion batteries present a number of disadvantages; these may be overcome in part by changes to the basic components, but this often results in a loss in terms of other performance aspects. First, these batteries are expensive, around 50% dearer than their NI-Cd equivalents. This cost is unlikely to fall, due to the rarity of lithium and the scale of demand. Moreover, these components are chemically fragile, particularly at high temperatures, and precautions need to be taken while charging; this slows down the charging process, which follows a specific progression, starting with a constant current then progressively reducing this current until the battery is fully charged. The aging process for Li-ion batteries remains hard to control, and their lifecycle is too short (barely more than 500 charge/discharge cycles, and 2–4 years). They are subject to irreversible damage if the charge drops too low, and if the voltage falls below approximately 2 V. Li-ion batteries, therefore, need to be controlled by a specific circuit, which cuts off service before full discharge is reached. These circuits are often integrated into the battery casing itself, and are known as battery management systems (BMS). BMS are particularly important in the context of high-performance batteries. Finally, it should be noted that lithium is a combustible metal. Batteries may, therefore, catch fire or explode if the discharge current exceeds a certain level, for example due to an increase in temperature. The transportation of lithium batteries is subject to limitations, notably in the context of air transport; this constraint does not only apply to industrial contexts, as airlines also specify conditions for transporting rechargeable batteries in personal luggage[21].

Research into the improvement of lithium batteries is ongoing [HOR 14a], with a particular focus on cathode efficiency via the choice of active materials, including cobalt and nickel oxides; these new products are identified as $LiCoO_2$, $LiNiO_2$ or, more generally, $LiM_xNi_{-x}O_2$, where M is a metal such as cobalt, magnesium, manganese and aluminum. Lithium-iron-phosphate (Li-$FePO_4$) batteries, for example, have a longer life expectancy and are safer to use, but offer a lower energy/weight ratio. Other projects concern the replacement of the liquid phase electrolyte by a polymer electrolyte, in order to decrease the risks associated with housing rupture. These projects have yet to result in the creation of commercial products, but

21 See IATA: Lithium Battery Guidance Document - 2014 - 05/11/2013 - APCS/Cargo.

the term "lithium-polymer" (Li-Po) has already been used to designate batteries with a polymer-based housing. These batteries present interesting mechanical properties (notably the fact that they may take the form of very fine plates), but do not constitute significant progress from an energy perspective in relation to the classic Li-ion technology (they also have a relatively low life expectancy). Further work has been carried out with the aim of reducing charging time. Very good results have been obtained by replacing the graphite at the anode by nanoparticles of tin oxide, SnO_2, which have no other effects on battery performance [ELA 14].

Finally, over a much longer term, researchers are considering methods whereby current is produced by reduction of ambient oxygen at the cathode and oxidation of lithium at the cathode, creating a Li-air (lithium-air) battery. This technology potentially offers an energy/weight ratio 10 times higher than that of Li-ion cells; these solutions are intended for use in vehicles rather than cameras, but progress made in this increasingly significant sector is sure to have a knock-on effect in areas such as photography.

10

Photographic Software

Evidently, the range of software developed for the purpose of treating photographic images is too broad to cover in its entirety here; interested readers may wish to consult specialist publications [BOV 00, MAI 08a]. In this chapter, we will consider:

– programs integrated into cameras, which, in addition to the treatments required when taking photographs, are used in certain models to provide additional functions: facial tracking, motion detection, the creation of panoramas, etc. (section 10.1);

– programs which may be loaded onto camera hardware in order to increase performance beyond the levels intended by the manufacturer (section 10.2). Distributed by specialist suppliers, these programs either replace or supplement existing camera software. While programs of this type remain relatively rare at the time of writing (2015), they are likely to become more widespread due to the increasing synergy between photography and telecommunications;

– finally, programs which are generally installed on a computer and used for postprocessing, in order to derive full benefit from the properties of the recorded signal and to provide additional functions: focus across the whole field, very high dynamics, etc. Many of these programs are still at the experimental stage, and can be used to test functions which may, 1 day, be integrated into camera hardware (section 10.3).

10.1. Integrated software

The software integrated into a camera is intrinsically linked with a given processor. As we have seen, many types of camera may share the same material architecture. Manufacturers, therefore, use software in order to establish limits regarding the use of the processor. Resources which manufacturers do not wish to make available to users are not disclosed in the software, allowing manufacturers to limit their use to other, often costlier, products.

The software installed during the manufacturing process is not set in stone; updates, supplied by the manufacturer, may be downloaded when the camera is connected to a network. In this way, users benefit from feedback received by the manufacturer, and from improvements made since the original program was released.

The basic programs used to reformat signals have been discussed at length in the chapters on image quality, demosaicing and coding, and will not be covered further here. In this section, we will consider those programs which are most widely used for processing reconstructed images in modern cameras.

10.1.1. *Noise reduction*

Noise reduction is an important function in a camera. It is applied automatically in most cases at the same time as demosaicing. In some cases, users have a choice regarding the level of filtering (for example, whether or not to apply noise reduction in the case of long exposure times), but generally speaking, the procedure is fixed, and internal measures are used to determine filtering levels. These elements are of three types:

– ISO sensitivity, which determines the amplifier gain;

– the duration of exposure;

– a dark current, generally measured at unexposed photosites on the edge of the field which are reserved for this use. This current is dependent on sensor temperature, and hence on the conditions of use.

We have not yet touched on the noise reduction filters used within camera hardware; we will now consider the state-of-the-art in this area, as discussed in image processing journals, with a particular focus on those recent developments which offer the best performance. Note that in many cases, however, these types of filtering must be applied later using an external processor, as they are too costly for application by the integrated processor without incurring unreasonable delays.

10.1.2. *Classic approaches*

In the simplest cases, the noise b is taken to be Gaussian and independent of the signal i; these hypotheses are not unreasonable for a first approximation, but do not fully reflect the complexity of the problem. The noisy image i_b is written as:

$$i_b(x, y) = i(x, y) + b(x, y) \qquad [10.1]$$

The simplest form of filtering consists of averaging the signal over a window $V(x, y)$; the size of this window is determined by our knowledge of the scale of the noise.

$$i_{ave}(x, y) = \frac{1}{\alpha} \sum_{(x',y') \in V(x,y)} i_b(x', y') \quad \text{with} \quad \alpha = ||V(x, y)|| \qquad [10.2]$$

This form of linear filtering reduces noise (its variance is inversely proportional to the square root of the size $||V(x, y)||$ of the window), but causes unacceptable degradation to any high-signal frequencies present in the window. For this reason, it is better to use slightly more complex (and nonlinear) filters, such as the median filter, which offers better contour preservation and preserves structures with dimensions of at least half the window size. Another option is the inverse gradient filter, which averages the pixels in the window and weights them using the inverse of their deviation from the central pixel. A final possibility is the bilateral filter, which involves more complex weighting of the pixels in the window (for example, using Gaussian weighting) [PAR 08].

The inverse gradient filter is written as:

$$i_{inv}(x, y) = \frac{1}{\alpha'} \sum_{(x',y') \in V(x,y)} \frac{i_b(x', y')}{max(1, |i_b(x, y) - i_b(x', y')|)} \qquad [10.3]$$

Bilateral filters are written as a function of two functions ψ and ϕ, which are both positive and decreasing; ψ removes values which are too different to

$i_b(x, y)$, and ϕ attenuates the influence of pixels which are too far away:

$$i_{bil}(x,y) = \frac{1}{\alpha''} \sum_{(x',y') \in V(x,y)} i_b(x',y') \psi \left[|i_b(x,y) - i_b(x',y')| \right] \times \phi \left[|x - x'|, |y - y'| \right] \quad [10.4]$$

It is also possible to filter noise in a transformed space. A generic outline of these methods is given in [DON 94], using wavelet representations, and the hypothesis that the signal and noise exist in separate frequency domains. Another form of this technique exploits the architecture of the signal processor integrated into the camera in order to reduce noise in the high-frequency coefficients of the discrete cosine transform (DCT) [YU 11]. More complex approaches involve local adaptation of the image decomposition to the structure of the signal using bases of curvelets, which are wavelets adapted to the contours in the image [STA 02].

These approaches are all based on seeking a signal redundancy to enable noise elimination within an analysis window around the pixel in question. If the noise is strong, the dimensions of the window required become prohibitive. In these cases, noise reduction almost always results in a loss of fine details.

10.1.3. *Iterative methods*

The difficulty of noise filtering lies in our limited knowledge of the noise affecting an image. If the noise is distinguished from the signal by rapid variations, its level can be determined in low-contrast zones of the image; however, in this case, the noise will be harder to detect in zones of rapid signal variation. In these zones, it is easier to detect noise by looking for decorrelation: the signal is generally correlated, while the noise is not. The distinction is, therefore, clearer in a space where the image is decorrelated, such as a wavelet base, Fourier transform, principal components, etc. One approach is to use progressive filtering, alternating between filtering in the image domain (for example, using a bilateral filter) and filtering in a transformed domain (wavelets or DCT).

The dual domain image denoising (DDID) method operates in this way, and is highly effective, while remaining relatively simple [KNA 13]. This approach is based on filtering in both domains (the image domain, using bilateral filtering and wavelets). The progressive image denoising (PID) method, proposed by the same authors, optimizes these treatments by deterministic annealing [KNA 14].

10.1.4. *Non-local approaches*

Another approach has been put forward using a different principle, that of non-local (NL) filtering [BUA 08]. Using this method, multiple pixels are still involved in noise reduction, but instead of these pixels being selected around a specific point of treatment, we look for similar configurations of pixels within an image (for example, a window, or patch, of 5×5 pixels). Filtering is then carried out using the most similar configurations. A number of questions need to be considered:

– where should we look for similar patches? Around the pixel in question, in the whole image, or in a set of images obtained under the same conditions?

– how should we compare patches?

– how do we define the final value attributed to the filtered pixel?

Let $V(x, y)$ be a patch centered on (x, y) and $W(x, y)$ the set of patches used to filter the pixel (x, y). The filtering process is written as:

$$i_{NL}(x, y) = \frac{1}{\alpha'''} \sum_{V'(x',y') \in W(x,y)} \psi \left[d(V'(x', y'), V(x, y)) \right] i_b(x', y') \quad [10.5]$$

where the term $d(V, V')$ expresses the distance between the two patches V and V' and function ψ is a positive function, decreasing toward zero in the case of strong differences between patches.

A variety of responses to these questions have been suggested, depending on the context (image type, noise source and available processing time). Current NL methods are particularly powerful, and compare favorably with the best denoising techniques. They make good use of the high levels of similarity often observed at many points in any image, for example along a contour or within a textured area.

We will now consider a number of particularly effective NL restoration methods in greater detail.

10.1.4.1. *NL-means*

Taking the Euclidean distance as the distance d between patches, and retaining only patches at a given distance from $V(x, y)$ using function ψ, we calculate the mean of all similar patches [BUA 08]. These techniques have been extended for use with color images [BUA 09].

To improve efficiency, ψ may take the form: $\psi = \exp\left[-(d^2 - \sigma^2)/h^2\right]$, where h is a parameter determined by the user and σ is the variance of the noise.

10.1.4.2. *BM3D*

Instead of operating within the image space, we may filter the pile of the most similar patches after wavelet decomposition, carried out using the 3D block matching method (BM3D) [DAB 07] (see section 10.1.6).

10.1.4.3. *NL-Bayes*

To estimate the value of the new pixel, we add the calculation of the covariance matrix for each group of similar patches to the calculation of the mean, in order to fully determine a Gaussian vectorial process. From these parameters, we obtain a first estimation of the desired pixel (in terms of mean square error) by inversion of the correlation matrix. This denoised image is then used to recalculate a more precise correlation matrix for the denoised image, and this is reintroduced in the final stage of the filtering process [LEB 13, LEB 15].

Similar techniques use different optimization methods for the pile of similar patches: singular value decomposition, principal component decomposition or sparsity constraints. Other methods focus on learning, using predefined dictionaries of adapted patches (learned simultaneous sparse coding method (LSSC)) [MAI 09].

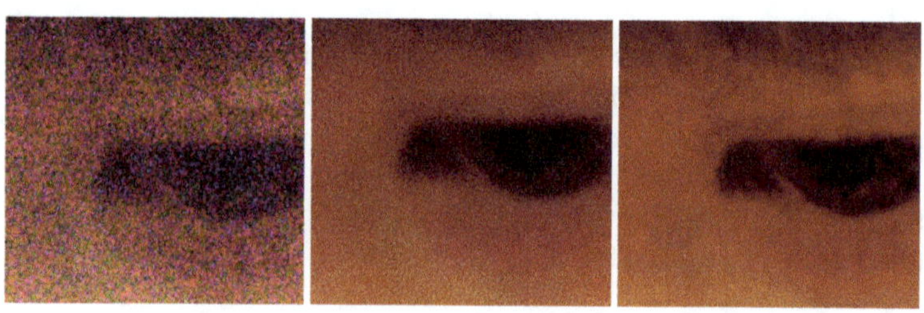

Figure 10.1. *Denoising an image. Left: detail of an original image, with noise. Center: image after denoising using a bilateral filter. Right: image after denoising using non-local Bayesian filtering. For a color version of this figure, see www.iste.co.uk/maitre/pixel.zip*

These filtering approaches perform remarkably well (see Figure 10.1); in simulated conditions, they are able to produce results which are very close to the theoretical limits. However, they are relatively costly in terms of machine time and processing resources (searches within the image, patch archiving, etc.); for this reason, these methods are generally only applied using a separate computer. They are included in a number of image processing programs (NeatImage, DxO Lab, etc.).

10.1.5. *Facial detection*

Facial detection and tracking has come to represent a key aspect of integrated camera software. Manufacturers have focused on the development of this technique for a number of reasons:

– portraiture is a key element of photography; even before the development of photography, it constituted a specific artistic domain in its own right, through mediums such as drawing, painting and sculpture. Academic rules associated with portrait creation affect a variety of key aspects of photography: framing, focus, depth of field, lighting and exposure time. These elements may be controlled by the camera on the condition that the face can be identified and localized. Careful automation of these processes can improve the quality of the photographs produced, especially by non-expert users;

– even in photographs which are not taken as portraits, faces constitute important elements in a scene, and rendering levels are more important for faces than for other components of a scene. A good level of focus on all of the faces in a scene, and ensuring these details are correctly exposed, provides a good basis for default image parameters producing high-quality results;

– while people often form an important part of a scene, they also move. Detecting faces and tracking their movement, maintaining focus throughout the capture process, is an extremely helpful camera feature, particularly in difficult conditions: photographing sports, children playing, events, theatrical representations, etc.

As we have already noted, capture support functions are particularly important for non-expert photographers. For this reason, they were first integrated into compact cameras, then telephones. These platforms often have significant limitations in terms of processing power and energy; this constraint led to the development of particularly efficient algorithms, with their origins in the field of biometry, notably in response to investments in the defense and security sectors [NAI 12b].

Three distinct families of algorithms are used: masking methods, characteristic point methods and methods using Eigenface databases for decomposition purposes.

10.1.5.1. *Masking methods*

These methods, made popular by Viola and Jones [VIO 04], use masks to systematically search for configurations similar to those found in faces. These masks initially took the form of pseudo-Haar functions, made up of black and white zones arranged following a very simple horizontal or vertical geometry. Later versions added windows with different orientations; more complex masks are now used, which make greater use of contours (histograms of oriented gradients (HOGs) [DAL 05]), in order to overcome issues in relation to lighting.

These methods involve systematic movement of a small window (typically 24×24 pixels), through the whole pyramid of representations of the image at various resolutions (these representations are easy to obtain, for example using the JPEG coder, in the case of both DCT (JPEG and MPEG) and wavelet (JPEG 2000) approaches) – see Chapter 8. Window values are then compared to thresholds, determined through a learning process using a large base of examples, in order to determine whether or not a face is present.

Two types of optimization may be applied to ensure that this search process is carried out quickly:

1) Window values for each position are precalculated by the digital signal processor (DSP) pipeline in a highly efficient manner (for pseudo-Haar masks, this takes the form of a precalculation of integral images [VIO 04]).

2) The recognition process for each position is accelerated using a decision tree, where the earliest nodes relate to the clearest discriminating criteria; non-candidate windows are, therefore, rapidly eliminated from the process. The decision tree is the product of a statistical learning process, constructing reliable classifiers on the basis of other reliable classifiers, for example using the AdaBoost method [SHA 99].

Masking techniques were the first to produce results for multiple facial detection in real-time video sequences.

10.1.5.2. *Characteristic points*

Characteristic point methods use very general shape detection techniques which draw on statistical learning tools.

This is carried out by detecting notable or characteristic points in an image, representing each of these points using a descriptor. We then look for identical configurations of characteristic points in a database. If a set of characteristic points is found which is very similar to configurations identified as belonging to faces, then the method concludes that a face is present. The characteristic points are recorded, which allows us to track them over time.

Characteristic points are always local maxima of contrast, which may be calculated either by derivative measures, or by applying template matching [MIK 04, MIK 05].

The choice of characteristic points was initially carried out using Harris points, before moving on to size invariant feature transform (SIFT) detectors.

Harris points are obtained from the matrix M of the second derivatives of the filtered image. A criterion H is calculated using $H = \det(M) - \kappa \text{Trace}^2(M)$, where κ is a parameter affecting the sensitivity of the detector. We then seek local maxima of H after successive applications of filtering. Filtering is carried out using a low-pass filter supplied by the DSP pipeline.

SIFT detectors also look for local contrast maxima (measured using the Laplacian), but they act in the scale space. Each point is subjected to Gaussian image filtering, with Gaussians of standard deviation σ_i which increases following a geometric progression of factor 2. A point is retained as a characteristic point if it constitutes an extremum both in terms of space (neighborhood of 3×3) and scale (neighborhood of $3 \times 3 \times 3$) [LOW 04, MIK 05]. Interpolation is then used to determine the precise position of the point.

Generally, several tens of characteristic points are available for an image.

As far as possible, these points use representations which are robust in relation to the deformations applied to images: not only translations, but also rotations and affinities. The neighborhood of a characteristic point is described by variations in the signal in various directions, and often at various ranges. Each point is, therefore, described by a vector with between 10 and 100 components.

As an image may include faces of very different sizes, the search process is applied on multiple scales. This is generally carried out using a pyramid decomposition, as before, where each level is subject to a specific search process.

10.1.5.3. *Eigenfaces*

These methods are more commonly used for facial recognition than for facial detection within images [NAI 12a]. They are based on a learning stage using a database of faces, all reduced to the same size and placed in the same position (generally using the height of the ocular retinas, and the end of the nose or the gap between the lips). Let $v_k(i,j)$ be the kth face in the base, expressed as a vector of size N^2 by developing the image: $w_k(l = i + Nj) = v_k(i,j)$. A matrix W, of size $N^2 \times K$, is then created by juxtaposing the K vectors w_k.

This databased is decomposed using principal component analysis (PCA). To do this, we determine the Eigenvectors ν_k of the matrix $\mathcal{M} = WW^t$, which are then placed in decreasing order using the Eigenvalues. The truncated base of the first Eigenvectors is used to represent any new face using a small number of components. If the truncated representation is very similar to the original, a face is said to be present; otherwise, the hypothesis is rejected. The value of the coefficients also provides information regarding the similarity of the detected face to precalculated reference faces.

10.1.6. *Motion tracking*

As we have seen, the need to detect movement between two images or two zones of images is something which occurs on a regular basis. This applies not only to focal tracking, but also to a number of image treatment processes, such as the creation of high-dynamic images (section 10.3.1) or images with a large depth of field (section 10.3.2) from multiple images captured in burst mode. Motion detection also constitutes an important step in video coding (section 8.8.2), as movement prediction leads to a significant reduction in bandwidth requirements.

Using two successive images, $i(x,y,t)$ and $i(x,y,t+dt)$, one very intuitive approach consists of seeking the displacement (dx,dy) which locally minimizes the following difference:

$$\Delta i_t = \sum_{x,y \in \mathcal{D}} [i(x+dx, y+dy, t+dt) - i(x,y,t)]^2 \qquad [10.6]$$

Domain $\mathcal{D}$ is chosen by the user; this quantity must be sufficiently large to guarantee the quality of the estimation, and sufficiently small to take account of fine details in the image, which may move in very different ways. For fixed images $i(x,y,t)$ and $i(x,y,t+dt)$, the minimization of Δi_t is very similar to maximization of the kth crossed cross-term obtained by developing equation [10.6], which then gives the correlation between the two images (block-matching technique):

$$\min_{dx,dy} \Delta i_t \approx \max_{dx,dy} \sum_{x,y \in \mathcal{D}} i(x+dx, y+dy, t+dt) i(x,y,t) \qquad [10.7]$$

The correlation function can then be calculated either by using equation [10.7] directly, or using a direct product following a Fourier transformation of the two functions.

A less intuitive approach involves considering that the luminance is invariant over time ($\exists(dx,dy) : \Delta i_t = 0$ (equation [10.6]). This gives us the following variational equation [KON 00, BEN 02]:

$$\frac{\partial i}{\partial x} dx + \frac{\partial i}{\partial y} dy + \frac{\partial i}{\partial t} dt = 0 \qquad [10.8]$$

which is known as the optical flow equation. This equation is defined for each pixel (x,y), with each measure including two unknown quantities (dx,dy).

Another family of solutions involves searching for the best field of speed $\mathbf{v} = (dx/dt, dy/dt)$ over the image as a whole (or over the subdomain $\mathcal{D}'$ under consideration), subject to regularity constraints. The first work using this principle [HOR 81] involved a simple regularization term, minimizing:

$$\int_{\mathcal{D}'} \left[\left(\frac{dx}{dt}\frac{\partial i}{\partial x} + \frac{dy}{dt}\frac{\partial i}{\partial y} + \frac{\partial i}{\partial t} \right)^2 + \alpha \left(\left[\frac{\partial \mathbf{v}}{\partial x} \right]^2 + \left[\frac{\partial \mathbf{v}}{\partial y} \right]^2 \right) \right] dxdy \qquad [10.9]$$

This minimization produces very large systems, which are then solved using multi-grid or overrelaxation approaches [WEI 06]. However, these systems are often too large, and iterative solutions are needed after linearization. The regularization term in equation [10.9] takes a very simple form, but other variants have been put forward, using different norms or functionals, and exploiting particular properties of the speed field (for example, in the presence of a contour, we can only obtain the component of the speed which is normal to this contour). These improvements generally produce considerably better performances, but are also more costly in terms of processing power [WEI 06].

10.1.7. *Image rotation*

Rotating an image by $\pm$ 90° is an essential function for cameras, used to enable portrait or landscape display independently of the orientation of the camera. This function is provided by the display processor. As vertical, horizontal or point symmetry (which is extremely rare) are exact transformations, these rotations do not affect the quality of the recorded image, and are fully reversible. Rotation is implemented simply by means of pixel addressing.

Rotations by any other given angle α, on the other hand, are often complex and irreversible; for this reason, they are generally not carried out by the camera itself, but using offline software. However, the angle of vertical lines and the horizontality of the horizon are often important in determining image quality, and many cameras provide tools, such as grids and frames, to assist users in this area.

In spite of the precautions taken during the capture process, it may still be necessary to correct the angle of images in postprocessing; this must be carried out efficiently, minimizing the number of operations, and precisely, to minimize distortion.

The rotation of an image $i(x, y)$ by an angle α (different to $k\pi/2$) is specifically expressed using a decomposition of the rotation matrix into a product of two terms:

$$\mathcal{R}(\alpha) = \begin{bmatrix} \cos\alpha & -\sin\alpha \\ \sin\alpha & \cos\alpha \end{bmatrix} = \begin{bmatrix} 1 & 0 \\ \tan\alpha & 1/\cos\alpha \end{bmatrix} \times \begin{bmatrix} \cos\alpha & -\sin\alpha \\ 0 & 1 \end{bmatrix} \quad [10.10]$$

The rotation process is thus transformed into two unidimensional operations (lines and columns) involving the compression and expansion of the signal; these operations are easy to carry out using a DSP architecture. An even more efficient form can be obtained using three products, eliminating the compression element:

$$\mathcal{R}(\alpha) = \begin{bmatrix} 1 & -\tan(\alpha/2) \\ 0 & 1 \end{bmatrix} \times \begin{bmatrix} 1 & 0 \\ \sin\alpha & 1 \end{bmatrix} \times \begin{bmatrix} 1 & -\tan(\alpha/2) \\ 0 & 1 \end{bmatrix} \quad [10.11]$$

This second approach, which can be applied simply using three successive shifts of N lines, has been shown to conserve high frequencies more effectively and produces more precise calculation results [UNS 95b]. Precision is determined by the size of the convolution node. For low orders (less than 7), it is best to carry out this convolution by direct calculation. When using longer filters, it may be advisable to replace the convolution by a product of Fourier transforms [COW 99].

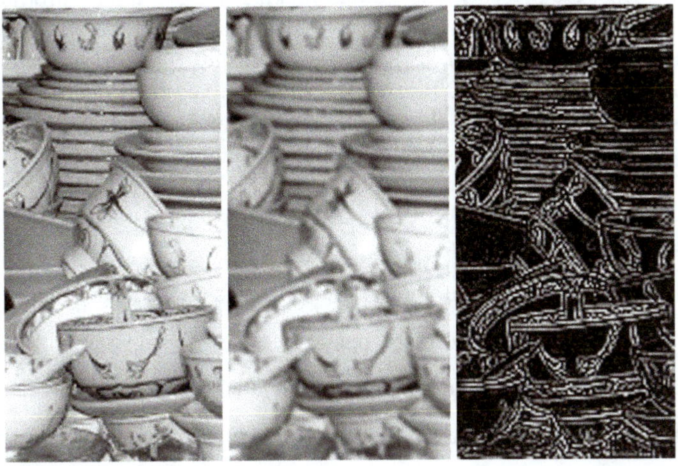

Figure 10.2. *Applying 10 successive rotations of $36°$, the image returns to its original position, but presents faults along the contours, due to successive approximations. Left: detail from the original image. Center: detail from the image after 10 rotations. Right: illustration of the differences (the contrast level has been multiplied by 10)*

Note, however, that any rotation which is not a multiple of $90°$ will degrade the image (see Figure 10.2). It is, therefore, best to avoid cumulating successive rotations when seeking the best orientation; in these cases, each test should be carried out using the original image. Furthermore, a rotation by an angle α followed by a rotation of $-\alpha$ will not give us the original image, unless a reversible rotation process, such as that proposed in [CON 08], is used; in this approach, convolution is carried out using fractional delay filters.

10.1.8. *Panoramas*

Many modern cameras allow users to assemble multiple images in order to create a single image with a wider field than that permitted by the camera lens assembly. Initially, this took the form of a function to assist users in taking a sequence of pictures which could then be reassembled later using computer software. These programs often made use of capture parameters specified in the EXIF file (particularly focal distance) for geometric-matching purposes. Other tools were developed to allow the creation of panoramas from any set of images with significant levels of overlap (if a certain number of properties were present), but without making use of capture information[1]. More recently, cameras have come to include integrated software enabling real-time panorama creation, using series of images obtained either in burst or, more often, video mode (i.e. with a higher, fixed-time interval between images). We will examine these products in greater detail below, with a description of some of the treatments which they apply.

Panoramas raise issues of two types: geometric and photometric. Geometric problems include:

– the choice of the final framework used to present the panorama;

– identification of the geometric operations used for registration;

– the treatment of the line separating individual images (the "seam");

– the treatment of mobile or deformable objects.

The main photometric issues concern the homogeneity of radiometric measures (luminance, contrast and white balance) and concealment of the joining line.

10.1.8.1. *Geometric problems: framework selection*

In a small number of cases, the projection framework is predetermined, for example in remote sensing, where convention dictates that all pixels must be placed onto the terrestrial geoid, or a flat local approximation, such as a Lambert and Mercator projection. This is also the case when representing a painting, or the facade of a building. In most situations, however, the projection surface is chosen in order to provide the best possible reconstitution of a scene, minimizing distortion of certain details; these include the relative size of objects, independently of their position in the mosaic, the quality of alignments, the parallel or orthogonal nature of contours, etc. These requirements are mutually exclusive. In practice, while a number of options exist, only three projections are used: onto a plane, onto a cylinder or onto a sphere (see Figure 10.3). The first option is used when representing distant objects viewed at a relatively small angle. Cylindrical

---

1 See Helmut Dersch's Website, *http://webuser.hs − furtwangen.de/ dersch/*, which provides open access to a set of "Panorama" tools. An updated, enriched version of this content is now available at *http://www.ptgui.com/panotools.html*.

projection is better for objects with low vertical range but a wide horizontal angle of observation. Finally, spherical projection is particularly suited to objects captured at close range, or with wide horizontal and vertical ranges (Figures 10.4–10.6).

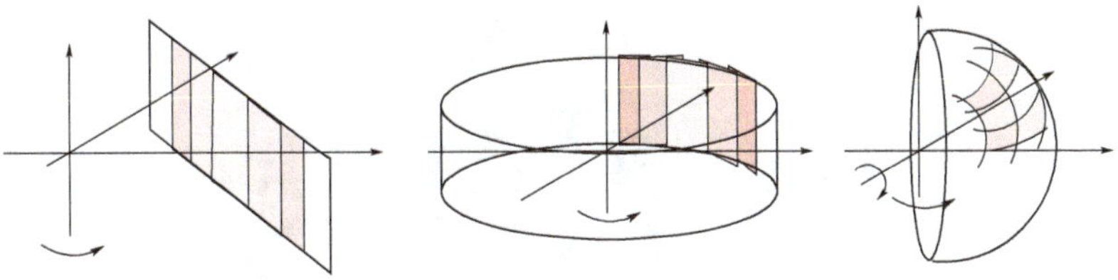

Figure 10.3. *Three geometric forms used in reconstructing panoramas: projection onto a plane (left), a cylinder (center) and a sphere (right)*

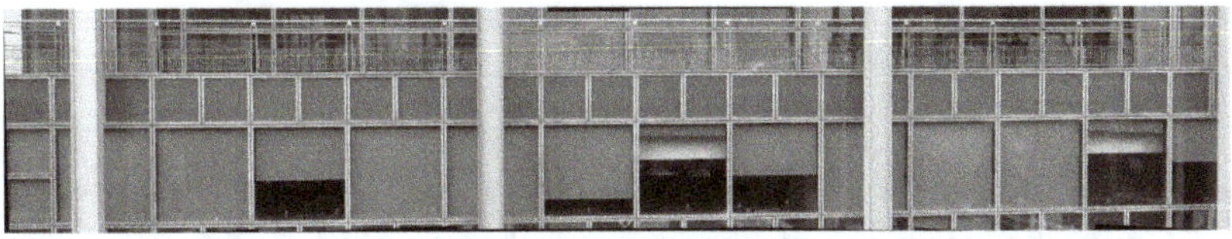

Figure 10.4. *Panorama constructed from six images of a facade, each with a field of $17° \times 13°$. The reconstruction has been carried out using a cylinder (with a total range of $60°$), prioritizing the linear and parallel representation of horizontal lines*

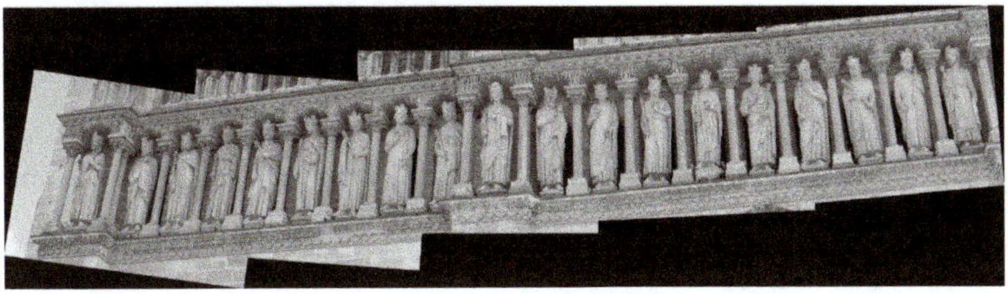

Figure 10.5. *Panorama constructed from 5 images, each with a field of $17° \times 13°$. The reconstruction has been carried out using a cylinder (with a total range of $82°$), prioritizing the linear and parallel representation of horizontal lines*

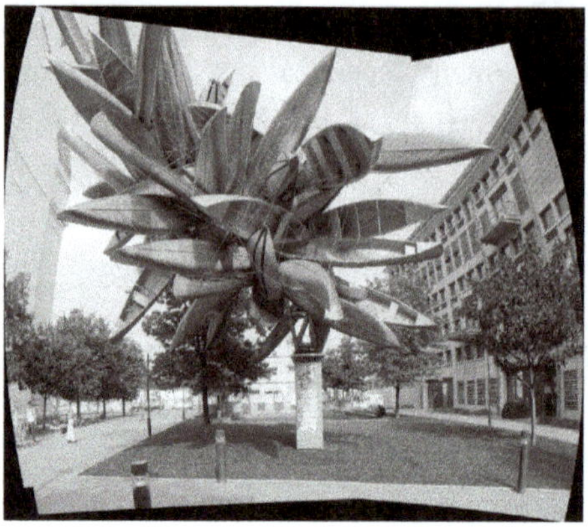

Figure 10.6. *Panorama constructed from nine images, each with a field of $32° \times 24°$. The reconstruction has been carried out using a sphere (with a total range of $93° \times 81°$). In this case, the size and appearance of the components of the monument takes priority, independently of their position in the image. The vertical lines present a strong curve toward the pole*

Given the geometry of the photographs taken by a camera, exact reconstruction of a scene is only possible in a very limited number of cases:

– when the camera lens assembly is rotated around its optical center;

– when the movement of the camera does not reveal elements which were hidden in the reference scene. This scenario occurs when a scene extends to infinity, and does not include any significant reliefs (as in spatial remote sensing); this also applies to flat or quasi-flat objects at finite distance (photographs of paintings, facades, etc.).

Representing a pixel in an image, of dimensions $\delta x \times \delta y$, by its indices (i, j) in relation to the projection center (see Figure 10.7), the point in the reconstructed space will have the following coordinates:

– using a flat projection: (i, j);

– using a spherical projection: $\left(\frac{p}{\delta x} \tan^{-1}(\frac{i\delta x}{p'}), \frac{p}{\delta y} \tan^{-1}(\frac{j\delta y}{p'})\right)$;

– using a cylindrical projection: $\left(\frac{p}{\delta x} \tan^{-1}(\frac{i\delta x}{p'}), j\right)$;

where $p'/p \sim f/p$ is the transversal zoom of the photograph.

10.1.8.2. *Mosaics*

The first steps taken in the field of image assembly concerned the creation of image mosaics for satellite applications [MIL 77]. In this case, the image assembly problem is relatively simple, as all points in the scene are at a more or less infinite distance in relation to the sensor, and scenes include very few concealed elements. Work in this area led to the definition of the key steps involved in any image assembly processes:

1) Image registration: this is sometimes based on precise knowledge of capture conditions, including all of the calibration parameters (see section 2.7.1); in other cases, points on the ground or points shared between images are used to define the transformations to apply to the geometry of the images, or to determine unknown parameters. This process is often followed by a correlation process around characteristic points (as described in section 10.1.6).

2) Geometric transformations are very often applied using a precise equation, which guarantees that the result will present the desired geometric properties: homography inversion and consideration of the curvature of the Earth's surface or a digital terrain model, so that each pixel is associated with a precise terrestrial latitude and longitude.

3) Choice of a representation of the image in the overlap zone: this may involve the selection of one of the two possible images (this gives a high level of precision in terms of details, but can result in abrupt transitions if the registration is not perfect), or the choice of a mixture of the two images, giving a seamless transition between the two (although this may result in blurring in the overlap zone). Elegant solutions have been proposed which create images using low frequencies from a mixture of the two sources (i.e. a broad neighborhood) and high frequencies from the dominant image (i.e. the image in which the point in question is closest to the center) alone [BUR 87].

4) Choice of a joining line between the two images. For reasons of simplicity, straight lines are often used; however, wherever possible, it is better to choose zones of very high contrast, which help to conceal minor radiometric or geometric errors.

5) The final stage involves radiometric or colorimetric homogenization on either side of the joining line, or more generally across the whole of the two images to avoid variations in appearance between the two sections of the image [RAB 11, FAR 14].

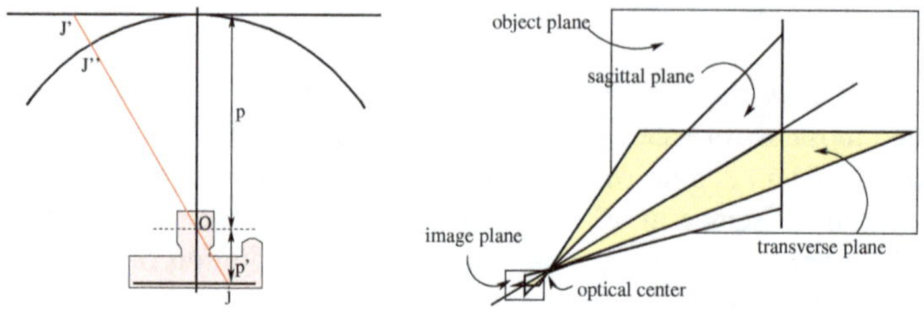

Figure 10.7. *Left: construction of an image using flat projection (point J'), or cylindrical and spherical projection (point J"). View of the transverse plane alone. Right: the transverse and sagittal reference planes used*

10.1.8.3. *Multimedia panoramas*

In more general multimedia domains, little information is available concerning the absolute orientation of the cameras used to capture the images requiring adjustment (unless test cards were used for calibration beforehand), and no information is available regarding the geometry of points in the scene. In these cases, relative calibration between images is used, with one image serving as a reference for the others (Figure 10.6), profiting from the number of available images in order to determine the unknown values in the problem (see the discussion on beam adjustment in section 2.7.1).

To do this, we look for matching points in overlap zones, taking images as pairs. We then carry out approximate registration of the images two-by-two in order to initialize the transformation. As the focal length is generally fixed, if the movement of the optical center (camera rotation) is small, then the intrinsic calibration parameters are fixed, and the calibration equations [2.47] are simplified; this is also the case if the camera movement constitutes a pure translation. In the absence of aberrations[2], points can be transported by homography; this may sometimes be simplified into an affinity by a first-order development of equation [2.47] if the depth of field is low. Finally, transformations are optimized globally, taking account of all the images used, by a beam adjustment which optimizes the whole registration process [BRO 07].

---

2 It is, therefore, best to correct geometric aberrations before attempting to create a mosaic (see section 2.8).

Open-access programs are now available, including AutoStitch[3], Hugin[4] and Image Composite Editor[5].

10.1.8.4. *Creating panoramas from a video flux*

Creating a panorama from a dense video flux, with regular but unmonitored sensor movement and a scene which is both mobile and unknown beforehand, requires the use of multiple hypotheses which allow us to obtain satisfactory solutions in real time; however, the properties of these solutions are often difficult to control. The basic principle used in these approaches was outlined in [PEL 00], but each manufacturer has made his own adjustments, meaning that each product is somewhat different.

The basic idea is to select a small area of each image, generally a narrow strip, to be used in the final image; this band is attached precisely to the previous image on one side, and to the following image on the other side (see Figure 10.8). The form of this area is defined based on the movement of the sensor, and is thus determined either using the optical flow measured throughout the sequence, or using the gyrometer and accelerometer included in the camera (see section 9.6).

Optical flow analysis is simple to implement, and provides a precise measure of movement at the important points of interest, expressed directly in pixels, which makes the mosaicing process easier. However, this method is easily influenced by moving objects, if multiple objects of this type are present in sensitive areas. Measures produced by movement sensors, however, are only linked to the motion of the camera and are not affected by image content. However, it needs to be transformed as we move across the image using capture parameters, including the focal length, focusing distance and, where applicable, the distance from the scene being observed.

Movement calculations are rarely continued up to the point of precise solution of the colinearity equations [2.47] or [2.49], and simplifying hypotheses are generally used. In the case of sensor rotation around a vertical axis (angle β in Figure 2.29), we presume, for example, that all of the

3 *http://cs.bath.ac.uk/brown/autostitch/autostitch.html*: AutoStitch repository by M. Brown, made available by Industrial Light & Magic (ILM).

4 *http://hugin.sourceforge.net/*: HUGIN repository based on work by H. Dersch, uses Panorama software.

5 *http://research.microsoft.com/en-us/um/redmond/groups/ivm/ice/*, produced by Microsoft.

observed points are located at a constant and very large distance Z, that the displacement of the optical center is null and that the tilt rotation (angle α) is also null. If non-null terms R_{12} and R_{21} appear during the analysis (equation [2.48]), they are often assigned to a rotation around the optical axis (angle γ); this rotation is corrected and the estimation is then reiterated in order to determine angle β, the unknown quantity in the problem.

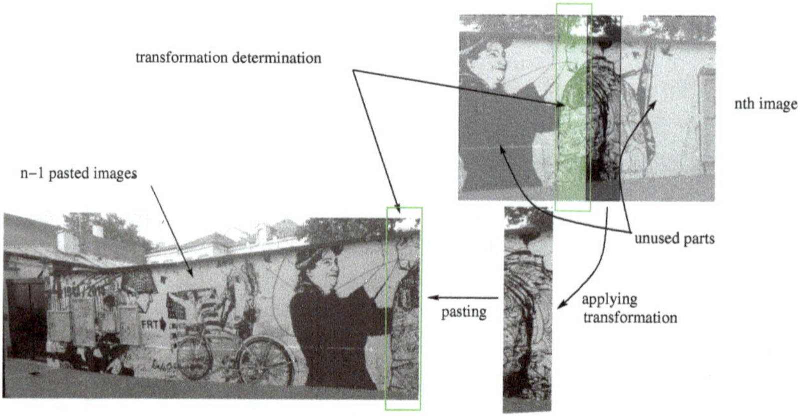

Figure 10.8. *Principle of panorama construction from a video flux. A strip is defined in image n. The width of this strip is determined by the magnitude of the sweeping movement, and its position is fixed in the center of the sensor. The geometric transformation to apply is determined by the strip immediately to the left*

Once the movement has been identified, we determine the zone in image n which will be included in the panorama. This is generally a strip with a width corresponding to the magnitude of the movement, with a form which is as long as possible perpendicular to the direction of movement. For a horizontal sweeping movement, this will, therefore, be a vertical strip covering the whole image. If the camera has moved closer to the object, we use a ring around the vanishing point. The selected strip may remain the same throughout the whole sequence, or the width may vary depending on the speed of movement. Generally, either one edge of the strip has a fixed position in relation to the sensor, or the strip is located around the central axis of the sensor. It is best to situate the strip close to the center of the sensor in order to avoid geometric distortion and focusing errors. The strip is deformed geometrically in order to fit precisely with the previous image along one edge (for example, the left side for left-to-right horizontal movement), while integrating geometric elements resulting from movement analysis. In the absence of full image calibration, and due to the low levels of movement affecting the sensor, a very simple model of this movement is often selected: in the very simplest cases, we may use homothety, affinity, or a combination of homothety, plane rotation and translation. The presence of a large overlap zone (the strip itself and a portion of the image to the left of the strip) is required in order to successfully determine the transformation (by block matching or the identification of characteristic points) and to guarantee high levels of continuity between images. The joining line is often a fixed geometric line, i.e. the edge of the strip. However, a high-contrast line close to this edge, and which also traverses the whole of the shared area, may also be used. Photometry is adapted on an image-by-image basis over the course of the capture process, using an estimation based on the overlap zone (adjustment of the mean level and the variance). If there are photometric variations during the capture process, graduations will be applied from the level of the first image to that of the final image (linear interpolations of the mean and square, applied as an offset and a gain).

In cases of irregular sensor movement, or if the scene is subject to distortion, points in a given strip may not exist in image n. In this case, linear interpolation is applied between neighboring points in order to fill in the hole. This technique notably allows us to avoid "rips" in cases of excessively fast movement.

In the case of left to right horizontal movement, the final image includes the whole of the first image up to the join with the second image, followed by strips from the second and subsequent images; the section of the final image to the right of the selected strip will also be included in its entirety (Figure 10.8).

In the absence of specific precautions during the capture process and when selecting geometric transformations, the geometry of the obtained images will be mediocre and the reconstructed image may be barely usable (Figure 10.9). Fronto-parallel translations, translations along the target line and rotations perpendicular to the optical axis offer the only means of controlling the geometry of the obtained image and generating something similar to a single image of the same scene obtained using a sensor with a different lens configuration. Mass-market image processing tools generally do not respect these constraints. Entertainment applications have even been developed which specialize in creating dramatic distortions in images, either by moving the camera or based on the movements of individual subjects during the capture process.

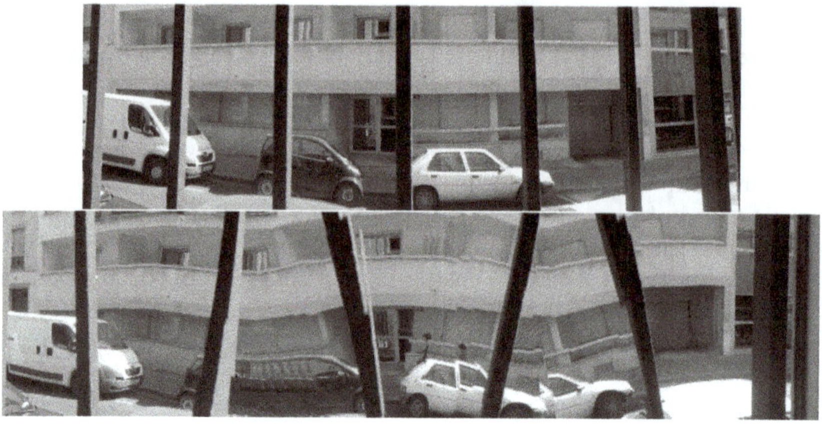

Figure 10.9. *Top: extract from a real-time automatic panorama of 25 images, obtained by rotating the sensor in a horizontal plane. The use of vertical strips produces a high-quality reconstruction. Bottom: the same panorama, but with a parasite rotation in a plane vertical to the sensor. The mosaicing method used to match up vertical strips is unable to identify overlap zones, provides a poor estimation of distortions and does not produce a satisfactory representation of the scene*

10.2. Imported software

As we have seen, modern cameras are increasingly similar to generic computer systems, connected to networks, and able to integrate additional software not included by the manufacturer. This tendency has been reinforced by the use of generic material platforms (such as the TMS320) and universal operating systems (such as the Android platform). Increased accessibility via USB or wireless technology, the ease with which users can now exchange images, and the need for regular software updates have also encouraged this tendency, promoted both by networks of aficionados and by professionals, working with large companies, who adapt mass-market products for specific research, cinematographic or defense purposes. Note that a certain number of open platforms are already available, such as the Franken camera [ADA 10], designed essentially for scientific research purposes.

Importing software in this way is something most manufacturers attempt to discourage, and these operations are only supported in specific, highly regulated circumstances. Camera warranties generally become invalid if any new software is added. This rather conservative policy may not last indefinitely; if manufacturers do continue along these lines, service providers are likely to emerge in order to profit from this promising new market. The services currently available in this area are of two types: the improvement of functions already included in the camera and the creation of new functions.

10.2.1. *Improving existing functions*

Improvements to camera performance partly result from *unlocking* functions which the processor is able to provide, but which are blocked by the existing software, and from the replacement of existing software by other, more powerful programs.

As we have seen, processors are often able to provide better performances than those offered by the camera. This is due to the fact that, for economic reasons, the same processor unit is used for whole ranges of cameras, but for marketing reasons, not all of the functions of this processor will be available using each platform. Classic examples of this type of performance improvement involve activating RAW and high-dynamic range (HDR) modes in basic cameras which are only meant to produce JPEG images.

More powerful programs can be used in certain cameras to replace the native software provided for demosaicing, white balance and/or facial tracking. Note that programs imported from outside often simply anticipate manufacturer updates, unless the product in question is considered to have reached the end of its useful lifespan, at which point manufacturers generally want users to switch to another model from their range. In these cases, the installation of up-to-date software gives users the means of extending the lifespan of their camera, but, as with computers, there is a risk that successive improvements will render the camera inefficient in terms of processing power and memory capacity. While identical components or processors may be included in models at opposite ends of a commercial range, manufacturers are more likely to select high-performance components for their top-end models. In these conditions, the processor included in a compact camera will not necessarily be able to support functions other than those proposed by the manufacturer for that specific model.

10.2.2. *Creating new functions*

The importation of functions not traditionally provided by cameras is particularly relevant in cases where the camera device is associated with powerful communication tools, notably cellular telephones and tablets. The generalization of communication abilities has led to an increase in their field of application.

The first group of treatments we will consider concerns shape recognition and automatic reading capacities, used to annotate files via the "user" fields of the EXIF file. This recognition process occurs outside of the camera itself, which sends a request to an online service. Printed text, in a variety of fonts, can now be recognized with very few errors if the capture conditions are suitable. This method is particularly effective for applications such as tourist guides, and has been widely used at sporting events, where the recognition function is used in tandem with a translation function; this example lies well outside of the traditional limits of photography. Performance in terms of handwriting recognition, however, is still extremely limited.

Symbol recognition applications, for example barcode and flashcode readers, operate in a very similar way; applications of this type are also used to recognize logos of various types as used in airports, train stations and public service buildings. As in the case of text recognition, systems which include even a very basic keyboard, such as telephones and tablets, offer a much wider range of possibilities; it seems likely that keyboard functions will be integrated into increasingly camera models in the near future, for example via touch screens. These applications could be particularly useful in assisting elderly or handicapped people, but would need to be used alongside a suitable acoustic or tactile output system, not generally included in cameras.

For many photographers, the most interesting new possibilities relate to the recognition of image content via the use of specialized sites, for example to identify fruits, flowers or plants [BEL 14], birds, stars, paintings or other works of art. Other applications cover manufactured objects, such as items of clothing. While these examples are currently the most widespread, many other domains are likely to be covered by similar approaches in the near future.

These applications are based on lists of particularly distinctive criteria found in images, such as color, morphology and contours, generally produced in collaboration with specialists from a specific domain. A database is then established to represent the domain, where each specimen is identified using values for these criteria. Automatic learning techniques are used to identify criteria variation domains for each subclass of the taxonomy of the domain. When a user submits an unknown sample, the same measures are calculated for the sample, and its position in the parameter space determines the class to which it will be assigned. Many applications also make use of additional information such as the capture date and precise location, if a GPS service is used.

The use of GPS, camera timeclocks and gyroscopes[6] has created new possibilities for extremely impressive applications, such as star identification and the localization of key summits in a mountain landscape.

New functions may also relate to a specific area of use, which requires certain special effects, for example chromatic effects, burst effects or flash effects. Certain functions require hardware adjustment; for instance, removal of antialiasing or infrared filters is now relatively common, as is the introduction of polarizers, chromatic filters and, in extreme cases, coded apertures.

10.3. External software

10.3.1. *High-dynamic images (HDR)*

The study of HDR imaging was initiated by Mann and Picard in 1995 [MAN 95b]. Based on the observation that it is very hard to represent real scenes featuring very dark and very luminous zones simultaneously using ordinary 8-bit image dynamics, the authors proposed the combination of two separate images, one focusing on the contrast in light zones, the other focusing on the contrast in darker areas. This issue was already well known in the context of film laboratory, where specific solutions were applied in laboratory settings from a very early date[7].

Initially limited to fixed scenes taken using a tripod, HDR imaging solutions were rapidly extended to hand-held photography, and then to dynamic scenes. HDR mode is now available on a number of cameras and cellular telephones, where it is used routinely for everyday photographs. In more traditional cameras, which do not include this feature, HDR photography is made considerably easier by the use of a bracketing function[8].

6 Gyroscopes permanently measure the orientation of a target axis in relation to the Earth. Gyrometers (as described in section 9.6.1) measure variations in orientation over time, without referring to a particular framework. Many cameras include gyrometers (for view stabilization purposes), and some models now include gyroscopes.

7 Two photographs taken by Gustave Le Gray in 1856–1857, *Brick au clair de lune* and *La grande vague, Sète*, use a technique known as combination printing to represent both strong lighting in the sky and dark shadows on the surface of the sea.

8 Bracketing allows users to take a series of three or five images, with different parameters, in a fraction of a second. Bracketing can thus be carried out on the basis of exposure time, aperture or focal distance. There is also nothing to prevent bracketing on the basis of film sensitivity or zoom levels.

HDR images can be created through bracketing on the basis of exposure time or aperture. Note, however, that bracketing using exposure time can be carried out electronically, and requires no specific mechanical action; in this case, shorter bursts can be used, and this represents a considerable advantage.

The discussion below is based on the work carried out by Aguerrebere [AGU 14b].

HDR techniques require the use of at least two low-dynamic photos: one for lighter tones, or (*high keys*), in which the darker areas are likely to be uniformly black (or "blocked"), and one for darker tones (*low keys*), in which the lighter areas will be saturated; however, more photos may be used. This results in an increase in the quality of the estimated signal, as more measures are used, but creates certain problems in relation to areas of movement. A number of authors have suggested taking all of the information required to reconstruct an HDR signal from a single image [NAY 00] using a modified sensor. These solutions will be examined in greater detail below.

Finally, note that the use of images in native format (RAW, see section 8.2) often reduces the scale of the HDR problem by using dynamics greater than 8 bits. If we wish to code an image using 1 byte, only part of the HDR problem is applicable, i.e. the rendering of high-contrast signals on terminals which are universally designed to only display limited ranges.

10.3.1.1. *Formalization*

A precise relationship between the value I_j^x of the image signal at pixel x during the exposure time τ_j and the radiance i^x of the observed scene is important in order to derive full benefit from the various measures obtained from an exposed pixel. Let Ψ_j denote this relationship (index j reminds us that this relationship is dependent on the exposure time τ_j):

$$I_j^x = \Psi_j(i^x) \qquad [10.12]$$

Models of this type, taking account of the various properties of sensors (photon noise, current conversions, electronic noise, etc.), were discussed in detail in Chapter 7, giving us equations [7.10] and [7.11].

The family of k values $\bar{i}_j^x, j = 1, \ldots, k$, obtained by inversion of equation [10.12] (on the condition, of course, that Ψ_j is invertible), generated by k captures, is used to obtain a single value $\tilde{i}^x$ judged to be most representative of the scene.

The model initially proposed by Mann [MAN 95b] involves the use of a weighted mean of the k available measures:

$$\tilde{i}^x_{mann} = \frac{\sum_j^k w_j^x \Psi_j^{-1}(I_j^x)/\tau_j}{\sum_j^k w_j^x} \qquad [10.13]$$

The authors selected weightings w_j^x which gave an advantage to pixels in the middle of the dynamic range, and accorded low confidence levels to both light and dark values. This choice works well for film cameras, as the intermediate zone is the most linear area.

Another option is to fix $w_j^x = \tau_j$: this gives us a simplified equation, as an identical treatment is then used for all pixels. In the case of Poisson's noise, this produces the following non-biased estimator of minimal variance:

$$\tilde{i}^x_{simpl} = \frac{\sum_j^k \Psi_j^{-1}(I_j^x)}{\sum_j^k \tau_j} \qquad [10.14]$$

A final possibility, often chosen for reasons of simplicity, is to take the highest value of i_j^x which is not saturated as the value of $\tilde{i}^x_m$, as this value is least affected by photon and thermal noise.

While these solutions are highly practical, they do not produce the best results, and further propositions have been made to improve performance.

10.3.1.2. *Criterion selection*

This problem may be approached in a number of ways.

First, we may use a full model of Ψ, which takes account of all the details of the acquisition system and derives full benefit from all of the available knowledge regarding this system (as in equations [7.10] and [7.11]). This approach is used in [GRA 10b] and [AGU 14b]. Alternatively, we may consider the situation at the end of the image production sequence, after corrections have been applied by the camera (as in [MAN 95b] and, more recently, [REI 10]), either because we trust the manufacturer to optimize noise processing, or because we do not have the information needed to correctly implement this step.

Methods are also distinguished by the optimization criterion used to deduce $\tilde{i}^x$ from the values of i_j^x: noise minimization, optimization of the signal-to-noise ratio, maximization of the likelihood of the estimation, etc.

The solution presented by Kirk [KIR 06] has proved to be particularly effective in practice [AGU 14b]. In this case, we look for the maximum likelihood $\tilde{i}_k^x$ based on the hypothesis of Poisson noise. Unfortunately, using this method, it is not possible to obtain the exact maximum likelihood, as we do not have the mean or the variance of the noise for each pixel; using the Poisson model, both quantities depend on the received energy. Kirk [KIR 06] proposes a two-step approximation process; this process was further improved by the use of iterative calculations in [GRA 10b].

Aguerrebere shows that, in practice, this estimator comes very close to the theoretical limit fixed by the Cramer–Rao bounds, and there is very little room for further improvement, notably due to the very small number of measures used to estimate the optimal value (in practice, only a few photos are used). However, the author suggests that taking account of saturated pixels in the estimation formula would result in minor performance improvements; these pixels are generally ignored in the decision process.

10.3.1.3. *HDR and moving scenes*

The results presented above were obtained on a pixel-by-pixel basis, presuming that the same signal always creates an image in the same pixel. This hypothesis does not hold true in two situations: first, when successive photographs are obtained in hand-held mode, and second, when the scene changes over the course of successive shots. In these cases, the situation is more complex. If it is possible to estimate the shift applied to an object between shots for each pixel, then the method used for fixed shots can still be applied. Registration techniques for very similar images, developed for both compression and stereovision applications, often make use of optical flow techniques [ROT 07, SUN 10, MOZ 13]. These methods are efficient and precise, but when used in addition to an estimation process, they do not allow us to optimize the process as a whole.

The solution proposed in [AGU 14a] uses an NL method, like those described in section 10.1.4, which simultaneously compensates for small shifts resulting from operator movement, allows us to disregard the movement of mobile objects, and estimates the value $\tilde{i}_j^x$ which is most compatible with the context of the image.

In the case of HDR reconstruction, a single image is used as a reference (and allows us to fix mobile objects in a single position). Candidate patches are identified by moving an equivalent patch across the other images, in the neighborhood of the central point of interest. The distance between the two patches is the mean square difference of the corresponding pixels, normalized by the variance of the patch. In order to obtain comparable values, these distances are not taken from the images I_j^x themselves, but from radiance images $\bar{i}_j^x$ which are reconstructed after inversion of the formation model (equation [10.12]) (the variances are estimated in a similar way, using model [7.10], for example). All candidate patches at a distance below a given threshold will give a candidate radiance value for use in filtering the center of the reference patch. The decision is, therefore, made in a statistical framework which is often more favorable than that used in previous methods, which only filtered pixels from images obtained in the burst.

The search area is determined by our knowledge of the movements in the scene and the stability of the camera during the shot. It generally covers a few tens of pixels in each direction. Note that the reference image plays an important role, as its saturated pixels, along with those concealed by noise, cannot be found. As this situation is generally impossible to avoid (and constitutes the reason for using HDR), it is best to begin by evaluating these pixels, for example using the conventional methods discussed above.

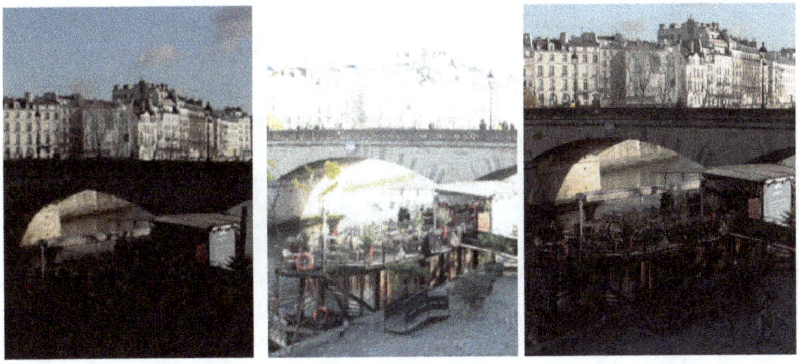

Figure 10.10. *HDR treatment of four images (only the two extreme images are presented, with the least exposed on the left and the most saturated in the center). The result of this treatment, for the shared zone of the images, accounting for movement and compressed onto 8 bits, is shown on the right. Taken from [AGU 14a]. For a color version of this figure, see www.iste.co.uk/maitre/pixel.zip*

10.3.1.4. *Single-image HDR*

The possibility of modifying camera hardware in order to obtain both low intensities and high intensities (for different pixels) in the same shot was mentioned in section 5.4.2; similarly, it is also possible to record the green channel for certain pixels and the red or blue channel for others. This idea allows us to avoid issues relating to operator movements and mobile objects. However, it also results in a loss of resolution, and requires a more complex sensor, necessitating changes within the camera hardware itself.

The first solution, presented in [NAY 00], consists of masking pixels using regular motifs (a block of 2×2 pixels using four different attenuation levels, with a progression of factor 2 or 4 depending on the dynamic of the scene). Reconstruction is then carried out by bicubic interpolation of measures, plane-by-plane, following the removal of values which are saturated or drowned in noise. This produces an image with the highest resolution permitted by the sensor.

Another solution involves the use of attenuation masks distributed randomly across the sensor in order to avoid the introduction of potentially troublesome motifs into the image.

The solution proposed by Fuji, described in section 5.4.2 (Figure 5.20), is based on the use of an especially developed photodetector known as the ESR; in HDR mode, this element groups pixels into pairs, one measuring high intensities, the other measuring low intensities.

10.3.2. *Plenoptic imaging: improving the depth of field*

Obtaining images in which all objects in the field are clearly defined is a key objective in photography. A variety of solutions have been proposed, known collectively as plenoptic imaging. These solutions are based on very different physical principles. In this section, we will describe three of the most successful approaches: multiple image combination, the use of microlenses and coded apertures. Other methods also produce good results, such as focal blur measurement [NAY 94, LAD 05, DEL 12], use of the achromatic element of lenses [BAN 08, GUI 09, TRO 12] and use of geometric aberrations [TAN 13].

10.3.2.1. *Plenoptic imaging by image combination*

As we have seen, depth of field is closely defined by the aperture of the lens assembly (section 2.2). In certain fixed scene situations, several shots may be taken using different focusing distances, in order to cover the full scene; these images may then be recombined to produce a single image with a very high depth of field, selecting the version of each pixel which contributes best to the final image.

To apply this technique, we must begin by ensuring that photos are taken from exactly the same point and in identical conditions. This constraint can be avoided by the use of registration techniques, such as those described in section 10.3.1.3, looking for similar zones in each image around the position of each pixel (NL patch technique). However, this adds a level of complexity to the process, and it is better to respect the constraint wherever possible. We thus consider that our images are fully stackable, as in the case of tripod photography. These images are obtained using the same focal distance but with variable focusing distances, adjusted manually or using the focus bracketing function on a camera. The aperture and exposure time parameters are kept constant, so that the image dynamics remain identical from one shot to the next. Once the photos have been taken, we check that the images are stackable, applying a fine registration process where necessary (see section 10.1.6). For each position (x, y), we thus have the same pixel $i_k(x, y)$, affected by a different focal impairment in each image, corresponding to the parameter f_k chosen for this image k.

The value of $i_k(x, y)$ which is most in focus, i.e. which maximizes a local measure $\mu_k(x, y)$, is then selected to be the output value $\bar{i}(x, y)$. The focus criterion μ is generally the same as that used for parameterization in automatic mode. It therefore constitutes a measure of local contrast in the image, along with its variance, Laplacian or total variation [SUB 93, NAY 94, KUM 13]. When this treatment is carried out offline, more time is available to determine the best image. In this case, the mean of the filtered local gradient is often used. The frequency selective weighted median (FSWM) measure is particularly effective in these situations [CHO 99]. This measure uses two crossed filters in x and y, for example:

$$\begin{aligned}\mu_x(x, y) = &\ \text{med}[i(x+1, y), i(x, y), i(x-1, y)] \\ &-\frac{1}{2}\text{med}[i(x+3, y), i(x+2, y), i(x+1, y)] \\ &-\frac{1}{2}\text{med}[i(x-3, y), i(x-2, y), i(x-1, y)]\end{aligned}$$

This approach is likely to be noise-sensitive in low-contrast areas, as each pixel is treated independently of its neighbors; for this reason, a hierarchical approach is often adopted, using a binary tree decomposition of the image [CAS 04].

The results obtained using these techniques were initially used in scientific applications, particularly in microscopy for biology. In photography, these methods have been most widely used in macro-photography [BRI 07], but very good results have also been obtained in other domains involving fixed scenes, such as mineral nature and architectural photography (Figure 10.11). Open-access programs are also available, such as CombineZ, by Hadley [HAD 10], and produce good results.

It is not currently possible to obtain these results directly using the camera, but models including this function are likely to be released onto the market in the near future, as in the case of dynamic improvement.

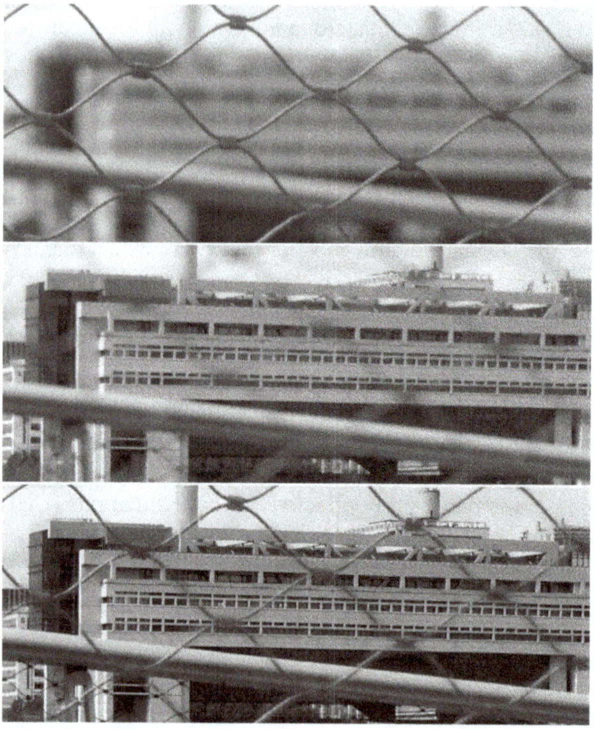

Figure 10.11. *Field depth extension using images with different focus levels. Top and center: the two extreme examples from the four-image stack. Bottom: the result of the field depth improvement process*

10.3.2.2. *Light field: physical considerations*

In the previous section, we discussed a solution to the depth of field problem based on the combination of multiple images taken using an ordinary camera. Now, we will consider two further techniques, which involve changes to the image formation process itself and thus require hardware modifications. Let us begin by using the propagation equations employed in both systems. The incoherent image formation equation presented in Chapter 2 is important in this context, specifically equation [2.35], which links the image i to the object o via the impulse response h_λ:

$$i_\lambda(x, y) = o_\lambda(x, y) * h_\lambda(x, y) \qquad [10.15]$$

where the impulse response h_λ is defined by its relation to the inverse Fourier transform of the pupil:

$$h_\lambda(x, y) = |k_\lambda(x, y)|^2 \qquad [10.16]$$

and

$$\begin{aligned} k_\lambda(x, y) &= \mathcal{TF}^{-1}\,(K(u, v)) \\ &= \iint_{lens} K(\eta, \zeta) \exp\left\{\frac{2j\pi(\eta x + \zeta y)}{\lambda f}\right\} d\eta d\zeta \end{aligned} \qquad [10.17]$$

In these equations, the expression of K is given as a function of the angles η and ζ created by the light ray (Figure 10.12).

Retaining only a small hole centered on point Θ from the pupil (of coordinates η and ζ), for each pixel, the image recorded by the sensor will only contain components emitted at angles η and ζ in relation to the non-deviant axial ray $M'M$. Given the quadruplet $\{x, y, \eta, \zeta\}$, it is then possible to obtain the single source point x', y', z'. Generalizing the approach, we can then reconstruct the whole of the incident light flow on the lens using the quadruplet x, y, η, ζ at each point of the image i; this is known as the light field. This light field allows us to find not only points seen in full (ordinary image points), but also those which are partially concealed, as in difficult partial masking configurations such as those presented in section 2.1.2 and illustrated in Figure 2.3.

There are two main ways of using these equations: first, using the approach presented above directly in conjunction with the notion of coded apertures, or, second, with the addition of spatial multiplexing, for example using a matrix of microlenses (discussed below) [NG 05, ISA 00].

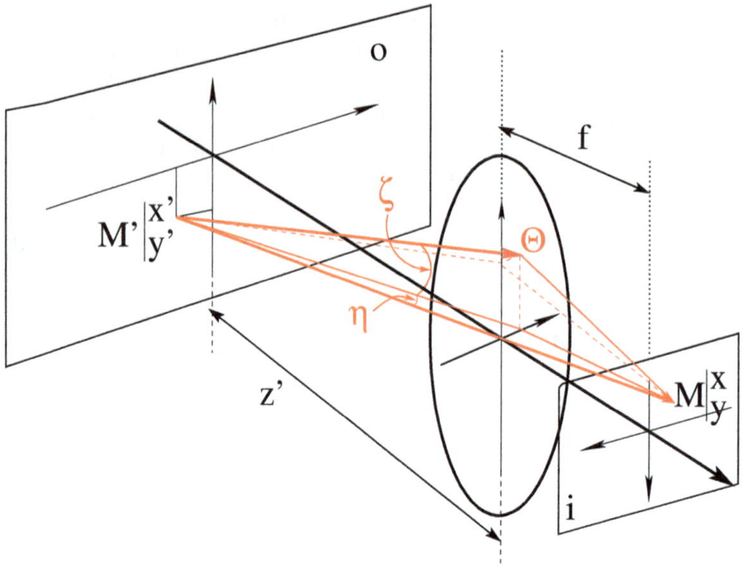

Figure 10.12. *Incoherent diffraction and image formation: a point Θ in the pupil codes the two incident angles (η and ζ) in relation to the axial ray of the beam onto the lens assembly, while point M in the image plane codes the direction of the source point M'*

10.3.2.3. *Coded apertures*

Coded apertures, along with the flutter shutter techniques which will be discussed below, are still a long way from inclusion in camera hardware. This technique consists of modifying the aperture of the photographic objective so that the lens assembly itself executes the functions described above. The image, or set of images created using different filters, produced in this way should allow the creation of new functions, notably concerning the distance of different objects in the scene: focus across the whole field, selection of a specific plane and three-dimensional mapping of the object space.

The principle behind coded apertures (widely used in astrophysics, particle physics and acoustic imaging) consists, in practice, of placing a binary filter, which either blocks or allows light to pass, either in the object nodal plane or in the image nodal plane of the lens assembly [GOT 07]. Each aperture at a point Θ_k defines a direction η_k, ζ_k, which, associated with the image point x, y, defines a point in the object space x', y', z'. The dimension of the aperture is inversely proportional to the precision of the object in terms of z', but it also determines the measured energy. A compromise is, therefore, required, which in practice results in the use of a small number of relatively large apertures (giving low depth resolution).

Rather than measuring the contribution of a single aperture Θ_k to each measure, it is also better to display several apertures simultaneously in order to reduce the number of images required to determine a full light field. To this end, the contributions of the various apertures must be easily separable during the reconstruction process. In this context, a disposition corresponding to two-dimensional Hadamard codes has been shown to be particularly efficient [WEN 05].

Masks are often created using bands of opaque material (paper or gelatin) which are used in successive positions where different directions are coded (see Figure 10.13). In automatic assemblies, electro-optical cells (Kerr cells or liquid crystals) are used within the lens assembly to allow rapid switching between different apertures [LIA 08]. The number of resolution cells remains relatively low for all cases, of around 100 apertures; this gives approximately the same number of 3D positions, although it would also be reasonable to interpolate the depth of field over continuous surfaces. The image resolution in terms of x, y, however, is excellent and is equal to that of the original image.

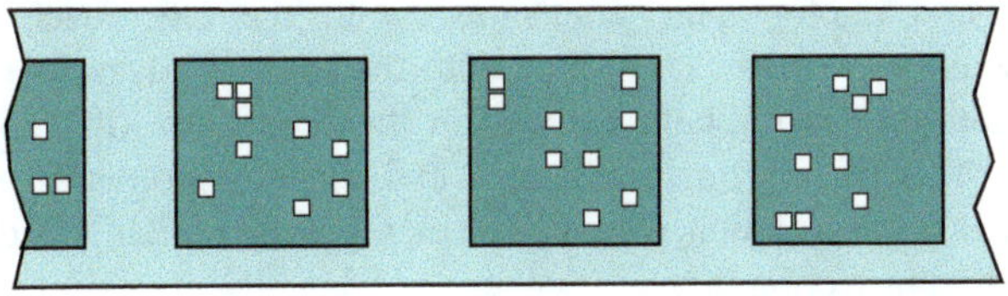

Figure 10.13. *Coded apertures: examples of masks (size: 8×8) placed successively within the optical system. Each mask allows us to code nine of the 64 possible directions simultaneously*

10.3.2.4. *Light fields using a microlens matrix*

The second way of using equations [10.15]–[10.17] consists of separating the beams behind the mask in order to record the images from different apertures Θ_k using different sensors. Solutions have been proposed involving the juxtaposition of cameras in the image plane after derivation of the various beams [KOL 02]; however, this possibility has yet to be fully explored. Other possibilities, which have been considered in greater detail, concern the use of a microlens matrix[9] placed in the image beam in contact with the captor, so that a ray with an orientation fixed by Θ will be received by a site neighboring which receives the beam passing through the optical center. This idea was first introduced by Lippman [LIP 08], and reexamined experimentally a century later [LIA 08, LEV 09a, LIN 13]. Solutions of this type involve spatial multiplexing of the angular coordinates η and ζ. The size and focal distance of the microlenses is chosen so that a single pixel will be distributed across, for example, 5×5 sites. A single shot thus generates 25 depth maps. The payoff for this result is a reduction in the number of pixels in the image by the same factor, i.e. 25.

Cameras using this principle are now available, providing either a depth value for each pixel or a representation of the scene. Examples include the Lytro camera [GEO 13], designed for mass-market applications, and the RayTrix [PER 12], intended for professional applications; other prototypes exist which are able to produce a full light field using a single camera.

10.3.3. *Improving resolution: super-resolution*

Improving resolution is another key concern in photography. Over the last century, the majority of work has concentrated on the improvement of lens assemblies and photographic emulsions; with the emergence of digital imaging technology, focus has shifted onto the sensor. Advances in the field of microphotolithography have resulted in the production of sensors which are smaller than the diffraction figure, and the compromises involved in obtaining the best resolution without sacrificing image quality in these cases were discussed in section 6.3.

The key aim of super-resolution is to produce resolutions even higher than those determined by the lens assembly.

More specifically, super-resolution involves the reconstruction of signal frequencies outside of the bandwidth determined by the acquisition system. This issue has been widely discussed in the image processing community, mostly in the context of astronomical and microscopic imaging, since the

9 Generally, a microlens matrix is placed in contact with the sensor (as shown in Figure 3.8) in order to derive full benefit from the light field. The step used is that of the photodetectors. To measure the light field, we generally replace this matrix by a matrix with a greater step (for example, covering $p \times p$ detectors). This gives us p^2 possible positions in terms of z for each pixel measured.

1950s–1960s [FRA 55, HAR 64, LUK 66]. The first investigations in this area concerned the improvement of a single image, making use of available knowledge regarding the solution and requiring the verification of a certain number of properties (finite range of the support of the desired object, bounded values for the signal and its derivatives, etc.). The proposed solutions were based on analytical extensions of the signal spectrum, projections onto adapted bases (prolate spheroidal functions) or alternating projections onto convex sets, all within spaces respecting certain precise constraints.

These early projects demonstrated the limits of analytical super-resolution for a single image, essentially in relation to the sensitivity of the proposed solutions to measurement noise. Subsequently, super-resolution evolved in different directions, considering the means of combining information from multiple images in order to improve the quality of each individual image. In this context, the diversity of representations of a scene appeared to offer the key to improved resolution. In the field of video technology, efforts focused on the movement of objects or the captor, but consideration was also given to the roles of perspective, lighting, focus, etc.

The emergence of solid-state captors modified the issues relating to super-resolution in two key ways:

– first, the use of sampling in precise positions and the use of photosites of finite size invalidated the linear and spatially invariant model used in Shannon–Nyquist filtering analysis and opened up new perspectives, as it is almost impossible to avoid aliasing, and thus the recording – generally considered parasitic, but in this case, beneficial – of frequencies outside of the bandwidth;

– second, it is now very easy to obtain multiple images, with very little change in terms of the viewpoint, meaning that the super-resolution issue can be considered within a much wider context involving multiple, similar but distinct, images of the scene under reconstruction.

We will consider the issue of super-resolution in this later context, starting with a first key question: given multiple images of the same scene, is it possible to create an image with a higher resolution than any of the originals? This discussion is based on the work of Traonmilin [TRA 14a][10].

10.3.3.1. *Formulation*

Consider N images, captured with the same camera settings, but from slightly different viewpoints, such as those obtained using burst mode in

10 Note that this publication is by no means limited to the results presented here.

hand-held photography. From an object $o(x, y)$, we thus obtain N images $i_k, k = 1, \ldots, N$ such that:

$$i_k = ShQ_k o_k + n_k \tag{10.18}$$

where n_k is the noise affecting image k, Q_k is its translation into the image reference frame, h is the pulsed response from the camera and S is the sampling on the sensor grid.

Let us combine the N images i_k to form a single image by concatenating the measures. We obtain:

$$i = Ao + n \qquad \text{with} \qquad A = (ShQ_k) \quad \text{for } k = 1, \ldots, N \tag{10.19}$$

Our aim is to invert this system in order to find as much detail concerning the unknown object o as possible, above and beyond the information provided by the i_k images on an individual basis.

Two spatial frequency limits are involved in this case:

– the limit established by the photodetector matrix and expressed by S; this is generally the inverse of the photosite step, $1/\delta_{sens}$;

– the limit imposed by the optical element and expressed by h; conventionally, this quantity is given by the inverse of the extent of the diffraction figure: $1/\delta_{diff}$.

Generally speaking[11] $\delta_{sens} > \delta_{diff}$; the aim of multi-image super-resolution is to come as close as possible to the bandwidth limit $1/\delta_{diff}$, greater than the bandwidth offered by the sensor. The improvement objective is thus determined by the limit $M = \delta_{sens}/\delta_{diff}$, or the nearest integer.

The reconstruction of an improved image $\tilde{i}$, close to o, is obtained using a very general inversion formulation of relationship [10.19]:

$$\tilde{i} = argmin_o \|Ao - i\|_p^p + \lambda R(o) \tag{10.20}$$

11 This is not always the case in telephones and ultra-compact cameras, which have very small photosites; in this context, the resolution limit is often defined by their poor impulse response.

where R is a regularization term which measures the difference between the solution obtained and our prior knowledge regarding this solution. Least squares minimization ($p = 2$) or robust regression minimization ($p = 1$), along with regularizations applied to the derivatives, minimization of order 2 (Tychonov regularization), or the absolute value (regularization by total variation) are the most common methods used in image processing [PAR 03, TIA 11].

10.3.3.2. *The noiseless case*

First, let us consider cases where the noise is negligible. This gives some interesting theoretical results.

If the movement of the sensor is purely a translation, we can show [AHU 06] that a perfect reconstruction is obtainable if we have $\nu = M^2$ images in the general position[12]. The problem is, therefore, reduced to irregular sampling of a two-dimensional signal [ALM 04], and equation [10.20] may be used without the addition of a regularization term.

If the movement is affine (often the case for a series of hand-held shots of a distant object), generally speaking, this still holds true [TRA 14a].

The most general case (homography covering all of the shots of a camera without aberrations) has been considered by a number of authors, but no general conclusions have yet been reached.

10.3.3.3. *The noisy case*

In cases where the measurement noise cannot be ignored, if we have $\nu > M^2$ images, it is possible to make the reconstruction error tend toward zero for increasing values of ν, on the condition that the algorithm used to solve equation [10.20] includes a regularization term. The speed of convergence depends on the precise form of equation [10.20], which is itself dependent not only on the noise, but also on the position of the shots. An optimal configuration is reached using a technique close to regular oversampling, very similar to that given by a sensor with a step value of δ_{sens}/M.

The calculation above presumes perfect knowledge of the movement parameters of the various photographs. If this information is not available, it must be deduced from the images themselves, and this constitutes a potential

12 "In the general position" implies that there is no degeneration in the sample network.

source of uncertainty. In the case of a simple translation, we only need to determine two unknowns per image; six are required in the affine case, and the methods involved are iterative techniques (gradient descent from an initial known value) which do not always converge well.

In real capture conditions, it is very hard to keep a scene constant during the burst period. Traonmilin [TRA 14a] has shown that, in cases where objects are moved during the capture process, zones requiring stronger regularization, or which should even be eliminated from the reconstruction process, can be identified during restoration.

Finally, Aguerrebere and Traonmilin have shown that HDR reconstructions (see section 10.3.1) and reconstructions in super-resolution can be beneficially combined in a single restoration operation without regularization [TRA 14b].

10.3.3.4. *Compressed sensing*

Finally, note that the considerable amount of work carried out in recent years on the topic of compressed sensing (discussed briefly in section 8.9) may well produce new results over the next few years, as, in principle, reconstructions carried out using this technique are not subject to the Shannon's limit [DON 06, BAR 07, CAN 08].

10.3.4. *Flutter-shutters*

The work presented below draws on the article [TEN 13].

10.3.4.1. *Motion blur*

One of the compromises to be made in photography concerns the choice between a long exposure time (enabling more photons to be received, but at the risk of increased motion blur) and high sensitivity (which allows short exposure times to be used, but results in increased noise). The motion blur may come from the scene, due to the presence of moving objects, or from the camera, which is not always fixed in place. Motion blur is often considered, as a first approximation, as the monodimensional convolution of the image by a rectangular function, with a width proportional to the exposure time. This approximation is valid if the observed scene is at a fixed distance from the camera, if the objects all move in the same way, and if this movement is relatively constant over the exposure period.

Using these hypotheses, representing the image along axes defined by the direction of movement, and leaving aside zoom and filtering effects due to the lens assembly, the recorded image may be written as:

$$i(x,y) = \int_{t=0}^{\delta t} o(x,y,t)dt \quad = \quad o'(x,y) * \mathrm{rect}(x/\delta x) \qquad [10.21]$$

where $o'(x,y)$ denotes the scene captured at instant $\delta t/2$ which we wish to reconstruct and δx is the distance traveled by a point in the image while moving.

Unfortunately, equation [10.21] cannot be inverted numerically for motion of more than 2 pixels as the default becomes singular; in this case, the zeros in the transfer function (a cardinal sine function) are located inside the bandwidth. Solutions which have been put forward to enable image reconstruction in these cases all involve significant regularizations, and the images in question are often of limited quality. Other solutions, using multiple images with different exposure times, allow us to use one image to retrieve the signal for frequencies lost in another image, but this solution is relatively unusual.

10.3.4.2. *The flutter-shutter solution*

Agrawal [AGR 07] proposes a new, elegant solution, in which the single, long exposure time used in obtaining an image is replaced by a series of very short micro-exposures within the exposure time frame. The sensor then adds the contributions from each microexposure to create a single image. For carefully selected exposure periods, the non-invertible cardinal sine function can be replaced by a regular function h, which is fully invertible (in the interval containing o', i.e. for a limited frequency bandwidth), leaving aside the noise:

$$i(x,y) = o'(x,y) * h(x,y) \quad \Longrightarrow \quad o'(x,y) = i(x,y) * k(x,y) [10.22]$$

with:

$$k(x,y) = \mathcal{TF}^{-1}\left[\mathcal{TF}\left(\frac{1}{h(x,y)}\right)\right] \qquad [10.23]$$

This approach, known as flutter-shutter, was optimized in [AGR 10]. It has been applied in practice, either via the addition of an extra shutter to the camera [SHI 09], or using stroboscopic lighting which illuminates the scene in an equivalent manner. The choice of random aperture sequences has been considered in some detail, covering much of the same ground studied in the context of compressed sensing (see section 8.9).

Another, seemingly different approach to photographing moving objects has been put forward, known as motion invariant photography [LEV 08, LEV 09b]. Using this approach, the camera is subjected to constant acceleration in the direction of movement over the course of the shot.

10.3.4.3. *Theoretical analysis of flutter-shutter*

Article [TEN 13] provides a detailed analysis of these different approaches and their level of optimality from two key angles: first, in terms of their ability to reverse motion blurring, and second in terms of their resistance to noise. To do this, the authors reformulated the problem in more general terms. Starting with a set of images $i_k(x, y)$ from observations at instant $t = k\delta t$, the flutter-shutter image is constructed using a weighted sum of these observed images[13].

$$i(x, y) = \frac{1}{N}\sum_{k}^{N} \alpha_k i_k(x, y) \qquad [10.24]$$

The sequence of values α_k is known as the flutter-shutter code, and another aim of the study was to determine the optimum code for given movement speeds and observation periods.

The authors identified three different configurations, according to the value of the coefficient α_k affecting the kth microinterval during reconstruction of the image using formula [10.22]:

1) The initial, basic flutter-shutter model proposed by Agarwal *et al.*, using discrete microintervals, separated over time and with a value of 0 or 1 ($\alpha_k \in \{0, 1\}$) depending on whether the shutter is open or closed.

2) An analog flutter-shutter model, controlled by continuous optical command in terms of both time and signal attenuation, and able to modulate light at very close intervals ($\alpha_k \in [0, 1]$) (in this case, attenuation may be carried out using a Kerr or Pockels cell, for example, using birefringence induced by a photo-electric effect in a material).

13 The advantage to carrying out this sum immediately, rather than transmitting the N images, lies in the fact that only one image then needs to be transmitted; this is highly beneficial in applications with limited bandwidth, such as satellite remote sensing, drone control, etc.

3) A digital flutter-shutter model, controlled after acquisition of the microsignal (coefficient $\alpha_k \in \mathbb{R}$), again using modulation over very close intervals, but with the ability to take negative values or values greater than 1 (in this case, however, it must be possible to create the function numerically using a series of finite, non-null intervals).

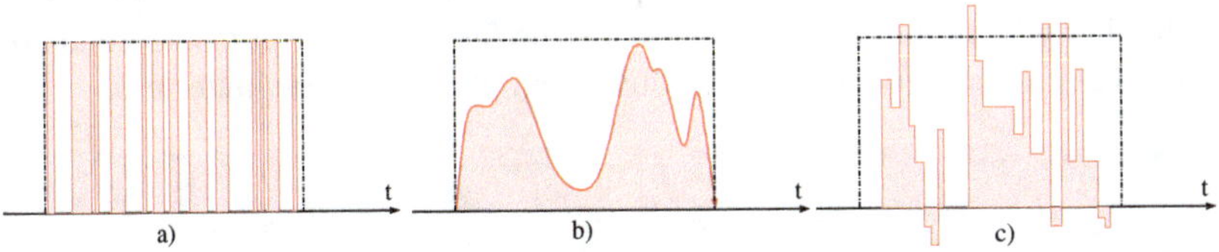

Figure 10.14. *Flutter-shutter: three different strategies may be used within the capture time window: a) The original flutter-shutter opens the shutter following a random sequence of binary microexposures; b) the analog flutter-shutter modulates the photon flux over the duration of the exposure period; c) the digital flutter-shutter freely associates measures taken during microexposures using a linear combination technique*

The authors also used a sensor model using Poisson's noise, similar to that presented in equation [7.10]. They reached the following conclusions:

– for any code, it is possible to determine the invertibility of the proposed system, and to predict the performance of the reconstructed image in terms of the signal-to-noise ratio, as a function of the speed of movement;

– for a given time period, flutter-shutter techniques always represent a better strategy than blurred shots, and the quality of the restored image is always better;

– digital flutter-shutters produce the best results in terms of quality, and the authors proposed an optimal code for a given speed of movement and observation period[14]. This code results in a regular numerical inversion (equation [10.22]) of the formation equation, taking account of equation [10.24];

– however, performances in terms of restoration remain limited, and the signal/noise ratio for images reconstructed using flutter-shutter is no more than 17 % higher than that of the best microimage involved in the observed sequence, for any observation period.

14 These codes perform better than those given in [AGR 10].

Finally, the authors show that the motion invariant photography approach may be considered as a specific instance of digital flutter-shutter, and that the codes proposed in [LEV 08, LEV 09b] are suboptimal in terms of the signal-to-noise ratio.

The authors conclude that this approach is extremely valuable, while also highlighting its limits; they refer to this as the flutter-shutter paradox. Tendero [TEN 12] provides codes and interactive examples of flutter-shutter use.

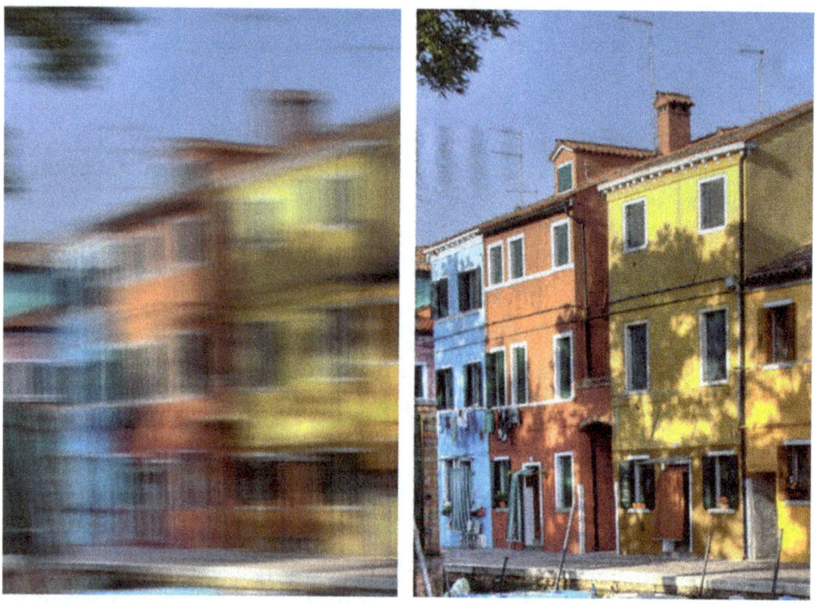

Figure 10.15. *Flutter-shutter: example of image processing, taken from [TEN 13]. Left: the image recorded using a flutter-shutter, with a total movement of 52 pixels. Right: the corrected image. This second image is very close to the original, with the exception of the small vertical ripples in the sky on the left-hand side. For a color version of this figure, see www.iste.co.uk/maitre/pixel.zip*

Bibliography

[ADA 10] ADAMS A., TALVALA E., PARK S. *et al.*, "The Franken camera: an experimental platform for computational photography", *ACM Transactions on Graphics*, vol. 29, no. 4, pp. 1–12, July 2010.

[ADO 12] ADOBE, Digital Negative (DNG) specification, version 1.4.0, Adobe Systems Inc., San Jose, June 2012.

[AGR 03] AGRANOV G., BEREZIN V., TSAI R., "Crosstalk and Microlens Study in a Color CMOS Image Sensor", *IEEE Transactions on Electron Devices*, vol. 50–(1), pp. 4–12, January 2003.

[AGR 07] AGRAWAL A., RASKAR R., "Resolving objects at higher resolution from a single motion-blurred image", *IEEE CVPR'07 Computer Vision and Pattern Recognition*, pp. 1–8, 2007.

[AGR 10] AGRAWAL A., GUPTA M., VEERARAGHAVAN A. *et al.*, "Optimal coded sampling for temporal super-resolution", *IEEE Computer Vision and Pattern Recognition 2010*, San Francisco, USA, pp. 599–606, 2010.

[AGU 12] AGUERREBERE C., DELON J., GOUSSEAU Y. *et al.*, Study of the digital camera acquisition process and statistical modeling of the sensor raw data, HAL-00733538-v3, 2012.

[AGU 14a] AGUERREBERE C., On the generation of high dynamic range images: Theory and practice from a statistical perspective., PhD, Télécom ParisTech, Paris, France, May 2014.

[AGU 14b] AGUERREBERE C., DELON J., GOUSSEAU Y. *et al.*, "Best algorithms for HDR image generation. A study of performance bounds", *SIAM Journal on Imaging Sciences*, vol. 1, pp. 1–34, 2014.

[AHM 74] AHMED N., NATARAJAN T., RAO K., "Discrete cosine transform", *IEEE transactions on Computers*, vol. C-23-(1), pp. 90–93, 1974.

[AHU 06] AHUJA N., BOSE N., "Multidimensional generalized sampling theorem for wavelet based image superresolution", *IEEE International conference on Image Processing*, pp. 1589–1592, October 2006.

[ALM 04] ALMANSA A., DURAND S., ROUGÉ B., "Measuring and improving image resolution by adaptation of reciprocal cell", *Journal of Mathematical Imaging and Vision*, vol. 21(3), pp. 235–279, 2004.

[ALV 99] ALVAREZ L., GOUSSEAU Y., MOREL J.-M., "The size of objects in natural and artificial images", *Avances in Imaging & Electron Physics*, vol. 111, pp. 167–242, 1999.

[AMI 14] AMIRSHAHI S., HAYN-LEICHSENRING G., DENZLER J. *et al.*, "Evaluating the rule of thirds in photographs and paintings", *Art & Perception*, vol. 2, pp. 163–182, 2014.

[ANS 48] ANSCOMBE F., "The transformation of Poisson, binomial and negative binomial data", *Biometrika*, vol. 35, no. 3-4, pp. 246–254, 1948.

[ARN 10] ARNHEIM R., *Art and Visual Perception: A Psychology of the Creative Eye*, University of California Press, Oakland, 2010.

[ATK 08] ATKINS B., "Image Stabilization-Body or Lens?", available at: www.bobatkins.com/ photography/digital/image_stabilization.html, 2008.

[AYD 15] AYDIN T. O., SMOLIC A., GROSS M., "Automated Aesthetic Analysis of Photographic Images", *IEEE Transactions on Visualization and Computer Graphics*, vol. 21, pp. 31–42, 2015.

[BAN 08] BANDO Y., CHEN B., NISHITA T., "Extracting depth and matter using a color filtered aperture", *ACM SIGGRAPH Asia 2008*, pp. 134, 2008.

[BAR 03a] BARLAUD M., ANTONINI M., "Transformées en ondelettes pour la compression d'images", in BARLAUD M., LABIT C., (eds.), *Compression et codage des images et des vidéos*, Hermès-Lavoisier, Paris, 2003.

[BAR 03b] BARLAUD M., LABIT C. (eds.), *Compression et codage des images et des vidéos*, Hermès-Lavoisier, Paris, 2003.

[BAR 07] BARANIUK C., "Compressive Sensing", *IEEE Signal Processing Magazine*, vol. 24, pp. 118–121, July 2007.

[BEL 14] BELHUMEUR P., JACOBS D., LOPEZ I., "Leafsnap, an electronic field guide", available at: http://leafsnap.com, 2014.

[BEN 69] BENSE M., *Einfüung in die informationstheoretische Ästhetik, Grundlegung und Andwendung in der Texttheorie*, Rowoldt Taschenbuch Verlag, 1969.

[BEN 02] BENOIS-PINEAU J., PATEUX S., BARBA D., "Modèles et représentation d'images vidéo et compression", in BARLAUD M., LABIT C., (eds.), *Compression et Codage des Images et des Vidéos*, Hermès–Lavoisier, Paris, 2002.

[BIG 14] BIGLER E., "Lumière, diaphragmes et pupilles, optiques épaisse, deuxième partie", available at: www.galerie-photo.com/pupilles-objectifs-photographie.html, February 2014.

[BIR 33] BIRKHOFF G., *Aesthetic Measure*, Harvard University Press, Cambridge, 1933.

[BLA 55] BLANC-LAPIERRE A., DUMONTET P., "La notion de cohérence en optique", *Revue de Physique Appliquée*, vol. 34, pp. 1–21, 1955.

[BOR 70] BORN M., WOLF E., *Principles of Optics*, 4th ed., Pergamon Press, Oxford, 1970.

[BOU 65] BOURDIEU P., *Un art moyen : essai sur les usages sociaux de la photographie*, Les Editions de Minuit, Paris, 1965.

[BOU 09] BOUILLOT R., *Cours de photographie numérique*, 3rd ed., Dunod, Paris, 2009.

[BOV 00] BOVIK A., *Image and Video Processing*, Academic Press, San Diego, 2000.

[BRA 99] BRADLEY A. P., "A wavelet visible difference predictor", *IEEE transactions on Image Processing*, vol. 8-5, pp. 717–730, 1999.

[BRI 07] BRIAN V., "ComposeZ", available at: www.dgrin.com/showthread.php?t=61316, May 2007.

[BRO 07] BROWN M., LOWE D., "Automatic panoramic image stitching using invariant features", *International Journal of Computer Vision*, vol. 74, no. 1, pp. 59–73, Aug. 2007.

[BUA 08] BUADES A., COLL B., MOREL J.-M., "Non-local image and movie denoising", *International Journal of Computer Vision*, vol. 76, no. 2, pp. 123–139, 2008.

[BUA 09] BUADES A., COLL B., MOREL J.-M., SBERT C., "Self-similarity driven color demosaicking", *IEEE transactions on Pattern Analysis and Machine Intelligence*, vol. 18 (6), pp. 1192–1202, 2009.

[BUC 80] BUCHSBAUM G., "A spatial processor model for object colour perception", *Journal of Franklin Institute*, vol. 310, pp. 1–26, July 1980.

[BUK 12] BUKSHTAB M., *Applied Photometry, Radiometry and Measurement of optical losses*, Springer Verlag, 2012.

[BUR 87] BURT P., ADELSON E., "A multiresolution spline with application to image mosaïcs", *ACM Transactions on Graphics*, vol. 2, no. 4, pp. 217–236, 1987.

[BUR 12] BURIE J., CHAMBAH M., TREUILLET S., "Color constancy", in FERNANDEZ-MALOIGNE C., ROBERT-INACIO F., MACAIRE L, (eds), *Digital Color Imaging*, ISTE Ltd, London and John Wiley & Sons, New York, 2012.

[CAL 98] CALLET P., *Couleur-lumière, couleur-matière - Interaction lumière-matière et synthèse d'images*, Diderot, Paris, 1998.

[CAL 12] CALDWELL B., BITTNER W., The SpeedBooster - a new type of optical attachement for increasing the speed of photographic lens, Report , Caldwell Photo Inc. & WB design, 2012.

[CAN 06] CANDÈS E.-J., ROMBERG J.-K., TAO T., "Stable signal recovery from incomplete ande inacurrate measurements", *Communications on Pure and Applied Mathematics*, vol. 58 (8), pp. 1207–1222, 2006.

[CAN 08] CANDÈS E.-J., WAKIN M.-B., "An introduction to compressive sampling", *IEEE Signal Processing Magazine*, vol. 25 (2), pp. 21–30, 2008.

[CAO 10a] CAO F., GUICHARD F., HORNUNG H., "Dead leaves model for measuring texture quality on a digital camera", *Digital Photography VI, Proceedings SPIE, Vol. 7537*, 2010.

[CAO 10b] CAO F., GUICHARD F., HORNUNG H., "Information capacity: a measure of potential image quality of a digital camera", *Proceedings SPIE Vol 7537 - Digital Potography VI*, vol. 7537, 2010.

[CAR 05] CARLSSON K., The Pinhole Camera Revisited or The Revenge of the Simple-Minded Engineer, Report, The Royal Institute of Technology, Stockholm, Sweden, available at: https://www.kth.se/social/files/542d2d2df276546ca71dffaa/Pinhole.pdf, 2005.

[CAR 08] CARTIER-BRESSON A., *Le vocabulaire technique de la photographie*, Marval, Paris, 2008.

[CAS 04] CASTORINA A., CAPRA A., CURTI S. *et al.*, "Extension of the depth of field using multi-focus input images", *IEEE International Symposium on Consumer Electronics*, pp. 146–150, 2004.

[CHA 07a] CHAIX DE LAVARÈNE B., L'échantillonnage spatio-chromatique dans la rétine humaine et les caméras numériques, PhD, Joseph Fourier University, Grenoble, 2007.

[CHA 07b] CHANDLER D., HEMAMI S., "VSNR: a wavelet based visual Signal-to-Noise Ratio for Natural Images", *IEEE Transactions on Image Processing*, vol. 16, no. 9, pp. 2284–2298, 2007.

[CHA 12a] CHAKRABARTI A., HIRAKAWA K., ZICKLER T., "Color constancy with spatio spectral statistics", *IEEE Transactions on Pattern Analysis and Machine Intelligence*, vol. 34, no. 8, pp. 1509–1520, 2012.

[CHA 12b] CHARRIER C., LEZORAY O., LEBRUN G., "Machine learning to design full-reference image quality assessment algorithm", *Signal Processing: Image Communications*, vol. 27-3, pp. 209–219, 2012.

[CHA 13] CHANDLER R., "Seven Challenges in image quality assessment: past, present and future research", *ISRN Signal Processing*, vol. 2013, pp. 1–53, 2013.

[CHA 14] CHALAMALA B., GUTTROMSON R., MASIELLO R., "Energy storage, Part I: Batteries and energy conversion systems", *Proceedings of the IEEE*, vol. 102, (6), no. 6, pp. 936–938, June 2014.

[CHE 12] CHEN J., CRANTON W., FIHN M., *Handbook of Visual Display Technology*, Springer, New York, 2012.

[CHO 99] CHOI K.-S., LEE J.-S., KO S.-J., "New autofocusing technique using the frequency selective weighted median filter for video cameras", *IEEE transactions on Consumer Electronics*, vol. 45, no. 3, pp. 620–827, 1999.

[CHU 06] CHUNG K., CHAN Y., "Color demosaicking using variance of color differences", *IEEE Transactions on Image Processing*, vol. 15 (10), pp. 2944–2955, 2006.

[CIE 06] CIE, Fundamental chromaticity diagram with physiological axes - Part 1, Report, 1, 2006.

[CLI 08] CLIFFORD J., KONSTANTATOS G., JOHNSTON K. *et al.*, "Fast, sensitive and spectrally tuneable colloidal quantum dot photodetectors", *Nature Nanotechnology Letters*, vol. NNANO-2008-313, 2008.

[COF 14] COFFIN D., "DCRAW", available at: http://cybercom.net/~dcoffin/dcraw, 2014.

[COH 92] COHEN A., DAUBECHIES I., FAUVEAU J.-C., "Biorthogonal bases of compacity supported wavelets", *Communications on Pure and Applied Mathematics*, vol. 45-5, pp. 485–560, June 1992.

[COH 93] COHEN M., WALLACE J., *Radiosity and realistic image synthesis*, Academic Press, 1993.

[CON 08] CONDAT L., VAN DE VILLE D., "Fully reversible image rotation by 1-D filtering", *IEEE Conference on Image Processing*, pp. 913–916, 2008.

[COO 67] COONS S., *Surfaces from Computer-aided Design of Space Forms*, MIT Press, 1967.

[COW 99] COW R., TONG R., "Two- and Three dimensional image rotation using the FFT", *IEEE Transactions on Image Processing*, vol. 8, no. 9, pp. 1297–1299, 1999.

[COX 13] COXETER H., *Projective Geometry*, Springer Verlag, 2013.

[DAB 07] DABOV K., FOI A., KRATKOVNIK V., EGIAZARIAN K., "Image denoising by sparse 3-D transform-domain collaborative filtering", *IEEE Transactions on Image Processing*, vol. 16-8, pp. 2080–2095, 2007.

[DAL 90] DALY J., "Application of noise-adaptive contrast sensitivity function to image data compression", *Optical Engineering*, vol. 29, pp. 977–987, 1990.

[DAL 05] DALAL N., TRIGGS B., "Histograms of oriented gradients for human detection", *CVPR Conference Computer Vision and Pattern Recognition*, San Diego USA, pp. 886–893, 2005.

[DEB 92] DEBRAY R., *Vie et mort de l'image, une histoire du regard en Occident*, Folio Essais, Paris, 1992.

[DEL 12] DELBRACIO M., ALMANSA A., MUSÉ P., MOREL J.-M., "Subpixel Point Spread Function Estimation from Two Photographs at Different Distances", *SIAM Journal on Imaging Science*, vol. 5 (4), pp. 1234–1260, November 2012.

[DEL 13] DELBRACIO-BENTANCOR M., Two problems of digital image formation, PhD, Ecole Normale Supérieure de Cachan, France, 2013.

[DEN 09] DENIS L., LORENTZ D., THIÉBAUT E., FOURNIER C., TREDE E., "Inline hologram reconstruction with sparsity constraint", *Optics Letters*, vol. 34(22), pp. 3475–3477, 2009.

[DER 09] DERICHE R., "Self calibration of video sensors", in DOHME M. (ed.), *Visual Perception Through Video Imagery*, ISTE Ltd, London and John Wiley & Sons, New York, 2009.

[DES 08] DESOLNEUX A., MOISAN L., MOREL J.-M., *From Gestalt Theory to Image Analysis: A Probabilistic Approach*, Springer-Verlag, 2008.

[DET 97] DETTWILLER L., *Les instruments d'Optique: étude théorique, expérimentale et pratique*, Ellipses, Paris, 1997.

[DHO 09] DHOME M. (ed.), *Visual Perception Through Video Imagery*, ISTE Ltd, London and John Wiley & Sons, New York, 2009.

[DON 94] DONOHO D., JOHNSTONE J., "Ideal spatial adaptation by wavelet shrinking", *Biometrika*, vol. 81-3, pp. 425–455, 1994.

[DON 06] DONOHO D., "Compressed sensing", *IEEE Transactions on Information Theory*, vol. 52 (4), pp. 1289–1306, 2006.

[DON 09] DONOHO D., TANNER J., "Observed universality of phase transitions in high-dimensionnal geometry with implications for modern data analysis and signal processing", *Philosophical Transactions of the Royal Society, A*, vol. 367, pp. 4273–4330, 2009.

[DUA 08] DUARTE M., DAVENPORT M., TAKHAR D. *et al.*, "Single-pixel imaging via compressive sampling", *IEEE Signal Processing Magazine*, vol. 25(2), pp. 83–91, March 2008.

[DUF 45] DUFFIEUX P., LANSRAUX G., "L'intégrale de Fourier et ses applications à l'Optique", *Revue d'Optique*, vol. 24, pp. 65, 151, 215, 1945.

[DUM 55] DUMONTET P., "Sur la correspondance objet-image en optique", *Optica Acta*, vol. 2, pp. 53–63, 1955.

[DUP 94] DUPONT B., TROTIGNON J., *Unités et grandeurs, Symboles et normalisation*, AFNOR-Nathan, Paris, 1994.

[ELA 14] ELACHERI V., SEISENBAEVA G. *et al.*, "Ordered network of interconnected SnO_2 nanoparticles for excellent Lithium-ion storage", *Advanced Energy Materials*, 2014.

[ELD 12] ELDAR Y., KUTYNIOK G., *Compressed Sensing: Theory and Applications*, Cambridge University Press, Cambridge, 2012.

[ELG 05] EL GAMAL A., ELTHOUKHY H., "CMOS Image Sensors", *IEEE Circuits & Devices Magazine*, vol. 05, pp. 6–20, May-June 2005.

[FAR 06] FARRELL J., XIAO F., KAVUSI S., "Resolution and light sensitivity tradeoff with pixel size", *Proceedings of SPIE Conference on Digital Photography*, vol. 6069, SPIE, pp. 211–218, 2006.

[FAR 14] FARIDUL H., STAUDER J., TRÉMEAU A., "Illuminant and device independant image stitching", *IEEE conference on Image Processing (ICIP'2014)*, pp. 56–60, 2014.

[FAU 93] FAUGERAS O., *Three-dimensional Computer Vision: a Geometric Viewpoint*, MIT Press, 1993.

[FER 09] FERZLI R., KARAM L., "A no-reference objective image sharpness metric based on the notion of Just Noticeable Blur (JNB)", *IEEE Transactions on Image Processing*, vol. 18-4, pp. 717–728, 2009.

[FEY 85] FEYNMAN R., *QED: The Strange Theory of Light and Matter*, Princeton Science Library, 1985.

[FIS 05] FISHER R., DAWSON-HOWE K., FITZGIBBON A. *et al.*, *Dictionary of Computer Vision and Image Processing*, John Wiley & Sons, Chichester, England, 2005.

[FLE 62] FLEURY P., MATHIEU J., *Images optiques*, 3rd ed., Eyrolles, Paris, 1962.

[FOR 90] FORSYTH D., "A novel algorithm for color constancy", *International Journal of Computer Vision*, vol. 5, no. 1, pp. 5–36, 1990.

[FOW 90] FOWLES G., *Introduction to modern Optics*, 2nd ed., Dover Books on Physics, 1990.

[FRA 55] TORALDO DI FRANCIA G., "Resolving power and information", *Journal Optical Society of America*, vol. 45, no. 7, pp. 497–501, July 1955.

[FRA 69] TORALDO DI FRANCIA G., "Degrees of freedom of an image", *Journal Optical Society of America*, vol. 59, no. 7, pp. 799–804, July 1969.

[FRI 68] FRIEDEN B., "How well can a lens system transmit entropy?", *Journal Optical Society of America*, vol. 58, pp. 1105–1112, 1968.

[FUJ 11] FUJIFILM C., "The debut of a new technology: EXR-cmos," available at: www.fujifilm.com/products/digital cameras/f/finepix f550exr/features/, 2011.

[GEO 13] GEORGIEV T., LUMSDAINE A., GOMA S., "Lytro camera technology: theory, algorithms, performance analysis", *Multimedia content and mobile devices, SPIE 8667*, 2013.

[GET 12] GETREUER P., "Image demosaicking with contour stencils", available at: www.ipol.im/pub/art/2012/g-dwcs/, March 2012.

[GIJ 11] GIJSENIJ A., GEVERS T., VAN DE WEIJER J., "Computational color constancy: survey and experiments", *IEEE Transactions on Image Processing*, vol. 20, no. 9, pp. 2475–2489, September 2011.

[GOL 08] GOLZ J., "The role of chromatic scene statistics in color constancy", *Journal of Vision*, vol. 8, no. 13, pp. 1–16, 2008.

[GOM 82] GOMBRICH E. H., *The image and the Eye*, Phaidon, 1982.

[GOO 76] GOODMAN J., "Some fundamental properties of speckle", *Journal Optical Society of America*, vol. 66, no. 11, pp. 1145–1150, 1976.

[GOT 07] GOTTESMAN S., "Coded apertures: past, present and future application and design", in CASASENT D., CLARK T., (eds.), *Adaptive Coded Aperture Imaging and Non-imaging Sensors*, 2007.

[GOU 07] GOUSSEAU Y., ROUEFF F., "Modeling Occlusion and Scaling in Natural Images", *SIAM Journal of Multiscale Modeling and Simulation*, vol. 6(1), pp. 105–134, 2007.

[GOY 08] GOYAL V., FLETCHER A., RAGAN S., "Compressive sampling and lossy compression", *IEEE Signal Processing Magazine*, vol. 25(2), pp. 48–56, March 2008.

[GRA 10a] GRAHAM D. J., FRIEDENBERG J. D., MCCANDLESS C. H. *et al.*, "Preference for art: similarity, statistics, and selling price.", *S&T/SPIE Electronic Imaging International Society for Optics and Photonics*, vol. 75271A-75271A, 2010.

[GRA 10b] GRANADOS M., ADJIN M., THEOBALT C. *et al.*, "Optimal HDR reconstuction with linear digital cameras", *Computer Vision and Pattern Recognition CVPR*, San Francisco, pp. 215–222, 2010.

[GUE 90] GUENTHER R., *Modern Optics*, John Wiley and Sons, Hoboken, 1990.

[GUI 03a] GUILLEMOT C., PATEUX S., "Eléments de théorie de l'information et de communication", in BARLAUD M., LABIT C., (eds.), *Compression et codage des images et des vidéos*, Hermès-Lavoisier, Paris, 2003.

[GUI 03b] GUILLOIS J., CHARRIER M., LAMBERT C., PAUCARD B., "Standards de compression d'images fixes", in BARLAUD M., LABIT C., (eds.), *Compression et codage des images et des vidéos*, Hermès-Lavoisier, Paris, 2003.

[GUI 09] GUICHARD F., NGUYEN H.-P., TESSIÈRES R. *et al.*, "Extended depth-of-field using sharpness transport across color channels", in RODRICKS B., SUSSTRUNK S., (eds.), *IST / SPIE Electronic Imaging*, 2009.

[HAD 10] HADLEY A., "CombineZ", www.hadleyweb.pwp.blueyonder.co.uk/, June 2010.

[HAG 14] HAGOLLE O., "The physics of optical remote sensing", in TUPIN F., NICOLAS J., INGLADA J. (eds), *Remote Sensing Imagery*, ISTE Ltd, London and John Wiley & Sons, New York, 2014.

[HAM 97] HAMILTON J. J., ADAMS J. J., Adaptive color plane interpolation in single sensor color electronic camera, US Patent n:5,629,734, May 1997.

[HAM 13] HAMAMATSU, "Learning Center in Digital Imaging : Charge Coupled Device CCD Linearity", www.hamamatsu.magnet.fsu.edu/articles/ccdlinearity.html, 2013.

[HAR 64] HARRIS J., "Resolving power and decision making", *Journal Optical Society of America*, vol. 54, pp. 606–611, 1964.

[HAR 98] HARDEBERG J., BRETTEL H., SCHMITT F., "Spectral characterization of electronic cameras", *Electronic Imaging: Processing, Printing and Publishing in Colo*, no. 3409, pp. 100–109, 1998.

[HEA 94] HEALEY G. E., KONDEPUDY R., "Radiometric CCD camera calibration and noise estimation", *IEEE Transactions on Pattern Analysis and Machine Intelligence*, vol. 16 (3), pp. 267–276, 1994.

[HEA 14] HEALPIX, "Hierarchical Equal Area isolatitude projection of a sphere", helpix.sourceforge.net, January 2014.

[HEM 10] HEMAMI S., REIBMAN A., "No-reference image and video quality estimation: Application and human-motivated design", *Signal Processing: Image Communications*, vol. 25, pp. 469–481, 2010.

[HEN 85] HENRY C., *Introduction à une esthétique scientifique*, Revue contemporaine, Paris, 1885.

[HOR 81] HORN B., SCHUNCK B., "Determining optical flow", *Artificial Intelligence*, vol. 17, pp. 181–203, 1981.

[HOR 14a] HORIBA T., "Lithium-Ion battery Systems", *Proceedings of IEEE*, vol. 102, (6), no. 6, pp. 1–12, June 2014.

[HOR 14b] HORVÀTH G., "RawTherapee", www.rawtherapee.com, 2014.

[HUG 10] HUGHES C., DENNY P., JONES E., GLAVIN M., "Accuracy of fish-eye lens models", *Applied Optics*, vol. 49-17, pp. 3338–3347, 2010.

[HUN 95] HUNT R., *The reproduction of Colour*, 5th ed., Fountain Press, London, 1995.

[HUO 10] HUO Y., FESENMAIER C., CATRYSSE P., "Microlens Performance Limits in sub 2μm pixel CMOS image sensors", *Optics Express*, vol. 18, no. 6, pp. 5861–5872, March 2010.

[IEE 14] IEEE-1431, "Standart Specification Format Guide and Test Procedure for Coriolis vibratory gyros", 2014.

[IPO 14] IPOL, "Image Processing On Line", www.ipol.im/, 2014.

[ISA 00] ISAKSEN A., MCMILLAN L., GORTLER S., "Dynamically reparametrized light fields", *SIGGRAPH 2000*, pp. 297–306, 2000.

[ISO 96] ISO, Isolation thermique, Transfert de chaleur par rayonnement, grandeurs physiques et définition, ISO Standard 9288, 1996.

[ISO 00] ISO, Photography – Electronic Still picture cameras – Resolution Measurements, ISO Standard 12233, 2000.

[ISO 02] ISO, Photography – Illuminants for sensitometry – Specifications for daylight, incandescent tungsten and printer, ISO Standard 7589, 2002.

[ISO 03] ISO, Photography microfilms - Determination of ISO Standard 9848 speed, 2003.

[ISO 06a] ISO, Photography - digital still cameras - determination of exposure index, ISO speed ratings, standard output sensitivity and recommended exposure index, ISO Standard 12232, 2006.

[ISO 06b] ISO, Graphics Technology and photography – Colour characterisation of digital still cameras (DSCs). Part 1 : Stimuli, metrology and test procedures, ISO Standard 17321, 2006.

[ISO 10] ISO, Electronic still-picture imaging – removable memory – Part 2: TIFF/EP Image Data Format, ISO Standard 12234, 2010.

[ISO 12] ISO, Photography – Electronic still picture imaging – Vocabulary, ISO Standard 12231, 2012.

[ITU 13] ITU-R, "Methodology for the subjective assessment of the quality of television pictures, Recommendation", 2013.

[JAC 62] JACKSON J., *Classical Electrodynamics*, John Wiley & Sons, New York, 1962.

[JIA 13] JIANG J., LIU A., GU J., SÜSSTRUNK S., "What is the space of spectral sensitivity functions for digital color cameras?", *2013 IEEE Workshop on Applications of Computer Vision (WACV)*, pp. 168–179, 2013.

[JOB 97a] JOBSON D., RAHMAN Z., WOODELL G., "A multiscale Retinex for bridging the gap between color images and the human observation of scenes", *IEEE Transactions on Image Processing*, vol. 6, pp. 965–976, 1997.

[JOB 97b] JOBSON D., RAHMAN Z., WOODELL G., "Properties and performance of a center/surround Retinex", *IEEE Transactions on Image Processing*, vol. 6, pp. 451–462, 1997.

[JOH 15] JOHANNSEN J., ULSTRUP S., CREPALDI A. *et al.*, "Tunable Carrier Multiplication and Cooling in Graphene", *Nano Letters*, vol. 15–1, pp. 326–331, 2015.

[JOS 08] JOSHI N., SZELISKI R., KRIEGMAN D., "PSF estimation using sharp edge prediction", *IEEE Conference on Computer Vision & Pattern Recognition (CVPR 2008)*, pp. 1–8, 2008.

[JUD 64] JUDD D., MACADAM D. L., WYSZECKI G., "Spectral distribution of typical daylight as a function of correlated color temperature", *Journal Optical Society of America, A*, vol. 54, no. 8, pp. 1031–1040, 1964.

[KEM 90] KEMP M., *The Science of Art, Optical themes in Western Art from Brunelleschi to Seurat*, Yale University Press, 1990.

[KIR 06] KIRK K., ANDERSEN A., "Noise Characterization of weighting schemes for combination of multiple exposures", *British Machine Vision Conference*, Edinburgh, pp. 1129–1138, September 2006.

[KNA 13] KNAUS C., ZWICKER M., "Dual-domain image denoising", *Proceedings IEEE Conference on Image Processing, ICIP'13*, pp. 440–444, 2013.

[KNA 14] KNAUS C., ZWICKER M., "Progressive Image Denoising", *IEEE Transactions on Image Processing*, vol. 23–7, pp. 3114–3125, 2014.

[KOE 30] KOEHLER W., *Gestalt Psychology*, Bell, New York, 1930.

[KOL 02] KOLMOGOROV V., ZABIH R., "Multicamera scene reconstruction via graph-cuts", *Proceedings European Conference on Computer Vision*, no. 3, pp. 82–96, 2002.

[KON 00] KONRAD J., "Motion detection and estimation", in BOVIK A., (ed.), *Handbook of Image and Video Processing*, Academic Press, San Diego, 2000.

[KOP 06] KOPOSOV S., BARTUNOV O., "Q3CQuad Tree Cube, the new sky-indexing concept for huge astronomical catalogues.", *Astronomical Data Analysis Softwares and Systems*, vol. 351, pp. 735, 2006.

[KOW 72] KOWALISKI P., *Applied Photographic Theory*, John Wiley, Hoboken, 1972.

[KOW 99] KOWALISKI P., *Vision et mesure de couleur*, Masson, Paris, 1999.

[KRA 12] KRAMES M., "Light Emitting Diodes: Fudamentals", in CHEN J., CRANTON W., FIHN M. (eds), *Handbook of Visual Display Technology*, Springer, New York, 2012.

[KUM 13] KUMAR A., AHUJA N., "A generative focus measurewith application to omnifocus images", *International Conference on Computational Photography, ICCP'13*, 2013.

[KUN 01] KUNSZT P. Z., SZALAY A. S., THAKAR A. R., "The hierarchical triangular mesh", in LEWIS J. (ed), *Mining the Sky*, Springer, Berlin Heidelberg, 2001.

[LAD 05] LADJAL S., Flou et quantification dans les images numériques, PhD, Télécom ParisTech, France, 2005.

[LAN 71] LAND E., MCCANN J., "Lightness and Retinex Theory", *Journal of the Optical Society of America*, vol. 61, pp. 1–11, 1971.

[LAN 93] LAND E. H., *Edwin H. Land's Essays*, McCann Society for Imaging Science and Technology, Springfield, USA, 1993.

[LAR 10] LARSON E., CHANDLER D., "Most apparent distorsion: Full-reference image quality and the role of strategy", *Journal of Electronic Imaging*, vol. 19–1, 2010.

[LAV 00] LAVEST J., DHOME M., "Comment calibrer les objectifs á très courte focale", *Conférence Reconnaissance des Formes et Intelligence Artificielle*, Paris, France, pp. 81–90, 2000.

[LAV 09] LAVEST J.-M., RIVES G., "Calibration of Vision Sensors", in DOHME M. (ed.), *Visual Perception through Video Imagery*, ISTE Ltd, London and John Wiley & Sons, New York, 2009.

[LEB 13] LEBRUN M., BUADES A., MOREL J.-M., "A non-local Bayesian image denoising algorithm", *SIAM Journal Imaging Science*, vol. 6–3, pp. 1665–1688, 2013.

[LEB 15] LEBRUN M., COLOM M., MOREL J.-M., "The Noise Clinic: a Blind Image Denoising Algorithm", *Image Processing On Line*, pp. 1–54, 2015.

[LEC 03] LE CALLET P., BARBA D., "A robust quality metric for image quality assessment", *Proceedings IEEE International conference on Image Processing (ICIP'03)*, pp. 437–440, September 2003.

[LEC 15] LECLAIRE A., MOISAN L., "No-reference image quality assessment and blind deblurring with sharpness metricsexploiting Fourier phase indformation", *Journal of Mathematical Imaging and Vision*, 2015.

[LEG 67] LEGRAND Y., MILLODOT M., *Form and Space Vision*, Indiana University Press, 1967.

[LEG 68] LEGRAND Y., HUNT R., *Light, Colour and Vision*, Chapman and Hall, 1968.

[LEM 12] LE MONTAGNER Y., ANGELINI E., OLIVO-MARIN J.-C., "Video Reconstruction using compressed sensing measurementsand 3D total variation regularization for bio-imaging applications", *IEEE International Conference on Image Processing ICIP-12*, Orlando (Florida), pp. 917–920, 2012.

[LEV 08] LEVIN A., SAND P., CHO T. *et al.*, "Motion Invariant Photography", *ACM transactions on Graphics*, vol. 27, Page 71, 2008.

[LEV 09a] LEVIN A., HASINOFF S., GREEN P. *et al.*, "4D frequency analysis of computational cameras for depth of field extension", *ACM SIGGRAPH*, 2009.

[LEV 09b] LEVIN A., SAND P., CHO T. *et al.*, Method and apparatus for motion invariant imaging, US Patent n: 20,090,244,300, 2009.

[LI 10] LI C., LOUI A., CHEN T., "Towards Aesthetics: a Photo Quality Assessment and Photo Selection System", *Conf. ACM on Multimedia*, 2010.

[LIA 08] LIANG C.-K., LIN T.-H., WONG B.-Y., LIU C., CHEN H., "Programmable aperture photography: Multiplexed light field acquisition", *ACM Transactions on Graphics, Proceedings of Siggraph*, vol. 27 (3), Page 55, 2008.

[LIN 13] LIN X., SUO J.-L. *et al.*, "Coded focal stack photography", *IEEE International Conference on Computational Photography, ICCP*, 2013.

[LIP 08] LIPPMAN G., "La photographie intégrale", *Comptes Rendus de l'Académie des Sciences*, vol. 146, pp. 446–551, 1908.

[LIU 14] LIU C.-H., CHANG Y.-C., NORRIS T. *et al.*, "Graphene photodetectors with ultra-broadband and high responsivity at room temperature", *Nature Nanotechnology Letters*, vol. NNANO.2014.31, 2014.

[LIV 02] LIVINGSTONE M., *Vision and Art: the Biology of Seeing*, Abrams. Publishers., New York, 2002.

[LO 12] LO K. Y., LIU K. H., CHEN C. S., "Assessment of photo aesthetics with efficiency.", *IEEE International Conference on Pattern Recognition (ICPR)*, pp. 2186–2189, 2012.

[LOH 96] LOHMANN A., DORSCH R., MENDLOVIC D. *et al.*, "Space-bandwidth product of optical signals and systems", *Journal Optical Society of America*, vol. 13, no. 3, pp. 470–473, March 1996.

[LOP 08] LOPES A., GARELLO R., LE HÉGARAT-MASCLE S., "Speckle Models", in MAITRE H. (ed.), *Processing of Synthetic Aperture Radar Images*, ISTE Ltd, London and John Wiley & Sons, New York, 2008.

[LOW 04] LOWE D., "Distinctive image features from scale invariant key-points", *International Journal of Computer Vision*, vol. 60, no. 2, pp. 91–110, 2004.

[LU 15] LU P., PENG X., LI R., WANG X., "Towards aesthetics of image: A Bayesian framework for color harmony modeling", *Signal Processing: Image Communication*, vol. 39, pp. 487–498, 2015.

[LUK 66] LUKOSZ W., "Optical systems with resolving power exceeding the classical limit", *Journal Optical Society of America*, vol. 56, pp. 1463–1472, 1966.

[LUS 08] LUSTIG M., DONOHO D., SANTOS J., PAULY J., "Compressed sensing MRI", *IEEE Signal Processing Magazine*, vol. 25(2), pp. 78–82, 2008.

[MA 12a] MA K., "Organic Light emitting Diodes (OLEDS)", in CHEN J., CRANTON W., FIHN M. (eds.), *Handbook of Visual Display Technology*, Springer, 2012.

[MA 12b] MA R., "Active Matrix for OLED display", in CHEN J., CRANTON W., FIHN M. (eds), *Handbook of Visual Display Technology*, Springer, New York, 2012.

[MAC 42] MACADAM D., "Visual sensitivities to coulor differences in daylight", *Journal of Optics*, vol. 32, no. 5, pp. 247–273, 1942.

[MAC 43] MACADAM D., "Specification of small chromaticity differences", *Journal of the Optical Society of America*, vol. 33, no. 1, pp. 18–26, 1943.

[MAI 08a] MAITRE H. (ed.), *Image Processing*, ISTE Ltd, London and John Wiley & Sons, New York, 2008.

[MAI 08b] MAITRE H., "Statistical properties of images", in MAITRE H. (ed.), *Image Processing*, ISTE Ltd, London and John Wiley & Sons, New York, 2008.

[MAI 09] MAIRAL J., BACH F., PONCE J. *et al.*, "Non-local sparse model for image restoration", *Proceedings IEEE International conference on Computer Vision, ICCV'09*, pp. 2272–2279, September 2009.

[MAL 93] MALLAT S., ZHANG Z., "Matching pursuits with time-frequencies dictionnaries", *IEEE Transactions on Signal Processing*, vol. 12, pp. 3397–3415, December 1993.

[MAN 74] MANNOS J., SAKRISON D., "The effects of visual fidelity criterion on the encoding of images", *IEEE Transactions on Information Theory*, vol. 20, no. 4, pp. 525–536, 1974.

[MAN 95a] MANDEL L., WOLF E., *Optical Coherence and Quantum Optics*, Cambridge University Press, Cambridge, 1995.

[MAN 95b] MANN S., PICARD R., "On being 'undigital' with digital cameras: extending dynamic range by combining differently exposed pictures", *48th Proceedings IS&T*, Cambridge MA, pp. 442–448, May 1995.

[MAR 10] MARTIN-GONTHIER P., Contribution à l'amélioration de la dynamique des capteurs d'image CMOS à réponse linéaire, PhD, University of Toulouse, ISAE, Toulouse, France, 2010.

[MAS 05] MASSOUD M., *Engineering Thermofluids, Thermodynamics, Fluid-Mechanics and Heat-transfer*, Springer Verlag, 2005.

[MAT 93] MATSUI Y., NARIAI K., *Fundamentals of Practical Aberration Theory*, World Scientific Publishing Co, Singapore, 1993.

[MAZ 12] MAZIN B., DELON J., GOUSSEAU Y., "Illuminant Estimation from Projections on the Planckian Locus", *CPCV workshop, ECCV 2012, LNCS 7584*, vol. II, pp. 370–379, 2012.

[MCE 10] MCELVAIN J., CAMPBELL S., MILLER J., "Texture-based measurement of spatial frequency response using the dead leaves target: extensions, and application to real camera systems", *Electronic Imaging Processing SPIE 7537, 75370D*, 2010.

[MEY 15] MEYNANTS G., "Global Shutter Image Sensors", *Laser+ Photonics*, vol. 1, pp. 44–48, January 2015.

[MIC 03] MICUSIK B., PAJDLA T., "Estimation of omnidirectional camera model from epipolar geometry", *Proceedings of Computer Vision and Pattern Recognition*, vol. 1, pp. 485–490, June 2003.

[MID 60] MIDDLETON D., *An Introduction to Statistical Communications*, McGraw Hill, Columbus, 1960.

[MIK 04] MIKOLAJCZYCK K., SCHMID C., "Scale and affine invariant point detectors", *International Journal of Computer Vision*, vol. 60, no. 1, pp. 63–86, 2004.

[MIK 05] MIKOLAJCZYCK K., SCHMID C., "A performance evaluation of local descriptors", *IEEE Transactions on Pattern Analysis and Machine Intelligence*, vol. 27, no. 10, pp. 1615–1630, 2005.

[MIL 77] MILGRAM D., "Adaptive techniques for photomosaicking", *IEEE transactions on Computers*, vol. 26, pp. 1175–1180, November 1977.

[MIL 99] MILLER H., "Color filter array for CCD and CMOS image sensors using a chemically amplified, thermally cured, pre-dyed, positive tone photoresist for 365 nm lithography", *Proceedings of SPIE*, vol. 3678–2, Bellingham, pp. 1083–1090, 1999.

[MIT 12] MITTAL A., MOORTHY A., BOVIK A., "No-reference image quality assessment in the spatial domain", *IEEE Transactions on Image Processing*, vol. 21 (12), pp. 4695–4708, 2012.

[MIT 13] MITTAL A., SOUNDARARAJAN R., BOVIK A., "Making a 'complete blind' image quality analyzer", *IEEE Signal Processing Letters*, vol. 20 (3), pp. 209–212, 2013.

[MIY 64] MIYIAMOTO K., "Fish Eye lens", *Journal of the Optical Society of America*, vol. 54, pp. 1060–1061, 1964.

[MOL 57] MOLES A. A., "Théorie de l'information et perception esthétique", *Revue Philosophique de la France et de l'Etranger*, pp. 233–242, 1957.

[MOO 11] MOORTHY A., BOVIK A., "Blind image quality assessment: From natural scene statistics to perceptual quality", *IEEE Transactions on Image Processing*, vol. 20 (12), pp. 3350–3364, 2011.

[MOZ 13] MOZEROV M., "Constrained optical flow estimation as a matching problem", *IEEE Transactions on Image Processing*, vol. 22, no. 5, pp. 2044–2055, May 2013.

[MUS 06] MUSÉ P., SUR F., CAO F. *et al.*, *Shape recognition based on an a contrario methodology. In Statistics and analysis of shapes*, Birkhauser, 2006.

[NAI 12a] NAIT-ALI A., CHERIFI D., "Introduction to 2D face recognition", in NAIT-ALI A., FOURNIER R., (eds.), *Signal and Image Processing for Biometrics*, ISTE Ltd, London and John Wiley & Sons, New York, 2012.

[NAI 12b] NAIT-ALI A., FOURNIER R. (eds.), *Signal and Image Processing for Biometrics*, ISTE Ltd, London and John Wiley & Sons, New York, 2012.

[NAY 94] NAYAR S. K., NAKAGAWA Y., "Shape from focus", *IEEE Transactions on Pattern Analysis and Machine Intelligence*, vol. 16 (8), pp. 824–831, 1994.

[NAY 00] NAYAR S., MITSUNAGA T., "Hig dynamic range imaging spatially varying pixel exposure", *IEEE International Conference on Computer Vision and Pattern Recognition (CVPR)*, vol. 1, Hilton Head, USA, pp. 472–479, June 2000.

[NEA 11] NEAMEN D., *Semiconductor Physics and Devices: Basic Principles*, Irwin, 2011.

[NEV 95] NEVEUX M., *Le nombre d'or, radiographie d'un mythe*, Le Seuil, Paris, 1995.

[NG 05] NG R., "Fourier Slice Photography", *ACM Transactions on Graphics, Proceedings of ACM SIGGRAPH*, vol. 24 (3), pp. 735–744, 2005.

[NIC 03] NICOLAS H., GUILLEMOT C., "Normes de compression vidéo", in BARLAUD M., LABIT C., (eds.), *Compression et codage des images et des vidéos*, Hermès-Lavoisier, Paris, 2003.

[PAL 99] PALMER S. E., *Vision Science, Photons to Phenomenology*, MIT Press, Cambridge, 1999.

[PAL 10] PALMER S. E., SCHLOSS K. B., "An ecological valence theory of human color preference", *Proceedings of the National Academy of Sciences, USA*, vol. 107, pp. 8877–8882, 2010.

[PAR 03] PARK S., PARK M., KANG M., "Superresolution image reconstruction: a technical overview", *IEEE Signal Processing ASSP Magazine*, vol. 20, no. 3, pp. 21–36, May 2003.

[PAR 06] PARMAR M., REEVES S., "Selection of Optimal Spectral Sensitivity Functions for Color Filter Arrays", *IEEE International Conference on Image Processing, 2006*, IEEE, pp. 1005–1008, 2006.

[PAR 08] PARIS S., KORNPROBST P., TUMBLIN J. *et al.*, "A Gentle Introduction to Bilateral Filtering and its Applications", *Siggraph-08, Lecture Notes*, 2008.

[PAR 14] PARK H., DAN H., SEO K. *et al.*, "Filter-Free Image Sensor Pixels Comprising Silicon Nanowires with Selective Color Absorption", *Nano Letters*, vol. 14, no. 4, pp. 1804–1809, March 2014.

[PEL 00] PELEG S., ROUSSO B., RAV-ACHA A. *et al.*, "Mosaicking on adaptive manifolds", *IEEE Transactions on Pattern Analysis and Machine Intelligence*, vol. 22, no. 10, pp. 1144–1154, October 2000.

[PER 94] PEREZ J.-P., *Optique géométrique, ondulatoire et polarisation*, 4th ed., Masson, 1994.

[PER 02] PEREIRA F., EBRAHIMI T., *The MPEG-4 book*, Prentice-Hall, Upper Saddle River, 2002.

[PER 12] PERWASS C., WIETZKE L., "Single Lens 3D camera with extended depth of field", *Human Vision and Electronic Imaging, SPIE 8291*, 2012.

[PET 14] PETRO A., SBERT C., MOREL J., "Multiscale Retinex", *Image Processing On Line*, vol. 4, pp. 71–88, 2014.

[PHA 10] PHARR M., HUMPHREYS G., *Physically Based Rendering*, 2nd ed., Morgan Kaufmann, Burlington, 2010.

[PRO 03] PROST R., "Compression sans perte", in BARLAUD M., LABIT C. (eds), *Compression et codage des images et des vidéos*, Hermès-Lavoisier, Paris, 2003.

[PRO 05] PROVENZI E., DE CARLI L., RIZZI A., "Mathematical definition and analysis of the Retinex algorithm", *Journal Optical Society of America*, vol. 22, 12, pp. 2613–2621, 2005.

[PRO 08] PROVENZI E., GATTA C., FIERRO M. *et al.*, "A spatially variant white-patch and gray-world method for color image enhancement driven by local contrast", *IEEE Transactions on Pattern Analysis and Machine Intelligence*, vol. 30, no. 10, pp. 1757–1771, October 2008.

[RAB 11] RABIN J., DELON J., GOUSSEAU Y., "Removing artefacts from color and contrast modifications", *IEEE Transactions on Image Processing*, vol. 20, pp. 3073–3085, 2011.

[RAF 14] RAFINAZARI M., DUBOIS E., "Demosaicking algorithm for the Fujifilm X-Trans color filter array", *Proceedings of the IEEE International Conference on Image Processing*, pp. 660–663, October 2014.

[REI 10] REINHARD E., WARD G., PATTANAIK S. *et al.*, *High dynamic range imaging - Acquisition, display and Image-Based lighting*, Morgan-Kaufmann, Burlington, 2010.

[RIB 03] RIBES A., SCHMITT F., "A fully automatic method for the reconstructing spectral reflectances curves by using mixture density networks", *Pattern Recognition Letters*, vol. 24, no. 11, pp. 1691–1701, July 2003.

[RIG 08] RIGAU J., FEIXAS M., SBERT M., "Information Aesthetic Measures", *IEEE Computer Graphics and Applications*, vol. 2, pp. 24–34, 2008.

[RIO 07] RIOUL O., *Théorie de l'information et du codage*, Hermès-Lavoisier, Paris, 2007.

[RIZ 03] RIZZI A., GATTA C., MARINI D., "A new algorithm for unsupervised global and local color correction", *Pattern Recognition Letters*, vol. 24, no. 11, pp. 1663–1677, 2003.

[ROT 07] ROTH S., BLACK M., "On the spatial statistics of optical flow", *International Journal of Computer Vision*, vol. 47, no. 1, pp. 1–10, January 2007.

[RUD 94] RUDERMAN D., "The statistics of natural images", *Network Computing Neural Systems*, vol. 5 (4), pp. 517–548, 1994.

[SAA 03] SAADANE A., BARBA D., "Modèles psychovisuels de représentation d'images", in BARLAUD M., LABIT C. (eds), *Compression et codage des images et des vidéos*, Hermès-Lavoisier, Paris, 2003.

[SAL 78] SALEH B., *Photoelectrons Statistics*, Springer, New York, 1978.

[SAL 07] SALEH B., TEICH M., *Fundamentals of Photonics*, John Wiley, Singapore, 2007.

[SEI 11] SEITZ P., "Fundamentals of Noise in Optoelectronics", in SEITZ P., THEUWISSEN J. (eds), *Single-Photons Imaging*, Springer, Berlin, 2011.

[SEO 11] SEO K., WOBER M., STEINVURZEL P. *et al.*, "Multicolored Vertical Silicon Nanowires", *Nano Letters*, vol. 11, pp. 1851–1856, 2011.

[SEV 96] SEVE R., *Physique de la couleur : de l'apparence colorée à la technique colorimétrique*, Masson, Paris, 1996.

[SHA 48] SHANNON C., "A mathematical theory of communications", *Bell System Technical Journal*, vol. 27, pp. 379–423, 1948.

[SHA 99] SHAPIRE R., SINGER Y., "Improving boosting algorithm using confidence rated predictions", *Machine Learning*, pp. 80–91, 1999.

[SHA 05] SHARMA G., WU W., DALAL E., "The CEIDE2000 color difference formula implementation notes, supplementary test data, and mathematical observations", *Color Research & Applications*, vol. 30, no. 1, pp. 21–30, Feb 2005.

[SHC 06] SHCHERBACK I., SEGAL R., BELENKY A. *et al.*, "Two-dimensional CMOS image sensor characterization", *International Symposium on Circuits and Systems (ISCAS 2006)*, Kos, Greece, pp. 3582–3585, May 2006.

[SHE 06] SHEIKH H., BOVIK A., "Image Information and Visual Quality", *IEEE Transactions on Image Processing*, vol. 15–2, pp. 430–444, 2006.

[SHI 09] SHI G., GAO D., LIU D. *et al.*, "High resolution image reconstruction. A new imager via movable random exposure", *IEEE Conference on Image Processing ICIP*, Cairo, Egypt, pp. 1177–1180, 2009.

[SHN 06] SHNAYDERMAN A., GUSEV A., ESKICIOGLU A., "An SVD-based grayscale image quality measure for local and global assessment", *IEEE Transactions on Image Processing*, vol. 15–2, pp. 422–430, 2006.

[SIL 94] SILLION F., PUECH C., *Radiosity and Global Illumination*, Morgan-Kaufmann, Burlington, 1994.

[SIM 01] SIMONCELLI E., OLSHAUSEN B., "Natural image statistics and neural representation", *Annual review of neuroscience*, vol. 24 (1), pp. 1193–1216, 2001.

[SMI 90] SMITH W. J., *Modern Optical Engineering*, Mc-Graw-Hill, New York, 1990.

[SON 12a] SONY-CORPORATION, "Sony Exmor Stacked CMOS image Sensor", www.sony.net/SonyInfo/News/Press/201208/12-107E/index.html, August 2012.

[SON 12b] SONY-CORPORATION, "Sony Stacked CMOS Image Sensors solves all existing problems in one stroke", www.sony.net/Products/SC-HP/cx_news/vol68/pdf/sideview_vol68.pdf, January 2012.

[STA 02] STARCK J.-L., CANDÈS E.-J., DONOHO D.-L., "The curvelet transform for image denoising", *IEEE Transactions on Image Processing*, vol. 1–6, pp. 670–684, 2002.

[STE 04] STERN A., JAVIDI B., "Shannon number and information capacity of three-dimentional integral imaging", *Journal of the Optical Society of America*, vol. 21, no. 9, pp. 1602–1612, September 2004.

[SUB 93] SUBBARAO M., CHOI T., NIKZAD A., "Focussing Techniques", *Optical Engineering*, vol. 32–11, pp. 2824–2836, 1993.

[SUN 10] SUN D., ROTH S., BLACK M., "Secrets of optical flow estimation and their principles", *IEEE Computer Vision Pattern Recognition Conference*, San Francisco, USA, pp. 2432–2439, June 2010.

[SWE 96] SWELDENS W., "The lifting scheme: a custom-design construction of biorthogonal wavelets", *Applied and Computational Harmonic Analysis*, vol. 3–2, pp. 186–200, April 1996.

[TAN 13] TANG H., KUTULAKOS K., "What does an aberrated photo tell us about the lens and the scene?", *IEEE International Conference on Computational Photography,* ICCP'13, 2013.

[TAU 00] TAUBMAN D., "High performance scalable image compression with EBCOT", *IEEE Transactions on Image Processing*, vol. 9–7, 2000.

[TEM 14] TEMPLIER F., *OLED Microdisplays: Technology and Applications*, ISTE Ltd, London and John Wiley & Sons, New York, 2014.

[TEN 12] TENDERO Y., "The flutter-shutter camera simulator", *Image Processing On Line*, vol. 2, pp. 225–242, 2012.

[TEN 13] TENDERO Y., MOREL J.-M., ROUGÉ B., "The Flutter Shutter Paradox", *SIAM Journal Imaging Science*, vol. 6 (2), pp. 813–845, 2013.

[THE 96] THEUWISSEN A.J.P., *Solid-state Imaging with Charge-coupled Devices*, Kluwer, 1996.

[THE 08] THEUWISSEN A.J.P., "CMOS Image sensors:State-of-the-Art", *Solid State Electronics*, vol. 52, pp. 1401–1406, 2008.

[THE 10] THEUWISSEN A.J.P., "Better pictures through physics: the state of the art of CMOS image sensors", *IEEE Solid-State Circuits Magazine*, pp. 22–28, 2010.

[TIA 11] TIAN J., MA K., "A survey on super-resolution imaging", *Signal, Image and Video Processing*, vol. 21, no. 5, pp. 329–342, September 2011.

[TIS 08] TISSE C., GUICHARD F., CAO F., "Does resolution increase image quality?", in DICARLO J., RODRICKS B. (eds), *Digital Photography IV*, vol. 6817 of *SPIE-IS & T*, 2008.

[TOR 03] TORRALBA A., OLIVA A., "Statistics of natural image categories", *Network computation in neural systems*, vol. 14 (3), pp. 391–412, 2003.

[TRA 14a] TRAONMILIN Y., Relations entre le modèle d'image et le nombre de mesures pour une super-résolution fidèle, PhD, Télécom ParisTech, Paris, July 2014.

[TRA 14b] TRAONMILIN Y., AGUERREBERE C., "Simultaneous high dynamic range and super-resolution imaging without regularisation", *SIAM Journal on Imaging Sciences*, 2014.

[TRO 12] TROUVÉ P., Conception conjointe optique/traitement pour un imageur compact à capacité 3D, PhD, Ecole Centrale, Nantes, France, 2012.

[TRU 00] TRUSSELL H., "Color and Multispectral image representation and display", in BOVIK A. (ed.), *Handbook of Image and Video processing*, Academic Press, San Diego, 2000.

[UNS 91] UNSER M., ALDROUBI A., EDEN M., "Fast B-spline transforms for continuous image Representation and Interpolation", *IEEE Transactions on Pattern Analysis and Machine Intelligence*, vol. 13–3, pp. 277–285, 1991.

[UNS 95a] UNSER M., ALDROUBI A., EDEN M., "Enlargement or reduction of digital images with minimum loss of information", *IEEE Transactions on Image Processing*, vol. 4, no. 3, pp. 247–258, March 1995.

[UNS 95b] UNSER M., THEVENAZ P., YAROSLAVSKY L., "Convolution-based interpolation for fast, high-quality rotation of images", *IEEE Transactions on Image Processing*, vol. 4, no. 10, pp. 1371–1381, July 1995.

[VAL 93] VALOIS R., VALOIS K., "A multi-stage color model", *Vision Research*, vol. 33, pp. 1053–1065, 1993.

[VET 14] VETTERLI M., KOVACVIĆ J., GOYAL V.K., *Foundations of Signal Processing*, Cambridge University Press, 2014.

[VIE 12] VIENOT F., LE ROHELLEC J., "Colorimetry and Physiology - the LMS specification", in FERNANDEZ-MALOIGNE C., ROBERT-INACIO F., MACAIRE L. (eds.), *Digital Color Imaging*, ISTE Ltd, London and John Wiley & Sons, New York, 2012.

[VIO 04] VIOLA P., JONES M., "Robust Real Times Face Detection", *International Journal of Computer Vision*, vol. 57, pp. 137–154, 2004.

[VOE 12] VOELKEL R., "Wafer-scale micro-optics fabrication", *Advanced Optical Technology*, vol. 1, pp. 135–150, 2012.

[VON 79] VONKRIES J., "Influence of adaptation on the effects produced by luminous stimuli", in MACADAM J., (ed.), *Sources of Color Vision*, MIT Press, Cambridge, 1979.

[WAN 02] WANG Z., BOVIK A., "A universal Image quality index", *IEEE Signal Processing Letters*, vol. 9–3, pp. 81–84, 2002.

[WAN 11] WANG Z., LI Q., "Perception and quality models for image and video", *IEEE Transactions on Image Processing*, vol. 20–5, pp. 1185–1198, May 2011.

[WAN 13] WANG C., HSU T., CHEN C. *et al.*, "Nanotechnology development for CMOS image sensor applications", *IEEE Nanotechnology Materials and Devices Conference (IEEE NMDC 2013)*, October 2013.

[WAT 97] WATSON A., SOLOMON J., "A model of visual contrast, gain control and pattern masking", *Journal of the Optical Society of America*, vol. A–14, pp. 2378–2390, 1997.

[WEI 06] WEICKERT J., BRUHN A., BROX T. *et al.*, "A survey on variational optical flow methods for small displacements", *Mathematical Models of Registration and Applied Medical Imaging*, vol. 10, pp. 103–136, 2006.

[WEN 05] WENGER A., GARDNER A., TCHOU C. *et al.*, "Performance relighting and reflectance transformation with time-multiplexed illumination", *ACM SIGGRAPH'05*, New York, pp. 756–764, 2005.

[WES 82] WEST W., BRILL M., "Necessary and sufficient conditions for von Kries chromatic adaptation to give color constancy", *Journal of Mathematical Biology*, vol. 15, no. 2, pp. 249–258, 1982.

[WOO 11] WOODS J. W., *Multidimensional Signal, Image and Video Processing and Coding*, 2nd ed., Academic Press, San Diego, USA, 2011.

[WYS 82] WYSZECKI G., STILES W., *Color Sciences Concepts and Methods, Quantitative Data and Formulae*, 2nd ed., John Wiley & Sons, New York, 1982.

[XIA 05] XIAO F., FARREL J., WANDELL B., "Psychophysical thresholds and digital camera sensitivity: The thousand photon limit", *Proc SPIE*, no. 5678, SPIE, pp. 75–84, 2005.

[YEH 11] YEH J.-A., TSAI C.-G., YANG C.-C. *et al.*, "Smart Optical components based on liquid-liquid interface manipulation by dielectric forces", *Proc. IEEE International Conference on NanoMicro Engineered and Molecular Systems*, Kasjsiung, Taiwan, pp. 1192–1195, February 2011.

[YOS 09] YOSHIO M., BRODD R., KOZAWA A., *Lithium-Ion Batteries*, Springer-Verlag, New York, 2009.

[YU 11] YU G., SAPIRO G., "DCT image denoising, a simple and effective image denoising algorithm", *Image Processing On Line*, vol. 1, 2011.

[ZEK 99] ZEKI S., *Inner vision: An exploration of art and the brain*, Oxford University Press, 1999.

[ZHA 05] ZHANG L., WU X., "Color demosaicking via directional linear m.s.e. estimation", *IEEE Transactions on Image Processing*, vol. 14 (12), pp. 2167–2178, 2005.

[ZHA 06] ZHAO T., WANG R., LIU Y., YU F., "Characteristic Analysis of Optical Low Pass Filter Used in Digital Camera", *Optical Design and Fabrication, ICO20, Proceedings of SPIE*, vol. 6034, 2006.

Index

E, F

G, H, I

J, K, L

M, N

O, P

Q, R

S, T

U, V, W, X, Y, Z

Other titles from

ISTE

in

Digital Signal and Image Processing

2016

CESCHI Roger, GAUTIER Jean-Luc
Fourier Analysis

CHARBIT Maurice
Digital Signal Processing with Python Programming

FEMMAM Smain
Fundamentals of Signals and Control Systems

FEMMAM Smain
Signals and Control Systems: Application for Home Health Monitoring

2015

BLANCHET Gérard, CHARBIT Maurice
Digital Signal and Image Processing using MATLAB®
Volume 2 – Advances and Applications: The Deterministic Case – 2nd edition
Volume 3 – Advances and Applications: The Stochastic Case – 2nd edition

CLARYSSE Patrick, FRIBOULET Denis
Multi-modality Cardiac Imaging

GIOVANNELLI Jean-François, IDIER Jérôme
Regularization and Bayesian Methods for Inverse Problems in Signal and Image Processing

MAÎTRE Henri
From Photon to Pixel: The Digital Camera Handbook

2014

AUGER François
Signal Processing with Free Software: Practical Experiments

BLANCHET Gérard, CHARBIT Maurice
Digital Signal and Image Processing using MATLAB®
Volume 1 – Fundamentals – 2nd edition

DUBUISSON Séverine
Tracking with Particle Filter for High-dimensional Observation and State Spaces

ELL Todd A., LE BIHAN Nicolas, SANGWINE Stephen J.
Quaternion Fourier Transforms for Signal and Image Processing

FANET Hervé
Medical Imaging Based on Magnetic Fields and Ultrasounds

MOUKADEM Ali, OULD Abdeslam Djaffar, DIETERLEN Alain
Time-Frequency Domain for Segmentation and Classification of Non-stationary Signals: The Stockwell Transform Applied on Bio-signals and Electric Signals

NDAGIJIMANA Fabien
Signal Integrity: From High Speed to Radiofrequency Applications

PINOLI Jean-Charles
Mathematical Foundations of Image Processing and Analysis
Volumes 1 and 2

TUPIN Florence, INGLADA Jordi, NICOLAS Jean-Marie
Remote Sensing Imagery

VLADEANU Calin, EL ASSAD Safwan
Nonlinear Digital Encoders for Data Communications

2013

GOVAERT Gérard, NADIF Mohamed
Co-Clustering

DAROLLES Serge, DUVAUT Patrick, JAY Emmanuelle
Multi-factor Models and Signal Processing Techniques: Application to Quantitative Finance

LUCAS Laurent, LOSCOS Céline, REMION Yannick
3D Video: From Capture to Diffusion

MOREAU Eric, ADALI Tulay
Blind Identification and Separation of Complex-valued Signals

PERRIN Vincent
MRI Techniques

WAGNER Kevin, DOROSLOVACKI Milos
Proportionate-type Normalized Least Mean Square Algorithms

FERNANDEZ Christine, MACAIRE Ludovic, ROBERT-INACIO Frédérique
Digital Color Imaging

FERNANDEZ Christine, MACAIRE Ludovic, ROBERT-INACIO Frédérique
Digital Color: Acquisition, Perception, Coding and Rendering

NAIT-ALI Amine, FOURNIER Régis
Signal and Image Processing for Biometrics

OUAHABI Abdeljalil
Signal and Image Multiresolution Analysis

2011

CASTANIÉ Francis
Digital Spectral Analysis: Parametric, Non-parametric and Advanced Methods

DESCOMBES Xavier
Stochastic Geometry for Image Analysis

FANET Hervé
Photon-based Medical Imagery

MOREAU Nicolas
Tools for Signal Compression

2010

NAJMAN Laurent, TALBOT Hugues
Mathematical Morphology

2009

BERTEIN Jean-Claude, CESCHI Roger
Discrete Stochastic Processes and Optimal Filtering / 2nd edition

CHANUSSOT Jocelyn *et al.*
Multivariate Image Processing

DHOME Michel
Visual Perception through Video Imagery

GOVAERT Gérard
Data Analysis

GRANGEAT Pierre
Tomography

MOHAMAD-DJAFARI Ali
Inverse Problems in Vision and 3D Tomography

SIARRY Patrick
Optimization in Signal and Image Processing

2008

ABRY Patrice *et al.*
Scaling, Fractals and Wavelets

GARELLO René
Two-dimensional Signal Analysis

HLAWATSCH Franz *et al.*
Time-Frequency Analysis

IDIER Jérôme
Bayesian Approach to Inverse Problems

MAÎTRE Henri
Processing of Synthetic Aperture Radar (SAR) Images

MAÎTRE Henri
Image Processing

NAIT-ALI Amine, CAVARO-MENARD Christine
Compression of Biomedical Images and Signals

NAJIM Mohamed
Modeling, Estimation and Optimal Filtration in Signal Processing

QUINQUIS André
Digital Signal Processing Using Matlab

2007

BLOCH Isabelle
Information Fusion in Signal and Image Processing

GLAVIEUX Alain
Channel Coding in Communication Networks

OPPENHEIM Georges *et al.*
Wavelets and their Applications

2006

CASTANIÉ Francis
Spectral Analysis

NAJIM Mohamed
Digital Filters Design for Signal and Image Processing

www.ingramcontent.com/pod-product-compliance
Lightning Source LLC
LaVergne TN
LVHW020421301125
826312LV00002B/15
* 9 7 8 1 7 8 6 3 0 1 3 7 6 *